T0349833

Graduate Texts in Mathematics 133

Springer
New York
Berlin
Heidelberg
Barcelona
Budapest
Hong Kong
London
Milan
Paris
Santa Clara
Singapore
Tokyo

Graduate Texts in Mathematics

1 TAKEUTI/ZARING. Introduction to Axiomatic Set Theory. 2nd ed.
2 OXTOBY. Measure and Category. 2nd ed.
3 SCHAEFER. Topological Vector Spaces.
4 HILTON/STAMMBACH. A Course in Homological Algebra. 2nd ed.
5 MAC LANE. Categories for the Working Mathematician. 2nd ed.
6 HUGHES/PIPER. Projective Planes.
7 SERRE. A Course in Arithmetic.
8 TAKEUTI/ZARING. Axiomatic Set Theory.
9 HUMPHREYS. Introduction to Lie Algebras and Representation Theory.
10 COHEN. A Course in Simple Homotopy Theory.
11 CONWAY. Functions of One Complex Variable I. 2nd ed.
12 BEALS. Advanced Mathematical Analysis.
13 ANDERSON/FULLER. Rings and Categories of Modules. 2nd ed.
14 GOLUBITSKY/GUILLEMIN. Stable Mappings and Their Singularities.
15 BERBERIAN. Lectures in Functional Analysis and Operator Theory.
16 WINTER. The Structure of Fields.
17 ROSENBLATT. Random Processes. 2nd ed.
18 HALMOS. Measure Theory.
19 HALMOS. A Hilbert Space Problem Book. 2nd ed.
20 HUSEMOLLER. Fibre Bundles. 3rd ed.
21 HUMPHREYS. Linear Algebraic Groups.
22 BARNES/MACK. An Algebraic Introduction to Mathematical Logic.
23 GREUB. Linear Algebra. 4th ed.
24 HOLMES. Geometric Functional Analysis and Its Applications.
25 HEWITT/STROMBERG. Real and Abstract Analysis.
26 MANES. Algebraic Theories.
27 KELLEY. General Topology.
28 ZARISKI/SAMUEL. Commutative Algebra. Vol.I.
29 ZARISKI/SAMUEL. Commutative Algebra. Vol.II.
30 JACOBSON. Lectures in Abstract Algebra I. Basic Concepts.
31 JACOBSON. Lectures in Abstract Algebra II. Linear Algebra.
32 JACOBSON. Lectures in Abstract Algebra III. Theory of Fields and Galois Theory.
33 HIRSCH. Differential Topology.
34 SPITZER. Principles of Random Walk. 2nd ed.
35 ALEXANDER/WERMER. Several Complex Variables and Banach Algebras. 3rd ed.
36 KELLEY/NAMIOKA et al. Linear Topological Spaces.
37 MONK. Mathematical Logic.
38 GRAUERT/FRITZSCHE. Several Complex Variables.
39 ARVESON. An Invitation to C^*-Algebras.
40 KEMENY/SNELL/KNAPP. Denumerable Markov Chains. 2nd ed.
41 APOSTOL. Modular Functions and Dirichlet Series in Number Theory. 2nd ed.
42 SERRE. Linear Representations of Finite Groups.
43 GILLMAN/JERISON. Rings of Continuous Functions.
44 KENDIG. Elementary Algebraic Geometry.
45 LOÈVE. Probability Theory I. 4th ed.
46 LOÈVE. Probability Theory II. 4th ed.
47 MOISE. Geometric Topology in Dimensions 2 and 3.
48 SACHS/WU. General Relativity for Mathematicians.
49 GRUENBERG/WEIR. Linear Geometry. 2nd ed.
50 EDWARDS. Fermat's Last Theorem.
51 KLINGENBERG. A Course in Differential Geometry.
52 HARTSHORNE. Algebraic Geometry.
53 MANIN. A Course in Mathematical Logic.
54 GRAVER/WATKINS. Combinatorics with Emphasis on the Theory of Graphs.
55 BROWN/PEARCY. Introduction to Operator Theory I: Elements of Functional Analysis.
56 MASSEY. Algebraic Topology: An Introduction.
57 CROWELL/FOX. Introduction to Knot Theory.
58 KOBLITZ. p-adic Numbers, p-adic Analysis, and Zeta-Functions. 2nd ed.
59 LANG. Cyclotomic Fields.
60 ARNOLD. Mathematical Methods in Classical Mechanics. 2nd ed.

continued after index

Joe Harris

Algebraic Geometry

A First Course

With 83 Illustrations

Joe Harris
Department of Mathematics
Harvard University
Cambridge, MA 02138
USA

Mathematics Subject Classification: 14-01

Library of Congress Cataloging-in-Publication Data
Harris, Joe.
Algebraic geometry: a first course / Joe Harris.
p. cm.—(Graduate texts in mathematics; 133)
Includes bibliographical references and index.
ISBN 0-387-97716-3
1. Geometry, Algebraic. I. Title. II. Series.
QA564.H24 1992
516.3'5—dc20 91-33973

Printed on acid-free paper.

Typeset by Asco Trade Typesetting, North Point, Hong Kong.
Printed and bound by R.R. Donnelley & Sons, Harrisonburg, VA.
Printed in the United States of America.

9 8 7 6 5 4

ISBN 0-387-97716-3 Springer-Verlag New York Berlin Heidelberg
ISBN 3-540-97716-3 Springer-Verlag Berlin Heidelberg New York SPIN 10678499

For Diane, Liam, and Davey

Preface

This book is based on one-semester courses given at Harvard in 1984, at Brown in 1985, and at Harvard in 1988. It is intended to be, as the title suggests, a first introduction to the subject. Even so, a few words are in order about the purposes of the book.

Algebraic geometry has developed tremendously over the last century. During the 19th century, the subject was practiced on a relatively concrete, down-to-earth level; the main objects of study were projective varieties, and the techniques for the most part were grounded in geometric constructions. This approach flourished during the middle of the century and reached its culmination in the work of the Italian school around the end of the 19th and the beginning of the 20th centuries. Ultimately, the subject was pushed beyond the limits of its foundations: by the end of its period the Italian school had progressed to the point where the language and techniques of the subject could no longer serve to express or carry out the ideas of its best practitioners.

This was more than amply remedied in the course of several developments beginning early in this century. To begin with, there was the pioneering work of Zariski who, aided by the German school of abstract algebraists, succeeded in putting the subject on a firm algebraic foundation. Around the same time, Weil introduced the notion of abstract algebraic variety, in effect redefining the basic objects studied in the subject. Then in the 1950s came Serre's work, introducing the fundamental tool of sheaf theory. Finally (for now), in the 1960s, Grothendieck (aided and abetted by Artin, Mumford, and many others) introduced the concept of the scheme. This, more than anything else, transformed the subject, putting it on a radically new footing. As a result of these various developments much of the more advanced work of the Italian school could be put on a solid foundation and carried further; this has been happening over the last two decades simultaneously with the advent of new ideas made possible by the modern theory.

All this means that people studying algebraic geometry today are in the position of being given tools of remarkable power. At the same time, didactically it creates a dilemma: what is the best way to go about learning the subject? If your goal is simply to see what algebraic geometry is about—to get a sense of the basic objects considered, the questions asked about them and the sort of answers one can obtain—you might not want to start off with the more technical side of the subject. If, on the other hand, your ultimate goal is to work in the field of algebraic geometry it might seem that the best thing to do is to introduce the modern approach early on and develop the whole subject in these terms. Even in this case, though, you might be better motivated to learn the language of schemes, and better able to appreciate the insights offered by it, if you had some acquaintance with elementary algebraic geometry.

In the end, it is the subject itself that decided the issue for me. Classical algebraic geometry is simply a glorious subject, one with a beautifully intricate structure and yet a tremendous wealth of examples. It is full of enticing and easily posed problems, ranging from the tractable to the still unsolved. It is, in short, a joy both to teach and to learn. For all these reasons, it seemed to me that the best way to approach the subject is to spend some time introducing elementary algebraic geometry before going on to the modern theory. This book represents my attempt at such an introduction.

This motivation underlies many of the choices made in the contents of the book. For one thing, given that those who want to go on in algebraic geometry will be relearning the foundations in the modern language there is no point in introducing at this stage more than an absolute minimum of technical machinery. Likewise, I have for the most part avoided topics that I felt could be better dealt with from a more advanced perspective, focussing instead on those that to my mind are nearly as well understood classically as they are in modern language. (This is not absolute, of course; the reader who is familiar with the theory of schemes will find lots of places where we would all be much happier if I could just say the words "scheme-theoretic intersection" or "flat family".)

This decision as to content and level in turn influences a number of other questions of organization and style. For example, it seemed a good idea for the present purposes to stress examples throughout, with the theory developed concurrently as needed. Thus, Part I is concerned with introducing basic varieties and constructions; many fundamental notions such as dimension and degree are not formally defined until Part II. Likewise, there are a number of unproved assertions, theorems whose statements I thought might be illuminating, but whose proofs are beyond the scope of the techniques introduced here. Finally, I have tried to maintain an informal style throughout.

Acknowledgments

Many people have helped a great deal in the development of this manuscript. Benji Fisher, as a junior at Harvard, went to the course the first time it was given and took a wonderful set of notes; it was the quality of those notes that encouraged me to proceed with the book. Those who attended those courses provided many ideas, suggestions, and corrections, as did a number of people who read various versions of the book, including Paolo Aluffi, Dan Grayson, Zinovy Reichstein and John Tate. I have also enjoyed and benefited from conversations with many people including Fernando Cukierman, David Eisenbud, Noam Elkies, Rolfdieter Frank, Bill Fulton, Dick Gross and Kurt Mederer. I would also like to thank Benji Fisher, Seth Padowitz, David Patrick and Lyle Ramshew for pointing out errors in the first printing.

The references in this book are scant, and I apologize to those whose work I may have failed to cite properly. I have acquired much of my knowledge of this subject informally, and remain much less familiar with the literature than I should be. Certainly, the absence of a reference for any particular discussion should be taken simply as an indication of my ignorance in this regard, rather than as a claim of originality.

I would like to thank Harvard University, and in particular Deans Candace Corvey and A. Michael Spence, for their generosity in providing the computers on which this book was written.

Finally, two people in particular contributed enormously and deserve special mention. Bill Fulton and David Eisenbud read the next-to-final version of the manuscript with exceptional thoroughness and made extremely valuable comments on everything from typos to issues of mathematical completeness and accuracy. Moreover, in every case where they saw an issue, they proposed ways of dealing with it, most of which were far superior to those I could have come up with.

Joe Harris
Harvard University
Cambridge, MA
harris@zariski.harvard.edu

Using This Book

There is not much to say here, but I'll make a couple of obvious points.

First of all, a quick glance at the book will show that the logical skeleton of this book occupies relatively little of its volume: most of the bulk is taken up by examples and exercises. Most of these can be omitted, if they are not of interest, and gone back to later if desired. Indeed, while I clearly feel that these sorts of examples represent a good way to become familiar with the subject, I expect that only someone who was truly gluttonous, masochistic, or compulsive would read every single one on the first go-round. By way of example, one possible abbreviated tour of the book might omit (hyphens without numbers following mean "to end of lecture") 1.22–, 2.27–, 3.16–, 4.10–, 5.11–, 6.8–11, 7.19–21, 7.25–, 8.9–13, 8.32–39, 9.15–20, 10.12–17, 10.23–, 11.40–, 12.11–, 13.7–, 15.7–21, 16.9–11, 16.21–, 17.4–15, 19.11–, 20.4–6, 20.9–13 and all of 21.

By the same token, I would encourage the reader to jump around in the text. As noted, some basic topics are relegated to later in the book, but there is no reason not to go ahead and look at these lectures if you're curious. Likewise, most of the examples are dealt with several times: they are introduced early and reexamined in the light of each new development. If you would rather, you could use the index and follow each one through.

Lastly, a word about prerequisites (and post-requisites). I have tried to keep the former to a minimum: a reader should be able to get by with just some linear and multilinear algebra and a basic background in abstract algebra (definitions and basic properties of groups, rings, fields, etc.), especially with a copy of a user-friendly commutative algebra book such as Atiyah and MacDonald's [AM] or Eisenbud's [E] at hand.

At the other end, what to do if, after reading this book, you would like to learn some algebraic geometry? The next step would be to learn some sheaf theory, sheaf cohomology, and scheme theory (the latter two not necessarily in that order).

For sheaf theory in the context of algebraic geometry, Serre's paper [S] is the basic source. For the theory of schemes, Hartshorne's [H] classic book stands out as the canonical reference; as an introduction to the subject there is also Mumford's [M1] red book and the book by Eisenbud and Harris [EH]. Alternatively, for a discussion of some advanced topics in the setting of complex manifolds rather than schemes, see [GH].

Contents

PART I

EXAMPLES OF VARIETIES AND MAPS

LECTURE 1

Affine and Projective Varieties

A Note About Our Field

In this book we will be dealing with varieties over a field K, which we will take to be algebraically closed throughout. Algebraic geometry can certainly be done over arbitrary fields (or even more generally over rings), but not in so straightforward a fashion as we will do here; indeed, to work with varieties over nonalgebraically closed fields the best language to use is that of scheme theory. Classically, much of algebraic geometry was done over the complex numbers $\mathbb{C}$, and this remains the source of much of our geometric intuition; but where possible we will avoid assuming $K = \mathbb{C}$.

Affine Space and Affine Varieties

By *affine space* over the field K, we mean simply the vector space K^n; this is usually denoted $\mathbb{A}^n_K$ or just $\mathbb{A}^n$. (The main distinction between affine space and the vector space K^n is that the origin plays no special role in affine space.) By an *affine variety* $X \subset \mathbb{A}^n$, we mean the common zero locus of a collection of polynomials $f_\alpha \in K[z_1, \ldots, z_n]$.

Projective Space and Projective Varieties

By *projective space* over a field K, we will mean the set of one-dimensional subspaces of the vector space K^{n+1}; this is denoted $\mathbb{P}^n_K$, or more often just $\mathbb{P}^n$. Equivalently, $\mathbb{P}^n$ is the quotient of the complement $K^{n+1} - \{0\}$ of the origin in K^{n+1} by the action of the group K^* acting by scalar multiplication. Sometimes, we will want to refer to the projective space of one-dimensional subspaces of a vector space V over the field K without specifying an isomorphism of V with K^{n+1} (or perhaps without specifying the dimension of V); in this case we will denote it by $\mathbb{P}(V)$ or just $\mathbb{P}V$. A point of $\mathbb{P}^n$ is usually written as a homogeneous vector $[Z_0, \dots, Z_n]$, by which we mean the line spanned by $(Z_0, \dots, Z_n) \in K^{n+1}$; likewise, for any nonzero vector $v \in V$ we denote by $[v]$ the corresponding point in $\mathbb{P}V \cong \mathbb{P}^n$.

A polynomial $F \in K[Z_0, \dots, Z_n]$ on the vector space K^{n+1} does not define a function on $\mathbb{P}^n$. On the other hand, if F happens to be homogeneous of degree d, then since

$$F(\lambda Z_0, \dots, \lambda Z_n) = \lambda^d \cdot F(Z_0, \dots, Z_n)$$

it does make sense to talk about the zero locus of the polynomial F; we define a *projective variety* $X \subset \mathbb{P}^n$ to be the zero locus of a collection of homogeneous polynomials F_α. The group $\mathrm{PGL}_{n+1}K$ acts on the space $\mathbb{P}^n$ (we will see in Lecture 18 that these are all the automorphisms of $\mathbb{P}^n$), and we say that two varieties X, $Y \subset \mathbb{P}^n$ are *projectively equivalent* if they are congruent modulo this group.

We should make here a couple of remarks about terminology. First, the standard coordinates $Z_0, \dots, Z_n$ on K^{n+1} (or any linear combinations of them) are called *homogeneous coordinates* on $\mathbb{P}^n$, but this is misleading: they are not even functions on $\mathbb{P}^n$ (only their pairwise ratios Z_i/Z_j are functions, and only where the denominator is nonzero). Likewise, we will often refer to a homogeneous polynomial $F(Z_0, \dots, Z_n)$ of degree d as a polynomial on $\mathbb{P}^n$; again, this is not to suggest that F is actually a function. Note that if $\mathbb{P}^n = \mathbb{P}V$ is the projective space associated with a vector space V, the homogeneous coordinates on $\mathbb{P}V$ correspond to elements of the dual space V^*, and similarly the space of homogeneous polynomials of degree d on $\mathbb{P}V$ is naturally identified with the vector space $\mathrm{Sym}^d(V^*)$.

Let $U_i \subset \mathbb{P}^n$ be the subset of points $[Z_0, \dots, Z_n]$ with $Z_i \neq 0$. Then on U_i the ratios $z_j = Z_j/Z_i$ are well-defined and give a bijection

$$U_i \cong \mathbb{A}^n.$$

Geometrically, we can think of this map as associating to a line $L \subset K^{n+1}$ not contained in the hyperplane $(Z_i = 0)$ its point p of intersection with the affine plane $(Z_i = 1) \subset K^{n+1}$.

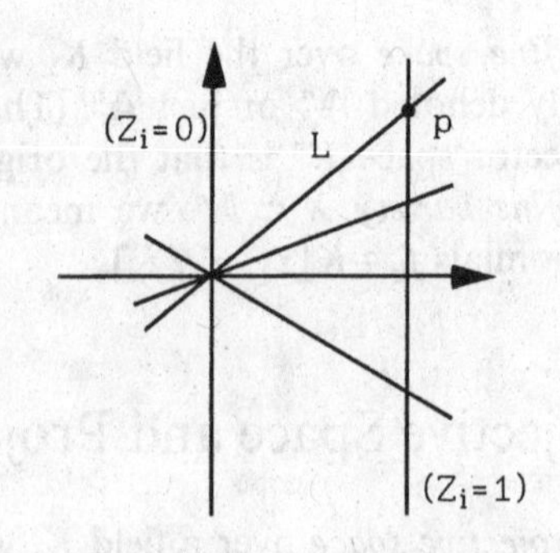

We may thus think of projective space as a compactification of affine space. The functions z_j on U_i are called *affine* or *Euclidean* coordinates on the projective space or on the open set U_i; the open sets U_i comprise what is called the standard cover of $\mathbb{P}^n$ by affine open sets.

If $X \subset \mathbb{P}^n$ is a variety, then the intersection $X_i = X \cap U_i$ is an affine variety: if X

is given by polynomials $F_\alpha \in K[Z_0, \ldots, Z_n]$, then X_0, for example, will be the zero locus of the polynomials

$$\begin{aligned} f_\alpha(z_1, \ldots, z_n) &= F_\alpha(Z_0, \ldots, Z_n)/Z_0^d \\ &= F_\alpha(1, z_1, \ldots, z_n) \end{aligned}$$

where $d = \deg(F_\alpha)$. Thus projective space is the union of affine spaces, and any projective variety is the union of affine varieties. Conversely, we may invert this process to see that any affine variety $X_0 \subset \mathbb{A}^n \cong U_0 \subset \mathbb{P}^n$ is the intersection of U_0 with a projective variety X: if X_0 is given by polynomials

$$f_\alpha(z_1, \ldots, z_n) = \sum a_{i_1, \ldots, i_n} \cdot z_1^{i_1} \cdot \ldots \cdot z_n^{i_n}$$

of degree d_α, then X may be given by the homogeneous polynomials

$$\begin{aligned} F_\alpha(Z_0, \ldots, Z_n) &= Z_0^{d_\alpha} \cdot f_\alpha(Z_1/Z_0, \ldots, Z_n/Z_0) \\ &= \sum a_{i_1, \ldots, i_n} \cdot Z_0^{d_\alpha - \sum i_1} \cdot Z_1^{i_1} \cdot \ldots \cdot Z_n^{i_n} \end{aligned}$$

Note in particular that a subset $X \subset \mathbb{P}^n$ is a projective variety if and only if its intersections $X_i = X \cap U_i$ are all affine varieties.

Example 1.1. Linear Spaces

An inclusion of vector spaces $W \cong K^{k+1} \hookrightarrow V \cong K^{n+1}$ induces a map $\mathbb{P}W \hookrightarrow \mathbb{P}V$; the image Λ of such a map is called a *linear subspace* of dimension k, or *k-plane*, in $\mathbb{P}V$. In case $k = n - 1$, we call Λ a *hyperplane*. In case $k = 1$ we call Λ a *line*; note that there is a unique line in $\mathbb{P}^n$ through any two distinct points. A linear subspace $\Lambda \cong \mathbb{P}^k \subset \mathbb{P}^n$ may also be described as the zero locus of $n - k$ homogeneous linear forms, so that it is a subvariety of $\mathbb{P}^n$; conversely, any variety defined by linear forms is a linear subspace.

The intersection of two linear subspaces of $\mathbb{P}^n$ is again a linear subspace, possibly empty. We can also talk about the span of two (or more) linear subspaces Λ, Λ'; if $\Lambda = \mathbb{P}W$, $\Lambda' = \mathbb{P}W'$, this is just the subspace associated to the sum $W + W' \subset K^{n+1}$, or equivalently the smallest linear subspace containing both Λ and Λ', and is denoted $\overline{\Lambda, \Lambda'}$. In general, for any pair of subsets Γ, $\Phi \subset \mathbb{P}^n$, we define the *span* of Γ and Φ, denoted $\overline{\Gamma \cup \Phi}$, to be the smallest linear subspace of $\mathbb{P}^n$ containing their union.

The dimension of the space $\overline{\Lambda, \Lambda'}$ is at most the sum of the dimensions plus one, with equality holding if and only if Λ and Λ' are disjoint; we have in general the relation

$$\dim(\overline{\Lambda, \Lambda'}) = \dim(\Lambda) + \dim(\Lambda') - \dim(\Lambda \cap \Lambda')$$

where we take the dimension of the empty set as a linear subspace to be -1. Note, in particular, one of the basic properties of projective space $\mathbb{P}^n$: whenever $k + l \geq n$, any two linear subspaces Λ, Λ' of dimensions k and l in $\mathbb{P}^n$ will intersect in a linear subspace of dimension at least $k + l - n$.

Note that the set of hyperplanes in a projective space $\mathbb{P}^n$ is again a projective space, called the *dual projective space* and denoted $\mathbb{P}^{n*}$. Intrinsically, if $\mathbb{P}^n = \mathbb{P}V$ is the projective space associated to a vector space V, the dual projective space $\mathbb{P}^{n*} = \mathbb{P}(V^*)$ is the projective space associated to the dual space V^*. More generally, if $\Lambda \cong \mathbb{P}^k \subset \mathbb{P}^n$ is a k-dimensional linear subspace, the set of $(k+1)$-planes containing Λ is a projective space $\mathbb{P}^{n-k-1}$, and the space of hyperplanes containing Λ is the dual projective space $(\mathbb{P}^{n-k-1})^*$. Intrinsically, if $\mathbb{P}^n = \mathbb{P}V$ and $\Lambda = \mathbb{P}W$ for some $(k+1)$-dimensional subspace $W \subset V$, then the space of $(k+1)$-planes containing Λ is the projective space $\mathbb{P}(V/W)$ associated to the quotient, and the set of hyperplanes containing Λ is naturally the projectivization $\mathbb{P}((V/W)^*) = \mathbb{P}(\mathrm{Ann}(W)) \subset \mathbb{P}(V^*)$ of the annihilator $\mathrm{Ann}(W) \subset V^*$ of W.

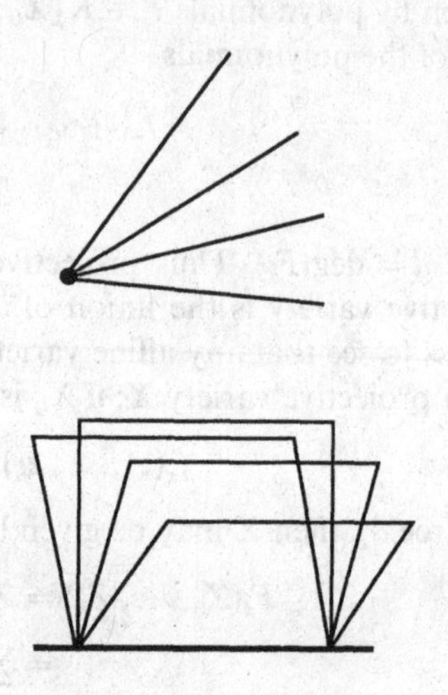

Example 1.2. Finite Sets

Any finite subset Γ of $\mathbb{P}^n$ is a variety: given any point $q \notin \Gamma$, we can find a polynomial on $\mathbb{P}^n$ vanishing on all the points p_i of Γ but not at q just by taking a product of homogeneous linear forms L_i where L_i vanishes at p_i but not at q. Thus, if Γ consists of d points (we say in this case that Γ has *degree d*), it may be described by polynomials of degree d and less.

In general, we may ask what are the smallest degree polynomials that suffice to describe a given variety $\Gamma \subset \mathbb{P}^n$. The bound given for finite sets is sharp, as may be seen from the example of d points lying on a line L; it's not hard to see that a polynomial $F(Z)$ of degree $d-1$ or less that vanishes on d points $p_i \in L$ will vanish identically on L. On the other hand, this is the only such example, as the following exercise shows.

Exercise 1.3. Show that if Γ consists of d points and is not contained in a line, then Γ may be described as the zero locus of polynomials of degree $d-1$ and less.

Note that the "and less" in Exercise 1.3 is unnecessary: if a variety $X \subset \mathbb{P}^n$ is the zero locus of polynomials F_α of degree $d_\alpha \leq m$, then it may also be represented as the zero locus of the polynomials $\{X^I \cdot F_\alpha\}$, where for each α the monomial X^I ranges over all monomials of degree $m - d_\alpha$.

Another direction in which we can go is to focus our attention on sets of points that satisfy no more linear relations than they have to. To be precise, we say that l points $p_i = [v_i] \in \mathbb{P}^n$ are *independent* if the corresponding vectors

v_i are; equivalently, if the span of the points is a subspace of dimension $l - 1$. Note that any $n + 2$ points in $\mathbb{P}^n$ are dependent, while $n + 1$ are dependent if and only if they lie in a hyperplane. We say that a finite set of points $\Gamma \subset \mathbb{P}^n$ is *in general position* if no $n + 1$ or fewer of them are dependent; if Γ contains $n + 1$ or more points this is the same as saying that no $n + 1$ of them lie in a hyperplane. We can then ask what is the smallest degree polynomial needed to cut out a set of d points in general position in $\mathbb{P}^n$; the following theorem and exercise represent one case of this.

Theorem 1.4. *If $\Gamma \subset \mathbb{P}^n$ is any collection of $d \leq 2n$ points in general position, then Γ may be described as the zero locus of quadratic polynomials.*

PROOF. We will do this for $\Gamma = \{p_1, \ldots, p_{2n}\}$ consisting of exactly $2n$ points; the general case will be easier. Suppose now that $q \in \mathbb{P}^n$ is any point such that every quadratic polynomial vanishing on Γ vanishes at q; we have to show that $q \in \Gamma$. To do this, observe first that by hypothesis, if $\Gamma = \Gamma_1 \cup \Gamma_2$ is any decomposition of Γ into sets of cardinality n, then each Γ_i will span a hyperplane $\Lambda_i \subset \mathbb{P}^n$; and since the union $\Lambda_1 \cup \Lambda_2$ is the zero locus of a quadratic polynomial vanishing on Γ we must have $q \in \Lambda_1 \cup \Lambda_2$. In particular, q must lie on at least one hyperplane spanned by points of Γ.

Now let $p_1, \ldots, p_k$ be any minimal subset of Γ such that q lies in their span; by the preceding, we can take $k \leq n$. Suppose that $\Sigma \subset \Gamma - \{p_1, \ldots, p_k\}$ is any subset of cardinality $n - k + 1$; then by the general position hypothesis, the hyperplane Λ spanned by the points $p_2, \ldots, p_k$ and Σ does not contain p_1. It follows that Λ cannot contain q, since the span of $p_2, \ldots, p_k$ and q contains p_1; thus q must lie on the hyperplane spanned by the remaining n points of Γ. In sum, then, q must lie on the span of p_1 and any $n - 1$ of the points $p_{k+1}, \ldots, p_{2n}$; since the intersection of all these hyperplanes is just p_1 itself, we conclude that $q = p_1$. □

Exercise 1.5. Show in general that for $k \geq 2$ any collection Γ of $d \leq kn$ points in general position may be described by polynomials of degree k and less (as we will see in Exercise 1.15, this is sharp).

As a final note on finite subsets of $\mathbb{P}^n$, we should mention (in the form of an exercise) a standard fact.

Exercise 1.6. Show that any two ordered subsets of $n + 2$ points in general position in $\mathbb{P}^n$ are projectively equivalent.

This in turn raises the question of when two ordered subsets of $d \geq n + 3$ points in general position in $\mathbb{P}^n$ are projectively equivalent. This question is answered in case $n = 1$ by the *cross-ratio*

$$\lambda(z_1, \ldots, z_4) = \frac{(z_1 - z_2)\cdot(z_3 - z_4)}{(z_1 - z_3)\cdot(z_2 - z_4)};$$

since $\lambda(z_1, \ldots, z_4)$ is the image of z_4 under the (unique) linear map of $\mathbb{P}^1$ to itself carrying z_1, z_2, and z_3 into 1, ∞, and 0 respectively, two subsets $z_1, \ldots, z_4$ and $z'_1, \ldots, z'_4 \in \mathbb{P}^1$ can be carried into one another in order if and only the cross-ratios $\lambda(z_1, \ldots, z_4) = \lambda(z'_1, \ldots, z'_4)$ (see for example [A]). We will give the answer to this in case $d = n + 3$ in Exercise 1.19; a similarly explicit answer for $d > n + 3$ is not known in general.

Example 1.7. Hypersurfaces

A *hypersurface* X is a subvariety of $\mathbb{P}^n$ described as the zero locus of a single homogeneous polynomial $F(Z_0, \ldots, Z_n)$—for example, a plane curve or a surface in 3-space $\mathbb{P}^3$. (We will see in Lecture 11 that in fact any variety of dimension $n - 1$ in $\mathbb{P}^n$ is a hypersurface in this sense—but that will have to wait until we have defined the notion of dimension at least; see the discussion at the end of this lecture.)

Note that any hypersurface X is the zero locus of a polynomial F without repeated prime factors. With this restriction F will be unique up to scalar multiplication. (To see this we need the Nullstellensatz (Theorem 5.1). In particular, it requires that our base field be algebraically closed; for example, we would not want to consider a single point like $(0, 0, \ldots, 0) \in \mathbb{R}^n$ to be a hypersurface, even though it is the zero locus of the single polynomial $\sum x_i^2$.) When this is done the degree of F is called the *degree* of the hypersurface X. In Lecture 15, we will define a notion of degree for an arbitrary variety $X \subset \mathbb{P}^n$ that generalizes the two cases mentioned so far; for the time being, note that our two definitions agree on the overlap of their domains, that is, hypersurfaces in $\mathbb{P}^1$.

Example 1.8. Analytic Subvarieties and Submanifolds

This is not so much an example as a theorem that we should mention (without proof, certainly) at this point. To begin with, observe that since polynomials $f(z_1, \ldots, z_n) \in \mathbb{C}[z_1, \ldots, z_n]$ with complex coefficients are holomorphic functions of their variables $z_1, \ldots, z_n$, an algebraic variety X in $\mathbb{A}^n_{\mathbb{C}}$ or $\mathbb{P}^n_{\mathbb{C}}$ will be in particular a complex analytic subvariety of these complex manifolds (i.e., a subset given locally as the zero locus of holomorphic functions). Notably, this gives us an a priori notion of the dimension of an algebraic variety $X \subset \mathbb{P}^n_{\mathbb{C}}$, and likewise of smooth and singular points of X. These are not satisfactory definitions of these concepts from the algebraic point of view, so we will not rely on them, but we will occasionally invoke them implicitly, as, for example, when we refer to a variety as a "curve."

The theorem we should quote here is the famous converse to this, in the case of subsets of projective space.

Theorem 1.9. (Chow's Theorem). *If $X \subset \mathbb{P}^n_{\mathbb{C}}$ is any complex analytic subvariety then X is an algebraic subvariety.*

Note that this is certainly false if we replace $\mathbb{P}^n_{\mathbb{C}}$ by $\mathbb{A}^n_{\mathbb{C}}$; for example, the subset $\mathbb{Z} \subset \mathbb{C} \cong \mathbb{A}^1_{\mathbb{C}}$ of integers is an analytic subvariety. See [S2] for a thorough discussion of this and related theorems.

Example 1.10. The Twisted Cubic

This is everybody's first example of a concrete variety that is not a hypersurface, linear space, or finite set of points. It is defined to be the image C of the map $\nu: \mathbb{P}^1 \to \mathbb{P}^3$ given in terms of affine coordinates on both spaces by

$$\nu: x \mapsto (x, x^2, x^3);$$

or, in terms of homogeneous coordinates on both, as

$$\nu: [X_0, X_1] \mapsto [X_0^3, X_0^2 X_1, X_0 X_1^2, X_1^3] = [Z_0, Z_1, Z_2, Z_3].$$

C lies on the 3 quadric surfaces Q_0, Q_1, and Q_2 given as the zero locus of the polynomials

$$F_0(Z) = Z_0 Z_2 - Z_1^2,$$

$$F_1(Z) = Z_0 Z_3 - Z_1 Z_2, \qquad \text{and}$$

$$F_2(Z) = Z_1 Z_3 - Z_2^2,$$

respectively, and is equal to their intersection: if we have a point $p \in \mathbb{P}^3$ with coordinates $[Z_0, Z_1, Z_2, Z_3]$ satisfying these three polynomials, either Z_0 or Z_3 must be nonzero; if the former we can write $p = \nu([Z_0, Z_1])$ and if the latter we can write $p = \nu([Z_2, Z_3])$. At the same time, C is not the intersection of any two of these quadrics: according to the following Exercise Q_i and Q_j will intersect in the union of C and a line l_{ij}.

Exercise 1.11. a. Show that for any $0 \leq i < j \leq 2$, the surfaces Q_i and Q_j intersect in the union of C and a line L.

b. More generally, for any $\lambda = [\lambda_0, \lambda_1, \lambda_2]$, let

$$F_\lambda = \lambda_0 \cdot F_0 + \lambda_1 \cdot F_1 + \lambda_2 \cdot F_2$$

and let Q_λ be the surface defined by F_λ. Show that for $\mu \neq \nu$, the quadrics Q_ν and Q_μ intersect in the union of C and a line $L_{\mu,\nu}$. (A slick way of doing this problem is described after Exercise 9.16; it is intended here to be done naively, though the computation is apt to get messy.)

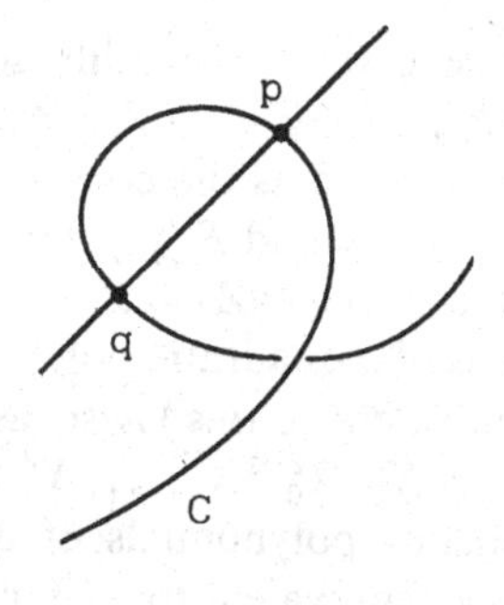

In fact, the lines that arise in this way form an interesting family. For example, we may observe that any chord of the curve C (that is, any line $\overline{pq}$ joining points of the curve) arises in this way. To see this, let $r \in \overline{pq}$ be any point other than p and q. In the three-dimensional vector space of polynomials F_λ vanishing on C there will be a two-dimensional subspace of those vanishing at r; say this subspace is spanned by F_μ and F_ν. But these quadrics all vanish at three points of the line $\overline{pq}$, and so vanish identically on this line; from Exercise 1.11 we deduce that $Q_\mu \cap Q_\nu = C \cup \overline{pq}$.

One point of terminology: while we often speak of the twisted cubic curve and have described a particular curve $C \subset \mathbb{P}^3$, in fact we call any curve $C' \subset \mathbb{P}^3$ projectively equivalent to C, that is, any curve given parametrically as the image of the map

$$[X] \mapsto [A_0(X), A_1(X), A_2(X), A_3(X)]$$

where A_0, A_1, A_2, A_3 form a basis for the space of homogeneous cubic polynomials in $X = [X_0, X_1]$, a twisted cubic.

Exercise 1.12. Show that any finite set of points on a twisted cubic curve are in general position, i.e., any four of them span $\mathbb{P}^3$.

In Theorem 1.18, we will see that given any six points in $\mathbb{P}^3$ in general position there is a unique twisted cubic containing all six.

Exercise 1.13. Show that if seven points $p_1, \ldots, p_7 \in \mathbb{P}^3$ lie on a twisted cubic, then the common zero locus of the quadratic polynomials vanishing at the p_i is that twisted cubic. (From this we see that the statement of Theorem 1.4 is sharp, at least in case $n = 3$.)

Example 1.14. Rational Normal Curves

These may be thought of as a generalization of twisted cubics; the rational normal curve $C \subset \mathbb{P}^d$ is defined to be the image of the map

$$v_d\colon \mathbb{P}^1 \to \mathbb{P}^d$$

given by

$$v_d\colon [X_0, X_1] \mapsto [X_0^d, X_0^{d-1}X_1, \ldots, X_1^d] = [Z_0, \ldots, Z_d].$$

The image $C \subset \mathbb{P}^d$ is readily seen to be the common zero locus of the polynomials $F_{i,j}(Z) = Z_i Z_j - Z_{i-1} Z_{j+1}$ for $1 \leq i \leq j \leq d - 1$. Note that for $d > 3$ it may also be expressed as the common zeros of a subset of these: the polynomials $F_{i,i}$, $1 = 1, \ldots, d - 1$ and $F_{1,d-1}$, for example. (Note also that in case $d = 2$ we get the plane conic curve $Z_0 Z_2 = Z_1^2$; in fact, it's not hard to see that any plane conic curve (zero locus of a quadratic polynomial on $\mathbb{P}^2$) other than a union of lines is projectively equivalent to this.) Also, as in the case of the twisted cubic, if we replace the monomials $X_0^d, X_0^{d-1}X_1, \ldots, X_1^d$ with an arbitrary basis $A_0, \ldots, A_d$ for the space of homogeneous polynomials of degree d on $\mathbb{P}^1$, we get a map whose image is projectively equivalent to $v_d(\mathbb{P}^1)$; we call any such curve a rational normal curve.

Note that any $d + 1$ points of a rational normal curve are linearly independent. This is tantamount to the fact that the Van der Monde determinant vanishes only if two of its rows coincide. We will see later that the rational normal curve is the unique curve with this property. (The weaker fact that no three points of a rational normal curve are collinear also follows from the fact that C is the zero locus of quadratic polynomials.)

Exercise 1.15. Show that if $p_1, \ldots, p_{kd+1}$ are any points on a rational normal curve in $\mathbb{P}^d$, then any polynomial F of degree k on $\mathbb{P}^d$ vanishing on the points p_i vanishes on C. Use this to show that the general statement given in Exercise 1.5 is sharp.

Example 1.16. Determinantal Representation of the Rational Normal Curve

One convenient (and significant) way to express the equations defining a rational normal curve is as the 2×2 minors of a matrix of homogeneous linear forms. Specifically, for any integer k between 1 and $d - 1$, the rational normal curve may be described as the locus of points $[Z_0, \ldots, Z_d] \in \mathbb{P}^d$ such that the rank of the matrix

$$\begin{pmatrix} Z_0 & Z_1 & Z_2 & \cdot & \cdot & Z_{k-1} & Z_k \\ Z_1 & Z_2 & \cdot & \cdot & \cdot & \cdot & Z_{k+1} \\ Z_2 & \cdot & \cdot & \cdot & \cdot & \cdot & \cdot \\ \cdot & \cdot & \cdot & \cdot & \cdot & \cdot & \cdot \\ \cdot & \cdot & \cdot & \cdot & \cdot & \cdot & Z_{d-1} \\ Z_{d-k} & \cdot & \cdot & \cdot & \cdot & Z_{d-1} & Z_d \end{pmatrix}$$

is 1. In general, a variety $X \subset \mathbb{P}^n$ whose equations can be represented in this way is called *determinantal*; we will see other examples of such varieties throughout these lectures and will study them in their own right in Lecture 9.

We will see in Exercise 1.25 that in case $k = 1$ or $d - 1$ we can replace the entries of this matrix with more general linear forms $L_{i,j}$, and unless these forms satisfy a nontrivial condition of linear dependence the resulting variety will again be a rational normal curve; but this is not the case for general k.

Example 1.17. Another Parametrization of the Rational Normal Curve

There is another way of representing a rational normal curve parametrically. It is based on the observation that if $G(X_0, X_1)$ is a homogeneous polynomial of degree $d + 1$, with distinct roots (i.e., $G(X_0, X_1) = \prod(\mu_i X_0 - \nu_i X_1)$ with $[\mu_i, \nu_i]$ distinct in $\mathbb{P}^1$), then the polynomials $H_i(X) = G(X)/(\mu_i X_0 - \nu_i X_1)$ form a basis for the space of homogeneous polynomials of degree d: if there were a linear relation $\sum a_i H_i(X_0, X_1) = 0$, then plugging in $(X_0, X_1) = (\nu_i, \mu_i)$ we could deduce that $a_i = 0$. Thus the map

$$\nu_d: [X_0, X_1] \mapsto [H_1(X_0, X_1), \ldots, H_{d+1}(X_0, X_1)]$$

has as its image a rational normal curve in $\mathbb{P}^d$. Dividing the homogeneous vector on the right by the polynomial G, we may write this map as

$$\nu_d: [X_0, X_1] \mapsto \left[\frac{1}{\mu_1 X_0 - \nu_1 X_1}, \ldots, \frac{1}{\mu_{d+1} X_0 - \nu_{d+1} X_1}\right].$$

Note that this rational normal curve passes through each of the coordinate points of $\mathbb{P}^d$, sending the zeros of G to these points. In addition, if all μ_i and v_i are nonzero the points 0 and ∞ (that is, [1, 0] and [0, 1]) go to the points $[\mu_1^{-1}, \ldots, \mu_{d+1}^{-1}]$ and $[v_1^{-1}, \ldots, v_{d+1}^{-1}]$, which may be any points not on the coordinate hyperplanes. Conversely, any rational normal curve passing through all $d + 1$ coordinate points may be written parametrically in this way. We may thus deduce the following theorem.

Theorem 1.18. *Through any $d + 3$ points in general position in $\mathbb{P}^d$ there passes a unique rational normal curve.*

We can use Theorem 1.18 to answer the question posed after Exercise 1.6, namely, when two subsets of $n + 3$ points in general position in $\mathbb{P}^n$ are projectively equivalent. The answer is straightforward, but cute: through $n + 3$ points $p_1, \ldots, p_{n+3} \in \mathbb{P}^n$ there passes a unique rational normal curve $v_n(\mathbb{P}^1)$, so that we can associate to the points $p_1, \ldots, p_{n+3} \in \mathbb{P}^n$ the set of $n + 3$ points $q_i = v_n^{-1}(p_i) \in \mathbb{P}^1$. We have then the following.

Exercise 1.19. Show that the points $p_i \in \mathbb{P}^n$ are projectively equivalent as an ordered set to another such collection $\{p'_1, \ldots, p'_{n+3}\}$ p'_i if and only if the corresponding ordered subsets $q_1, \ldots, q_{n+3}$ and $q'_1, \ldots, q'_{n+3} \in \mathbb{P}^1$ are projectively equivalent, that is, if and only if the cross-ratios $\lambda(q_1, q_2, q_3, q_i) = \lambda(q'_1, q'_2, q'_3, q'_i)$ for each $i = 4, \ldots, n + 3$. (You may use the characterization of the cross-ratio given on page 7.)

Example 1.20. The Family of Plane Conics

There is another way to see Theorem 1.18 in the special case $d = 2$. In this case, we observe that a rational normal curve C of degree 2 is specified by giving a homogeneous quadratic polynomial $Q(Z_0, Z_1, Z_2)$ (not a product of linear forms); Q is determined up to multiplication by scalars by C. Thus the set of such curves may be identified with a subset of the projective space $\mathbb{P}V = \mathbb{P}^5$ associated to the vector space

$$V = \{aZ_0^2 + bZ_1^2 + cZ_2^2 + dZ_0Z_1 + eZ_0Z_2 + fZ_1Z_2\}$$

of quadratic polynomials. In general, we call an element of this projective space a *plane conic curve*, or simply *conic*, and a rational normal curve—that is, a point of PV corresponding to an irreducible quadratic polynomial—a *smooth conic.* We then note that the subset of conics passing through a given point $p = [Z_0, Z_1, Z_2]$ is a hyperplane in $\mathbb{P}^5$, and since any five hyperplanes in $\mathbb{P}^5$ must have a common intersection (equivalently, five linear equations in six unknowns have a nonzero solution), there exists a conic curve through any five given points $p_1, \ldots, p_5$. If the points p_i are in general position, moreover, this cannot be a union of two lines.

Exercise 1.21. Check that the hyperplanes in $\mathbb{P}^5$ associated in this way to five points $p_1, \ldots, p_5$ with no four collinear are independent (i.e., meet in a single point), establishing uniqueness. (In classical language, $p_1, \ldots, p_5$ "impose independent conditions on conics.")

The description of the set of conics as a projective space $\mathbb{P}^5$ is the first example we will see of a *parameter space*, a notion ubiquitous in the subject; we will introduce parameter spaces first in Lecture 4 and discuss them in more detail in Lecture 21.

Example 1.22. A Synthetic Construction of the Rational Normal Curve

Finally, we should mention here a synthetic (and, on at least one occasion, useful) construction of rational normal curves. We start with the case of a conic, where the construction is quite simple. Let P and Q be two points in the plane $\mathbb{P}^2$. Then the lines passing through each point are naturally parameterized by $\mathbb{P}^1$ (e.g., if P is given as the zeros on two linear forms $L(Z) = M(Z) = 0$, the lines through P are of the form $\lambda L(Z) + \mu M(Z) = 0$ with $[\lambda, \mu] \in \mathbb{P}^1$). Thus the lines through P may be put in one-to-one correspondence with the lines through Q. Choose any bijection obtained in this way, subject only to the condition that the line $\overline{PQ}$ does not correspond to itself, so that corresponding lines will always intersect in a point. (Note that this rules out the simplest way of putting the lines through P and Q in correspondence: choosing an auxiliary line L and using it to parameterize the lines through both P and Q, that is, for each $R \in L$ making the lines $\overline{PR}$ and $\overline{QR}$ correspond.) We claim then that the locus of points of intersection of corresponding lines is a conic curve, and that conversely any conic may be obtained in this way.

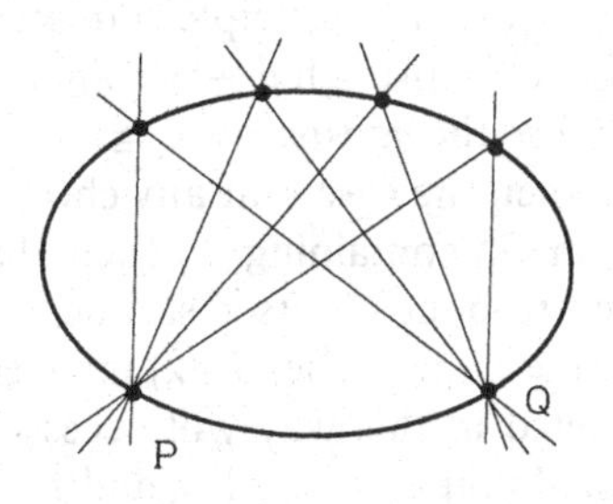

You could say this is not really a synthetic construction, inasmuch as the bijection between the families of lines through P and through Q was specified analytically. In the classical construction, the two families of lines were each parametrized by an auxiliary line in $\mathbb{P}^2$, which were then put in one-to-one correspondence by the family of lines through an auxiliary point. The construction was thus: choose points P, Q, and R not collinear, and distinct lines L and M not passing through any of these points in $\mathbb{P}^2$ and such that the point $L \cap M$ does not lie on the line $\overline{PQ}$. For every line N through R, let S_N be the point of intersection of the line L_N joining P to $N \cap L$ and the line M_N joining Q to $N \cap M$. Then the locus of the points S_N is a conic curve.

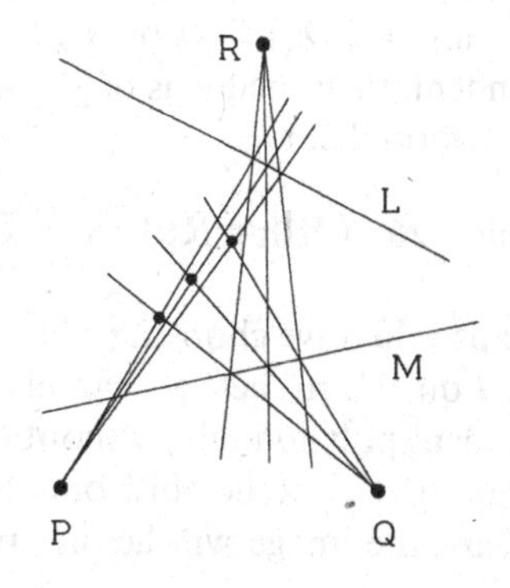

Exercise 1.23. Show that the locus constructed in this way is indeed a smooth conic curve, and that it does

pass through the points P, Q, and $L \cap M$, and through the points $\overline{RQ} \cap L$ and $\overline{RP} \cap M$. Using this, show one more time that through five points in the plane, no three collinear, there passes a unique conic curve.

As indicated, we can generalize this to a construction of rational normal curves in any projective space $\mathbb{P}^d$. Specifically, start by choosing d codimension two linear spaces $\Lambda_i \cong \mathbb{P}^{d-2} \subset \mathbb{P}^d$. The family $\{H_i(\lambda)\}$ of hyperplanes in $\mathbb{P}^d$ containing Λ_i is then parameterized by $\lambda \in \mathbb{P}^1$; choose such parameterizations, subject to the condition that for each λ the planes $H_1(\lambda), \ldots, H_d(\lambda)$ are independent, i.e., intersect in a point $p(\lambda)$. It is then the case that the locus of these points $p(\lambda)$ as λ varies in $\mathbb{P}^1$ is a rational normal curve.

Exercise 1.24. Verify the last statement.

We can use this description to see once again that there exists a unique rational normal curve through $d + 3$ points in $\mathbb{P}^d$ no $d + 1$ of which are dependent. To do this, choose the subspaces $\Lambda_i \cong \mathbb{P}^{d-2} \subset H$ be the span of the points $P_1, \ldots, \hat{P}_i, \ldots, P_d$. It is then the case that any choice of parametrizations of the families of hyperplanes in $\mathbb{P}^d$ containing Λ_i such that the hyperplane H spanned by $p_1, \ldots, p_d$ never corresponds to itself satisfies the independence condition, i.e., for each value $\lambda \in \mathbb{P}^1$, the hyperplanes $H_i(\lambda)$ intersect in a point $p(\lambda)$. The rational normal curve constructed in this way will necessarily contain the d points P_i; and given three additional points P_{d+1}, P_{d+2}, and P_{d+3} we can choose our parameterizations of the families of planes through the Λ_i so that the planes containing P_{d+1}, P_{d+2}, and P_{d+3} correspond to the values $\lambda = 0$, 1, and $\infty \in \mathbb{P}^1$, respectively.

Exercise 1.25. As we observed in Example 1.16, the rational normal curve $X \subset \mathbb{P}^d$ may be realized as the locus of points $[Z_0, \ldots, Z_d]$ such that the matrix

$$\begin{pmatrix} Z_0 & Z_1 & Z_2 & \cdot & \cdot & Z_{d-2} & Z_{d-1} \\ Z_1 & Z_2 & \cdot & \cdot & \cdot & \cdot & Z_d \end{pmatrix}$$

has rank 1. Interpret this as an example of the preceding construction: take Λ_i to be the plane $(Z_{i-1} = Z_i = 0)$, and $H_i(\lambda)$ the hyperplane $(\lambda_1 Z_{i-1} + \lambda_2 Z_i = 0)$. Generalize this to show that if $(L_{i,j})$ is any $2 \times d$ matrix of linear forms on $\mathbb{P}^d$ such that for any $(\lambda_1, \lambda_2) \neq (0, 0)$ the linear forms $\{\lambda_1 L_{1,j} + \lambda_2 L_{2,j}\}$, $j = 1, \ldots, d$ are independent, then the locus of $[Z] \in \mathbb{P}^d$ such that the matrix $L_{i,j}(Z)$ has rank 1 is a rational normal curve.

Example 1.26. Other Rational Curves

The maps v_d involve choosing a basis for the space of homogeneous polynomials of degree d on $\mathbb{P}^1$. In fact, we can also choose any collection $A_0, \ldots, A_m$ of linearly independent polynomials (without common zeros) and try to describe the image of the resulting map (if the polynomials we choose fail to be linearly independent, that just means the image will lie in a proper linear subspace of the target space $\mathbb{P}^m$).

For example, consider the case $d = 3$ and $m = 2$ and look at the maps

$$\mu, \nu: \mathbb{P}^1 \to \mathbb{P}^2$$

given by

$$\mu: [X_0, X_1] \mapsto [X_0^3, X_0X_1^2, X_1^3]$$

and

$$\nu: [X_0, X_1] \mapsto [X_0^3, X_0X_1^2 - X_0^3, X_1^3 - X_0^2X_1].$$

The images of these two maps are both cubic hypersurfaces in $\mathbb{P}^2$, given by the equations $Z_0Z_2^2 = Z_1^3$ and $Z_0Z_2^2 = Z_1^3 + Z_0Z_1^2$, respectively; in Euclidean coordinates, they are just the cuspidal cubic curve $y^2 = x^3$ and the nodal cubic $y^2 = x^3 + x^2$.

$y^2 = x^3$

Exercise 1.27. Show that the images of μ and ν are in fact given by these cubic polynomials.

$y^2 = x^3 + x^2$

Exercise 1.28. Show that the image of the map $\nu: \mathbb{P}^1 \to \mathbb{P}^2$ given by any triple of homogeneous cubic polynomials $A_i(X_0, X_1)$ without common zeroes satisfies a cubic polynomial. (In fact, any such image is projectively equivalent to one of the preceding two, a fact we will prove in Exercise 3.8 and again after Exercise 10.10.) (∗)

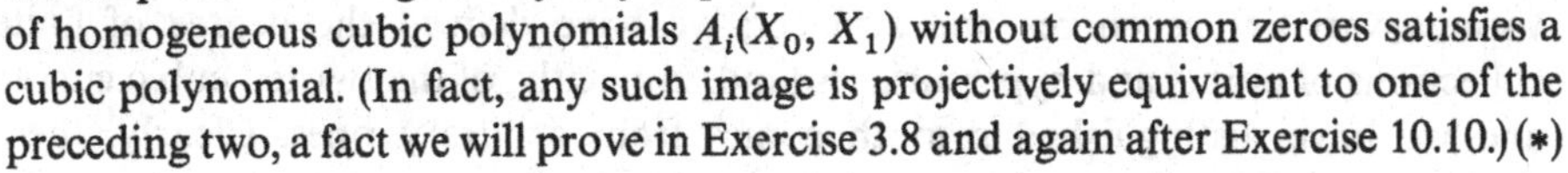

For another example, in which we will see (though we may not be able to prove that we have) a continuously varying family of non-projectively equivalent curves, consider the case $d = 4$ and $m = 3$ and look at the map

$$\nu_{\alpha,\beta}: \mathbb{P}^1 \to \mathbb{P}^3$$

given by

$$\nu_{\alpha,\beta}: [X_0, X_1]$$
$$\mapsto [X_0^4 - \beta X_0^3X_1, X_0^3X_1 - \beta X_0^2X_1^2, \alpha X_0^2X_1^2 - X_0X_1^3, \alpha X_0X_1^3 - X_1^4].$$

The images $C_{\alpha,\beta} \subset \mathbb{P}^3$ of these maps are called *rational quartic curves* in $\mathbb{P}^3$. The following exercise is probably hard to do purely naively, but will be easier after reading the next lecture.

Exercise 1.29. Show that $C_{\alpha,\beta}$ is indeed an algebraic variety, and that it may be described as the zero locus of one quadratic and two cubic polynomials.

We will see in Exercise 2.19 that the curves $C_{\alpha,\beta}$ give a continuously varying family of non-projectively equivalent curves.

Example 1.30. Varieties Defined Over Subfields of K

This is not really an example as much as it is a warning about terminology.

First of all, if $L \subset K$ is a subfield, we will write $\mathbb{A}^n(L)$ for the subset $L^n \subset K^n = \mathbb{A}^n_K$. Similarly, by $\mathbb{P}^n(L) \subset \mathbb{P}^n_K$ we will mean the subset of points $[Z_0, \ldots, Z_n]$ with $Z_i/Z_j \in L$ whenever defined—that is, points that may be written as $[Z_0, \ldots, Z_n]$ with $Z_i \in L$. All of what follows applies to projective varieties, but we will say it only in the context of affine ones.

We say that a subvariety $X \subset \mathbb{A}^n_K$ is *defined over* L if it is the zero locus of polynomials $f_\alpha(z_1, \ldots, z_n) \in L[z_1, \ldots, z_n]$. For such a variety X, the set of *points of* X *defined over* L is just the intersection $X \cap \mathbb{A}^n(L)$. We should not, however, confuse the set of points of X defined over L with X itself; for example, the variety in $\mathbb{A}^2_{\mathbb{C}}$ defined by the equation $x^2 + y^2 + 1 = 0$ is defined over $\mathbb{R}$ and has no points defined over $\mathbb{R}$, but it is not the empty variety.

A Note on Dimension, Smoothness, and Degree

We have, as we noted in the Introduction, a dilemma. Already in the first lecture we have encountered on a number of occasions references to three basic notions in algebraic geometry: dimension, degree, and smoothness. We have referred to various varieties as curves and surfaces; we have defined the degree of finite collections of points and hypersurfaces (and, implicitly, of the twisted cubic curve); and we have distinguished smooth conics from arbitrary ones. Clearly, these three ideas are fundamental to the subject; they give structure and focus to our analysis of varieties. Their formal definitions, however, have to be deferred until we have introduced a certain amount of technical apparatus, definitions, and foundational theorems. At the same time, I feel it is desirable to introduce as many examples as possible before or at the same time as the introduction of this apparatus.

The bottom line is that we have, to some degree, a vicious cycle: examples (by choice) come before definitions and foundational theorems, which come (of necessity) before the introduction and use of notions like dimension, smoothness, and degree, which in turn play a large role in the analysis of examples.

What do we do about this? First, we do have naive ideas of what notions like dimension and smoothness should represent, and I would ask the reader's forbearance if occasionally I refer to them. (It will not, I hope, upset the reader if I refer to the zero locus of a single polynomial in $\mathbb{P}^3$ as a "surface," even before the formal definition of dimension.) Secondly, when we introduce examples in Part I, I would encourage the reader, whenever interested, to skip around and look ahead to the analyses in Part 2 of their dimension, degree, smoothness, and/or tangent spaces.

LECTURE 2

Regular Functions and Maps

In the preceding lecture, we introduced the basic objects of the category we will be studying; we will now introduce the maps. As might be expected, this is extremely easy in the context of affine varieties and slightly trickier, at least at first, for projective ones.

The Zariski Topology

We begin with a piece of terminology that will be useful, if somewhat uncomfortable at first. The *Zariski topology* on a variety X is simply the topology whose closed sets are the subvarieties of X, i.e., the common zero loci of polynomials on X. Thus, for $X \subset \mathbb{A}^n$ affine, a base of open sets is given by the sets $U_f = \{p \in X: f(p) \neq 0\}$, where f ranges over polynomials; these are called the *distinguished* open subsets of X. Similarly, for $X \subset \mathbb{P}^n$ projective, a basis is given by the sets $U_F = \{p \in X: F(p) \neq 0\}$ for F a homogeneous polynomial; again, these open subsets are called distinguished.

This is the topology we will use on all the varieties with which we deal, so that if we refer to an open subset of a variety X without further specification, we will mean the complement of a subvariety. Implicit in our use of this topology is a fundamentally important fact: inasmuch as virtually all the constructions of algebraic geometry may be defined algebraically and make sense for varieties over any field, the ordinary topology on $\mathbb{P}^n_{\mathbb{C}}$ (or, as it's called, the *classical* or *analytic* topology) is not logically relevant. At the same time, we have to emphasize that the Zariski topology is primarily a formal construct; it is more a matter of terminology than a reflection of the geometry of varieties. For example, all plane curves given by irreducible polynomials over simply uncountable algebraically closed fields—

whether affine or projective and whatever their degree or the field in question—are homeomorphic; they are, as topological spaces, simply uncountable sets given the topology in which subsets are closed if and only if they are finite. More generally, note that the Zariski topology does not satisfy any of the usual separation axioms, inasmuch as any two open subsets of $\mathbb{P}^n$ will intersect. In sum, then, we will find it convenient to express most of the following in the language of the Zariski topology, but when we close our eyes and try to visualize an algebraic variety, it is probably the classical topology we should picture.

Note that the Zariski topology is what is called a *Noetherian topology*; that is, if $Y_1 \supset Y_2 \supset Y_3 \supset \cdots$ is any chain of closed subsets of a variety X, then for some m we have $Y_m = Y_{m+1} = \cdots$. This is equivalent to the statement that in the polynomial ring $K[Z_0, \ldots, Z_n]$ any ideal is finitely generated, a special case of the theorem that $K[Z_0, \ldots, Z_n]$ is a Noetherian ring (cf. [E], [AM]).

We should also mention at this point one further bit of terminology: an open subset $U \subset X$ of a projective variety $X \subset \mathbb{P}^n$ is called a *quasi-projective variety* (equivalently, a quasi-projective variety is a *locally closed* subset of $\mathbb{P}^n$ in the Zariski topology). The class of quasi-projective varieties includes both affine and projective varieties and is much larger (we will see in Exercise 2.3 the simplest example of a quasi-projective variety that is not isomorphic to either an affine or a projective variety); but in practice most of the varieties with which we actually deal will be either projective or affine.

By way of usage, when we speak of a *variety* X without further specification, we will mean a quasi-projective variety. When we speak of a *subvariety* X of a variety Y or of "a variety $X \subset Y$", however, we will always mean a closed subset.

It should be mentioned here that there is some disagreement in the literature over the definition of the terms "variety" and "subvariety": in many sources varieties are required to be irreducible (see Lecture 5) and in others a subvariety $X \subset Y$ is defined to be any locally closed subset.

Regular Functions on an Affine Variety

Let $X \subset \mathbb{A}^n$ be a variety. We define the *ideal* of X to be the ideal

$$I(X) = \{f \in K[z_1, \ldots, z_n]: f \equiv 0 \text{ on } X\}$$

of functions vanishing on X; and we define the *coordinate ring* of X to be the quotient

$$A(X) = K[z_1, \ldots, z_n]/I(X).$$

We now come to a key definition, that of a *regular function* on the variety X. Ultimately, we would like a regular function on X to be simply the restriction to X of a polynomial in $z_1, \ldots, z_n$, modulo those vanishing on X, that is, an element of the coordinate ring $A(X)$. We need, however, to give a local definition, so that we can at the same time describe the ring of functions on an open subset $U \subset X$. We therefore define the following.

Definition. Let $U \subset X$ be any open set and $p \in U$ any point. We say that a function f on U is *regular at* p if in some neighborhood V of p it is expressible as a quotient g/h, where g and $h \in K[z_1, \ldots, z_n]$ are polynomials and $h(p) \neq 0$. We say that f is *regular on* U if it is regular at every point of U.

That this definition behaves as we desire is the content of the following lemma.

Lemma 2.1. *The ring of functions regular at every point of* X *is the coordinate ring* $A(X)$. *More generally, if* $U = U_f$ *is a distinguished open subset, then the ring of regular functions on* U *is the localization* $A(X)[1/f]$.

The proof of this lemma requires some additional machinery. In particular, it is clear that in order to prove it we have to know that a polynomial $h(z_1, \ldots, z_n)$ that is nowhere zero on X is a unit in $A(X)$, which is part of the Nullstellensatz (Theorem 5.1); we will give its proof following the proof of the Nullstellensatz in Lecture 5. Note that it is essential in this definition and lemma that our base field K be algebraically closed; if it were not, we could have a function $f = g/h$ where h was nowhere zero on $\mathbb{A}^n_K$ but did have zeroes on $\mathbb{A}^n_{\bar{K}}$ (for example, the function $1/(x^2 + 1)$ on $\mathbb{A}^1(\mathbb{R})$) and we would not want to call such a function regular.

We should also warn that the conclusion of Lemma 2.1—that any function f regular on U is expressible as a quotient g/h with h nowhere zero on U—is false for general open sets $U \subset X$.

Note that the distinguished open subsets U_f are themselves naturally affine varieties: if $X \subset \mathbb{A}^n$ is the zero locus of polynomials $f_\alpha(z_1, \ldots, z_n)$, then the locus $\Sigma \subset \mathbb{A}^{n+1}$ given by the polynomials f_α—viewed formally as polynomials in $z_1, \ldots, z_{n+1}$—together with the polynomial

$$g(z_1, \ldots, z_{n+1}) = 1 - z_{n+1} \cdot f(z_1, \ldots, z_n)$$

is bijective to U_f. (Note that the coordinate ring of Σ is exactly the ring $A(X)[1/f]$ of regular functions on U_f.) Thus, for example, if $X = \mathbb{A}^1$ and U_f is the open subset $\mathbb{A}^1 - \{1, -1\}$, we may realize U_f as the subvariety $\Sigma \subset \mathbb{A}^2$ given by the equation $w(z^2 - 1) - 1 = 0$, as in the diagram.

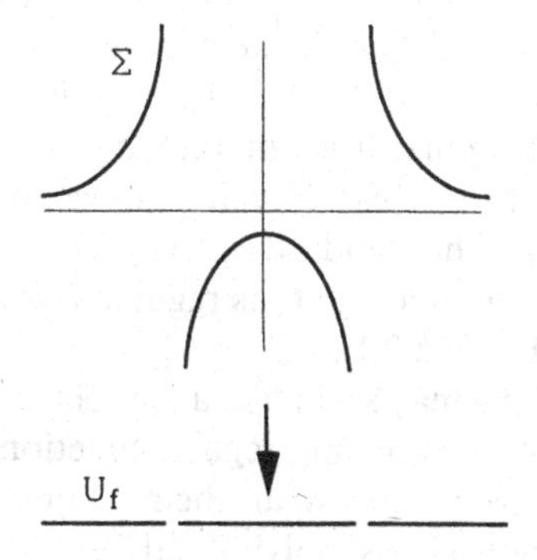

Exercise 2.2. What is the ring of regular functions on the complement $\mathbb{A}^2 - \{(0, 0)\}$ of the origin in $\mathbb{A}^2$?

We can recover an affine variety X (though not any particular embedding $X \hookrightarrow \mathbb{A}^n$) from its coordinate ring $A = A(X)$: just choose a collection of generators $x_1, \ldots, x_n$ for A over K, write

$$A = K[x_1, \ldots, x_n]/(f_1(x), \ldots, f_m(x)),$$

and take $X \subset \mathbb{A}^n$ as the zero locus of the polynomials f_α. More intrinsically, given the Nullstellensatz (Theorem 5.1), the points of X may be identified with the set of maximal ideals in the ring A: for any $p \in X$, the ideal $\mathfrak{m}_p \subset A$ of functions vanishing at p is maximal. Conversely, we will see in the proof of Proposition 5.18 that given any maximal ideal $\mathfrak{m}$ in the ring $A = K[x_1, \ldots, x_n]/(\{f_\alpha\})$, the quotient $A/\mathfrak{m}$ will be a field finitely generated as a K-algebra and hence isomorphic to K. If we then let a_i be the image of $x_i \in A$ under the quotient map

$$\varphi: A \to A/\mathfrak{m} \cong K,$$

the point $p = (a_1, \ldots, a_n)$ will lie on X, and $\mathfrak{m}$ will be the ideal of functions vanishing at p.

We will see as a consequence of the Nullstellensatz (Theorem 5.1) that in fact any finitely generated algebra over K will occur as the coordinate ring of an affine variety if and only if it has no nilpotent elements.

One further object we should introduce here is the *local ring of an affine variety* X at a point $p \in X$, denoted $\mathcal{O}_{X,p}$. This is defined to be the ring of germs of functions defined in some neighborhood of p and regular at p. This is the direct limit of the rings $A(U_f) = A(X)[1/f]$ where f ranges over all regular functions on X nonzero at p, or in other words the localization of the ring $A(X)$ with respect to the maximal ideal $\mathfrak{m}_p$. Note that if $Y \subset X$ is an open set containing p then $\mathcal{O}_{X,p} = \mathcal{O}_{Y,p}$.

Projective Varieties

There are analogous definitions for projective varieties $X \subset \mathbb{P}^n$. Again, we define the ideal of X to be the ideal of polynomials $F \in K[Z_0, \ldots, Z_n]$ vanishing on X; note that this is a homogeneous ideal, i.e., it is generated by homogeneous polynomials (equivalently, it is the direct sum of its homogeneous pieces). We likewise define the *homogeneous coordinate ring* $S(X)$ of X to be the quotient ring $K[Z_0, \ldots, Z_n]/I(X)$; this is again a graded ring.

A regular function on a quasi-projective variety $X \subset \mathbb{P}^n$—or more generally on an open subset $U \subset X$—is defined to be a function that is locally regular, i.e., if $\{U_i\}$ is the standard open cover of $\mathbb{P}^n$ by open sets $U_i \cong \mathbb{A}^n$, such that the restriction of f to each $U \cap U_i$ is regular. (Note that this is independent of the choice of cover by Lemma 2.1.)

This may seem like a cumbersome definition, and it is. In fact, there is a simpler way of expressing regular functions on an open subset of a projective variety: we can sometimes write them as quotients F/G, where F and $G \in K[Z_0, \ldots, Z_n]$ are homogeneous polynomials of the same degree with G nowhere zero in U. In particular, by an argument analogous to that for Lemma 2.1, if $U = U_G \subset X$ is the complement of the zero locus of the homogeneous polynomial G, then the ring of regular functions on U_G is exactly the 0th graded piece of the localization $S(X)[G^{-1}]$.

Finally, we may define the local ring $\mathcal{O}_{X,p}$ of a quasi-projective variety $X \subset \mathbb{P}^n$ at a point $p \in X$ just as we did in the affine case: as the ring of germs of functions regular in some neighborhood of X. Equivalently, if $\tilde{X}$ is any affine open subset of

X containing p, we may take $\mathcal{O}_{X,p} = \mathcal{O}_{\tilde{X},p}$, i.e., the localization of the coordinate ring $A(\tilde{X})$ with respect to the ideal of functions vanishing at p.

Regular Maps

Maps to affine varieties are simple to describe: a *regular map* from an arbitrary variety X to affine space $\mathbb{A}^n$ is a map given by an n-tuple of regular functions on X; and a map of X to an affine variety $Y \subset \mathbb{A}^n$ is a map to $\mathbb{A}^n$ with image contained in Y. Equivalently, such maps correspond bijectively to ring homomorphisms from the coordinate ring $A(Y)$ to the ring of regular functions on X.

This gives us a notion of isomorphism of affine varieties: two affine varieties X and Y are *isomorphic* if there exist regular maps $\eta: X \to Y$ and $\varphi: Y \to X$ inverse to one another in both directions, or equivalently if their coordinate rings $A(X) \cong A(Y)$ as algebras over the field K (in particular, the coordinate ring of an affine variety is an invariant of isomorphism).

Maps to projective space are naturally more complicated. To start with, we say that a map $\varphi: X \to \mathbb{P}^n$ is regular if it is regular locally, i.e., it is continuous and for each of the standard affine open subsets $U_i \cong \mathbb{A}^n \subset \mathbb{P}^n$ the restriction of φ to $\varphi^{-1}(U_i)$ is regular. Actually specifying a map to projective space by giving its restrictions to the inverse images of affine open subsets, however, is far too cumbersome. A better way of describing a map to projective space would be to specify an $(n + 1)$-tuple of regular functions, but this may not be possible if X is projective. If $X \subset \mathbb{P}^m$ is projective, we may specify a map of X to $\mathbb{P}^n$ by giving an $(n + 1)$-tuple of homogeneous polynomials of the same degree; as long as they are not simultaneously zero anywhere on X, this will determine a regular map. It happens, though, that this still does not suffice to describe all maps of projective varieties to projective space.

As an example of this, consider the variety $C \subset \mathbb{P}^2$ given by $X^2 + Y^2 - Z^2$, and the map φ of C to $\mathbb{P}^1$ given by

$$[X, Y, Z] \mapsto [X, Z - Y].$$

The map may be thought of as a stereographic projection from the point $p = [0, 1, 1]$: it sends a point $r \in C$ (other than p itself) to the point of intersection of the axis ($Y = 0$) with the line $\overline{pr}$. The two polynomials X and $Z - Y$ have a common zero on C at the point $p = [0, 1, 1]$, reflecting the fact that this assignment does not make sense at $r = p$; but the map is still regular (or rather extends to a regular map) at this point: we define $\varphi(p) = [1, 0]$ and observe that in terms of coordinates

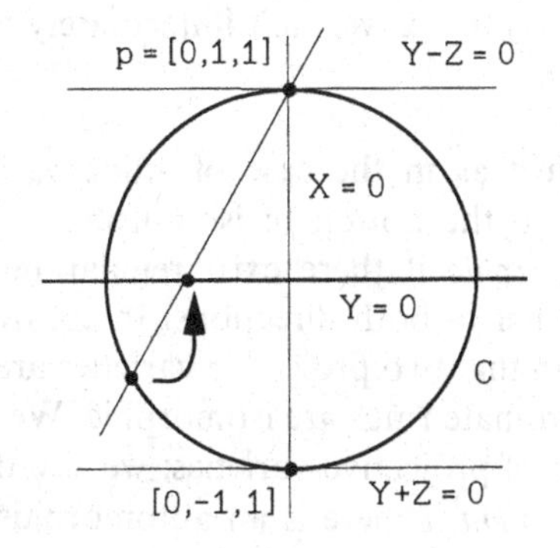

$[S, T]$ on $\mathbb{P}^1$ with affine opens $U_0\colon (S \neq 0)$ and $U_1\colon (T \neq 0)$ we have

$$\varphi^{-1}(U_0) = C - \{[0, -1, 1]\}$$

and

$$\varphi^{-1}(U_1) = C - \{[0, 1, 1]\}.$$

Now, on $\varphi^{-1}(U_1)$, the map φ is clearly regular; in terms of the coordinate $s = S/T$ on U_1, the restriction of φ is given by

$$[X, Y, Z] \mapsto \frac{X}{Z - Y},$$

which is clearly a regular function on $C - \{[0, 1, 1]\}$. On the other hand, on $\varphi^{-1}(U_0)$, we can write the map, in terms of the Euclidean coordinate $t = T/S$, as

$$[X, Y, Z] \mapsto \frac{Z - Y}{X}.$$

This may not appear to be regular at p, but we can write

$$\begin{aligned}\frac{Z-Y}{X} &= \frac{Z^2 - Y^2}{X(Y+Z)} \\ &= \frac{X^2}{X(Y+Z)} \\ &= \frac{X}{Y+Z},\end{aligned}$$

which is clearly regular on $C - \{[0, -1, 1]\}$. We note as well that the map $\varphi\colon C \to \mathbb{P}^1$ in fact cannot be given by a pair of homogeneous polynomials on $\mathbb{P}^2$ without common zeros on C.

This example is fairly representative: in practice, the most common way of specifying in coordinates a map $\varphi\colon X \to \mathbb{P}^n$ of a quasi-projective variety to projective space is by an $(n + 1)$-tuple of homogeneous polynomials of the same degree. The drawback of this is that we have to allow the possibility that these homogeneous polynomials have common zeros on X; and having written down such an $(n + 1)$-tuple, we can't immediately tell whether we have in fact defined a regular map.

Just as in the case of affine varieties, the definition of a regular map gives rise to the notion of isomorphism: two quasi-projective varieties X and Y are *isomorphic* if there exist regular maps $\eta\colon X \to Y$ and $\varphi\colon Y \to X$ inverse to one another in both directions. In contrast to the affine case, however, this does not mean that two projective varieties are isomorphic if and only if their homogeneous coordinate rings are isomorphic. We have, in other words, two notions of congruence of projective varieties: we say that two varieties $X, X' \subset \mathbb{P}^n$ are *projectively equivalent* if there is an automorphism $A \in \mathrm{PGL}_{n+1}K$ of $\mathbb{P}^n$ carrying X onto X', which is the same as saying that the homogeneous coordinate rings $S(X)$, $S(X')$ are isomorphic as graded K-algebras; while we say that they are isomorphic under the weaker condition that there is a biregular map between them. (We will see an explicit example where the two notions do not agree in Exercise 2.10.)

Exercise 2.3. Using the result of Exercise 2.2, show that for $n \geq 2$ the complement of the origin in $\mathbb{A}^n$ is not isomorphic to an affine variety.

Example 2.4. The Veronese Map

The construction of the rational normal curve can be further generalized: for any n and d, we define the *Veronese map* of degree d

$$v_d\colon \mathbb{P}^n \to \mathbb{P}^N$$

by sending

$$[X_0, \ldots, X_n] \mapsto [\ldots X^I \ldots],$$

where X^I ranges over all monomials of degree d in $X_0, \ldots, X_n$. As in the case of the rational normal curves, we will call the Veronese map any map differing from this by an automorphism of $\mathbb{P}^N$. Geometrically, the Veronese map is characterized by the property that the hypersurfaces of degree d in $\mathbb{P}^n$ are exactly the hyperplane sections of the image $v_d(\mathbb{P}^n) \subset \mathbb{P}^N$. It is not hard to see that the image of the Veronese map is an algebraic variety, often called a *Veronese variety.*

Exercise 2.5. Show that the number of monomials of degree d in $n + 1$ variables is the binomial coefficient $\binom{n+d}{d}$, so that the integer N is $\binom{n+d}{d} - 1$.

For example, in the simplest case other than the case $n = 1$ of the rational normal curve, the quadratic Veronese map

$$v_2\colon \mathbb{P}^2 \to \mathbb{P}^5$$

is given by

$$v_2\colon [X_0, X_1, X_2] \mapsto [X_0^2, X_1^2, X_2^2, X_0X_1, X_0X_2, X_1X_2].$$

The image of this map, often called simply the *Veronese surface,* is one variety we will encounter often in the course of this book.

The Veronese variety $v_d(\mathbb{P}^n)$ lies on a number of obvious quadric hypersurfaces: for every quadruple of multi-indices I, J, K, and L such that the corresponding monomials $X^I X^J = X^K X^L$, we have a quadratic relation on the image. In fact, it is not hard to check that the Veronese variety is exactly the zero locus of these quadratic polynomials.

Example 2.6. Determinantal Representation of Veronese Varieties

The Veronese surface, that is, the image of the map $v_2: \mathbb{P}^2 \to \mathbb{P}^5$, can also be described as the locus of points $[Z_0, \ldots, Z_5] \in \mathbb{P}^5$ such that the matrix

$$\begin{pmatrix} Z_0 & Z_3 & Z_4 \\ Z_3 & Z_1 & Z_5 \\ Z_4 & Z_5 & Z_2 \end{pmatrix}$$

has rank 1. In general, if we let $\{Z_{i,j}\}_{0 \le i \le j \le n}$ be the coordinates on the target space of the quadratic Veronese map

$$v_2: \mathbb{P}^n \to \mathbb{P}^{(n+1)(n+2)/2-1},$$

then we can represent the image of v_2 as the locus of the 2×2 minors of the $(n+1) \times (n+1)$ symmetric matrix with (i, j)th entry $Z_{i-1,j-1}$ for $i \le j$.

Example 2.7. Subvarieties of Veronese Varieties

The Veronese map may be applied not only to a projective space $\mathbb{P}^n$, but to any variety $X \subset \mathbb{P}^n$ by restriction. Observe in particular that if we restrict v_d to a linear subspace $\Lambda \cong \mathbb{P}^k \subset \mathbb{P}^n$, we get just the Veronese map of degree d on $\mathbb{P}^k$. For example, the images under the map $v_2: \mathbb{P}^2 \to \mathbb{P}^5$ of lines in $\mathbb{P}^2$ give a family of conic plane curves on the Veronese surface S, with one such conic passing through any two points of S.

More generally, we claim that the image of a variety $Y \subset \mathbb{P}^n$ under the Veronese map is a subvariety of $\mathbb{P}^N$. To see this, note first that homogeneous polynomials of degree k in the homogeneous coordinates Z on $\mathbb{P}^N$ pull back to give (all) polynomials of degree $d \cdot k$ in the variables X. Next, observe (as in the remark following Exercise 1.3) that the zero locus of a polynomial $F(X)$ of degree m is also the common zero locus of the polynomials $\{X_i F(X)\}$ of degree $m+1$. Thus a variety $Y \subset \mathbb{P}^n$ expressible as the common zero locus of polynomials of degree m and less may also be realized as the common zero locus of polynomials of degree exactly $k \cdot d$ for some k. It follows that its image $v_d(Y) \subset \mathbb{P}^N$ under the Veronese map is the intersection of the Veronese variety $v_d(\mathbb{P}^n)$—which we have already seen is a variety—with the common zero locus of polynomials of degree k.

For example, if $Y \subset \mathbb{P}^2$ is the curve given by the cubic polynomial $X_0^3 + X_1^3 + X_2^3$, then we can also write Y as the common locus of the quartics

$$X_0^4 + X_0X_1^3 + X_0X_2^3, \quad X_0^3X_1 + X_1^4 + X_1X_2^3, \quad \text{and} \quad X_0^3X_2 + X_1^3X_2 + X_2^4.$$

The image $v_2(Y) \subset \mathbb{P}^5$ is thus the intersection of the Veronese surface with the three quadric hypersurfaces

$$Z_0^2 + Z_1Z_3 + Z_2Z_4, \quad Z_0Z_3 + Z_1^2 + Z_2Z_5, \quad \text{and } Z_0Z_4 + Z_1Z_5 + Z_2^2.$$

In particular, it is the intersection of nine quadrics.

Exercise 2.8. Let $X \subset \mathbb{P}^n$ be a projective variety and $Y = v_d(X) \subset \mathbb{P}^N$ its image under the Veronese map. Show that X and Y are isomorphic, i.e., show that the inverse map is regular.

Exercise 2.9. Use the preceding analysis and exercise to deduce that any projective variety is isomorphic to an intersection of a Veronese variety with a linear space (and hence in particular that any projective variety is isomorphic to an intersection of quadrics).

Exercise 2.10. Let $X \subset \mathbb{P}^n$ be a projective variety and $Y = v_d(X) \subset \mathbb{P}^N$ its image under the Veronese map. What is the relation between the homogeneous coordinate rings of X and Y?

In case the field K has characteristic zero, Veronese map has a coordinate-free description that is worth bearing in mind. Briefly, if we view $\mathbb{P}^n = \mathbb{P}V$ as the space lines in a vector space V, then the Veronese map may be defined as the map

$$v_d: \mathbb{P}V \to \mathbb{P}(\mathrm{Sym}^d V)$$

to the projectivization of the dth symmetric power of V, given by

$$v_d: [v] \mapsto [v^d].$$

Equivalently, if we apply this to V^* rather than V, the image of the Veronese map may be viewed as the (projectivization of the) subset of the space $\mathrm{Sym}^d V^*$ of all polynomials on V consisting of dth powers of linear forms. Note that this is false for fields K of arbitrary characteristic: for example, if $\mathrm{char}(K) = p$, the locus in $\mathbb{P}(\mathrm{Sym}^p V)$ of pth powers of elements of V is not a rational normal curve, but a line. What is true in arbitrary characteristic is that the Veronese map v_d may be viewed as the map $\mathbb{P}V \to \mathbb{P}(\mathrm{Sym}^d V)$ sending a vector v to the linear functional on $\mathrm{Sym}^d V^*$ given by evaluation of polynomials at v.

Example 2.11. The Segre Maps

Another fundamental family of maps are the *Segre maps*

$$\sigma: \mathbb{P}^n \times \mathbb{P}^m \to \mathbb{P}^{(n+1)(m+1)-1}$$

defined by sending a pair $([X], [Y])$ to the point in $\mathbb{P}^{(n+1)(m+1)-1}$ whose coordinates are the pairwise products of the coordinates of $[X]$ and $[Y]$, i.e.,

$$\sigma: ([X_0, \ldots, X_n], [Y_0, \ldots, Y_m]) \mapsto [\ldots, X_i Y_j, \ldots],$$

where the coordinates in the target space range over all pairwise products of coordinates X_i and Y_j.

It is not hard to see that the image of the Segre map is an algebraic variety, called a *Segre variety*, and sometimes denoted $\Sigma_{n,m}$: if we label the coordinates on the target space as $Z_{i,j}$, we see that it is the common zero locus of the quadratic polynomials $Z_{i,j} \cdot Z_{k,l} - Z_{i,l} \cdot Z_{k,j}$. (In particular, the Segre variety is another example of a *determinantal variety*; it is the zero locus of the 2×2 minors of the matrix $(Z_{i,j})$.)

The first example of a Segre variety is the variety $\Sigma_{1,1} = \sigma(\mathbb{P}^1 \times \mathbb{P}^1) \subset \mathbb{P}^3$, that is, the image of the map

$$\sigma: ([X_0, X_1], [Y_0, Y_1]) \mapsto [X_0 Y_0, X_0 Y_1, X_1 Y_0, X_1 Y_1].$$

This is the locus of the single quadratic polynomial $Z_0 Z_3 - Z_1 Z_2$, that is, it is simply a quadric surface. Note that the fibers of the two projection maps from $\mathbb{P}^n \times \mathbb{P}^m$ to $\mathbb{P}^n$ and $\mathbb{P}^m$ are carried, under σ, into linear subspaces of $\mathbb{P}^{(n+1)(m+1)-1}$; in particular, the fibers of $\mathbb{P}^1 \times \mathbb{P}^1$ are carried into the families of lines $\{Z_1 = \lambda Z_0, Z_3 = \lambda Z_2\}$ and $\{Z_2 = \lambda Z_0, Z_3 = \lambda Z_1\}$. Note also that the description of the polynomial $Z_0 Z_3 - Z_1 Z_2$ as the determinant of the matrix

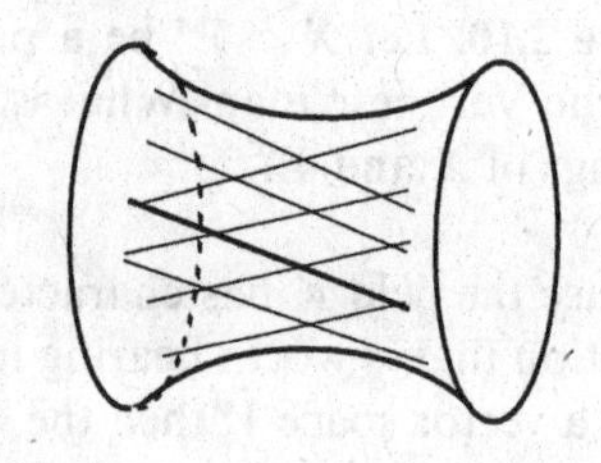

$$M = \begin{pmatrix} Z_0 & Z_1 \\ Z_2 & Z_3 \end{pmatrix}$$

displays the two families of lines nicely: one family consists of lines where the two columns satisfy a given linear relation, the other lines where the two rows satisfy a given linear relation.

Another common example of a Segre variety is the image

$$\Sigma_{2,1} = \sigma(\mathbb{P}^2 \times \mathbb{P}^1) \subset \mathbb{P}^5,$$

called the *Segre threefold*. We will encounter it again several times (for example, it is an example of a *rational normal scroll*, and as such is denoted $X_{1,1,1}$). For now, we mention the following facts.

Exercise 2.12. (i) Let L, M, and $N \subset \mathbb{P}^3$ be any three pairwise skew (i.e., disjoint) lines. Show that the union of the lines in $\mathbb{P}^3$ meeting all three lines is projectively equivalent to the Segre variety $\Sigma_{1,1} \subset \mathbb{P}^3$ and that this union is the unique Segre variety containing L, M, and N. (*)

(ii) More generally, suppose that L, M, and N are any three pairwise disjoint $(k-1)$-planes in $\mathbb{P}^{2k-1}$. Show that the union of all lines meeting L, M, and N is projectively equivalent to the Segre variety $\Sigma_{k-1,1} \subset \mathbb{P}^{2k-1}$ and that this union is the unique Segre variety containing L, M, and N. Is there an analogous description of Segre varieties $\Sigma_{a,b}$ with $a, b \geq 2$?

Exercise 2.13. Show that the twisted cubic curve $C \subset \mathbb{P}^3$ may be realized as the intersection of the Segre threefold with a three-plane $\mathbb{P}^3 \subset \mathbb{P}^5$.

Exercise 2.14. Show that any line $l \subset \Sigma_{2,1} \subset \mathbb{P}^5$ is contained in the image of a fiber of $\mathbb{P}^2 \times \mathbb{P}^1$ over $\mathbb{P}^2$ or $\mathbb{P}^1$ (*). (The same is true for any linear space contained in any Segre variety $\sigma(\mathbb{P}^n \times \mathbb{P}^m)$, but we will defer the most general statement until Theorem 9.22.)

Exercise 2.15. Show that the image of the diagonal $\Delta \subset \mathbb{P}^n \times \mathbb{P}^n$ under the Segre map is the Veronese variety $v_2(\mathbb{P}^n)$, lying in a subspace of $\mathbb{P}^{n^2+2n}$; deduce that in general the diagonal $\Delta_X \subset X \times X$ in the product of any variety with itself is a subvariety of that product, and likewise for all diagonals in the n fold product X^n.

Example 2.16. Subvarieties of Segre Varieties

Having given the product $\mathbb{P}^n \times \mathbb{P}^m$ the structure of a projective variety, a natural question to ask is how we may describe its subvarieties. A naive answer is immediate. To begin with, we say that a polynomial $F(Z_0, \ldots, Z_n, W_0, \ldots, W_m)$ in two sets of variables is *bihomogeneous of bidegree* (d, e) if it is simultaneously homogeneous of degree d in the first set of variables and of degree e in the second, that is, of the form

$$F(Z, W) = \sum_{\substack{I,J:\\ \Sigma i_\alpha = d,\\ \Sigma j_\beta = e}} a_{I,J} \cdot Z_0^{i_0} \ldots Z_n^{i_n} \cdot W_0^{j_0} \ldots W_m^{j_m}.$$

Now, since polynomials of degree d on the target projective space $\mathbb{P}^{(m+1)(n+1)-1}$ pull back to polynomials $F(Z, W)$ that are bihomogeneous of bidegree (d, d), the obvious answer is that subvarieties of $\mathbb{P}^n \times \mathbb{P}^m$ are simply the common zero loci of such polynomials (observe that the zero locus of any bihomogeneous polynomial is a well-defined subset of $\mathbb{P}^n \times \mathbb{P}^m$). At the same time, as in the discussion of subvarieties of the Veronese variety, we can see that the zero locus of a bihomogeneous polynomial $F(Z, W)$ of bidegree (d, e) is the common zero locus of the bihomogeneous polynomials of degree (d', e') divisible by it, for any $d' \geq d$ and $e' \geq e$; so that more generally we can say that the subvarieties of a Segre variety $\mathbb{P}^n \times \mathbb{P}^m$ are the zero loci of bihomogeneous polynomials of any bidegrees.

As an example, consider the twisted cubic $C \subset \mathbb{P}^3$ of Example 1.10 given as the image of the map

$$t \mapsto [1, t, t^2, t^3].$$

As we observed before, C lies on the quadric surface $Z_0Z_3 - Z_1Z_2 = 0$, which we now recognize as the Segre surface $S = \sigma(\mathbb{P}^1 \times \mathbb{P}^1) \subset \mathbb{P}^3$. Now, restrict to S the other two quadratic polynomials defining the twisted cubic. To begin with, the polynomial $Z_0Z_2 - Z_1^2$ on $\mathbb{P}^3$ pulls back to $X_0X_1Y_0^2 - X_0^2Y_1^2$, which factors into a product of X_0 and $F(X, Y) = X_1Y_0^2 - X_0Y_1^2$. The zero locus of this polynomial is thus the union of the twisted cubic with the line on S given by $X_0 = 0$ (or equivalently by $Z_0 = Z_1 = 0$). On the other hand, the polynomial $Z_1Z_3 - Z_2^2$ pulls back to $X_0X_1Y_1^2 - X_1^2Y_0^2$, which factors as $-X_1 \cdot F$; so its zero locus is the union of the curve C and the line $Z_2 = Z_3 = 0$. In sum, then, the twisted cubic curve is the zero locus of a single bihomogeneous polynomial $F(X, Y)$ of bidegree (1, 2) on the Segre surface $S = \sigma(\mathbb{P}^1 \times \mathbb{P}^1)$; the quadratic polynomials defining C restrict to S to give the bihomogeneous polynomials of bidegree (2, 2) divisible by F (equivalently, the quadric surfaces containing C cut on S the unions of C with the lines of one ruling of S).

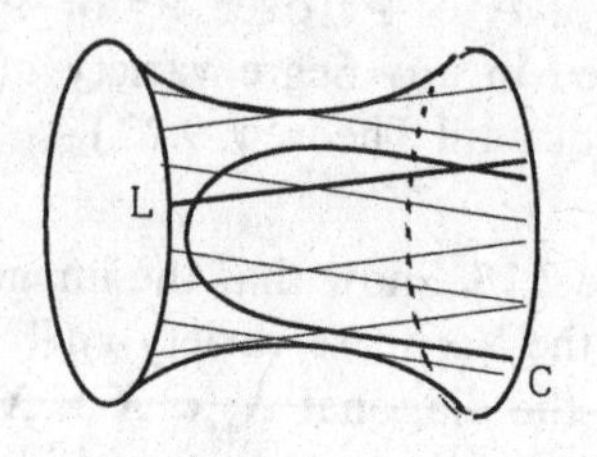

Exercise 2.17. Conversely, let $C \subset \mathbb{P}^1 \times \mathbb{P}^1$ be the zero locus of an irreducible bihomogeneous polynomial $F(X, Y)$ of bidegree (1, 2). Show that the image of C under the Segre map

$$\sigma: \mathbb{P}^1 \times \mathbb{P}^1 \to \mathbb{P}^3$$

is a twisted cubic curve.

Exercise 2.18. Now let $C = C_{\alpha,\beta} \subset \mathbb{P}^3$ be a rational quartic curve, as introduced in Example 1.26. Observe that C lies on the Segre surface S given by $Z_0Z_3 - Z_1Z_2 = 0$, that S is the unique quadric surface containing C, and that C is the zero locus of a bihomogeneous polynomial of bidegree (1, 3) on $S \cong \mathbb{P}^1 \times \mathbb{P}^1$. Use this to do Exercise 1.29.

Exercise 2.19. Use the preceding exercise to show in particular that there is a continuous family of curves $C_{\alpha,\beta}$ not projectively equivalent to one another. (*)

Exercise 2.20. a. Let $X \subset \mathbb{P}^n$ and $Y \subset \mathbb{P}^m$ be projective varieties. Show that the image $\sigma(X \times Y) \subset \sigma(\mathbb{P}^n \times \mathbb{P}^m) \subset \mathbb{P}^{nm+n+m}$ of the Segre map restricted to $X \times Y$ is a projective variety. b. Now suppose only that $X \subset \mathbb{P}^n$ and $Y \subset \mathbb{P}^m$ are quasi-projective. Show that $\sigma(X \times Y)$ is likewise quasi-projective, that is, locally closed in $\mathbb{P}^{nm+n+m}$.

Example 2.21. Products of Varieties

At the outset of Example 2.11, we referred to the product $\mathbb{P}^n \times \mathbb{P}^m$; we can only mean the product as a set. This space does not a priori have the structure of an algebraic variety. The Segre embedding, however, gives it one, which we will adopt as a definition of the product as a variety. In other words, when we talk about "the variety $\mathbb{P}^n \times \mathbb{P}^m$" we mean the image of the Segre map. Similarly, if $X \subset \mathbb{P}^n$ and $Y \subset \mathbb{P}^m$ are locally closed, according to Exercise 2.20, the image of the product $X \times Y \subset \mathbb{P}^n \times \mathbb{P}^m$ is a locally closed subset of $\mathbb{P}^{nm+n+m}$, which we will take as the definition of "the product $X \times Y$" as a variety.

A key point to be made in connection with this definition is that this is actually a *categorical product*, i.e., the projection maps $\pi_X: X \times Y \to X$ and $\pi_Y: X \times Y \to Y$ are regular and the variety $X \times Y$, together with these projection maps, satisfies the conditions for a product in the category of quasi-projective varieties and regular maps. What this means is that given an arbitrary variety Z and a pair of maps $\alpha: Z \to X$ and $\beta: Z \to Y$, there is a unique map $\alpha \times \beta: Z \to X \times Y$ whose compositions with the projections π_X and π_Y are α and β, respectively. It is not hard to see that this is the case: to begin with, the map $\alpha \times \beta$ is certainly uniquely determined by α and β; we have to check simply that it is regular. This we do locally: say $r_0 \in Z$ is any point, mapping via α and β to points $p \in X$ and $q \in Y$. Suppose p lies in the open set $Z_0 \neq 0$ in $\mathbb{P}^n$, so that in a neighborhood of r_0 the map α is given by

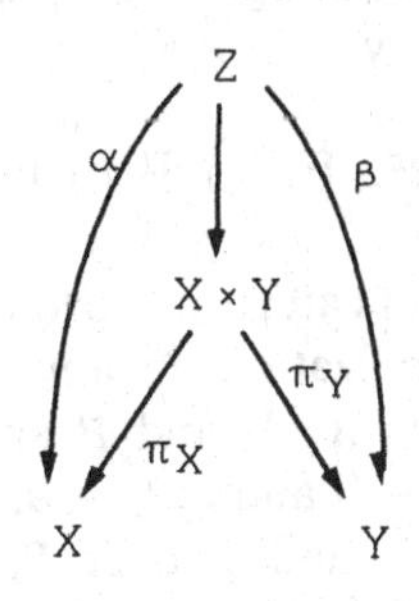

$$\alpha: r \mapsto [1, f_1(r), \ldots, f_n(r)],$$

with $f_1, \ldots, f_n$ regular functions of r on Z, and say β is given similarly by $\beta(r) = [1, g_1(r), \ldots, g_m(r)]$. Then in a neighborhood of r_0 the map $\alpha \times \beta: Z \to X \times Y \subset \mathbb{P}^{nm+n+m}$ is given by

$$\alpha \times \beta: r \mapsto [1, \ldots, f_i(r), \ldots, g_j(r), \ldots, f_i(r)g_j(r), \ldots]$$

and so is regular.

Exercise 2.22. Show that the Zariski topology on the product variety $X \times Y$ is *not* the product of the Zariski topologies on X and Y.

Example 2.23. Graphs

This is in some sense a subexample of Example 2.16, but it is important enough to warrant its own heading. The basic observation is contained in the following exercise.

Exercise 2.24. Let $X \subset \mathbb{P}^n$ be any projective variety and $\varphi: X \to \mathbb{P}^m$ any regular map. Show that the graph $\Gamma_\varphi \subset X \times \mathbb{P}^m \subset \mathbb{P}^n \times \mathbb{P}^m$ is a subvariety.

Note that it is not the case that a map $\varphi: X \to \mathbb{P}^m$ is regular if and only if the graph Γ_φ is a subvariety. For example, consider the map $\mu: \mathbb{P}^1 \to \mathbb{P}^2$ of Example 1.26 and denote by $X \subset \mathbb{P}^2$ its image, the cuspidal curve $Z_0 Z_2^2 = Z_1^3$. Inasmuch as the map μ is one to one, we can define a set-theoretic inverse map

$$\varphi: X \to \mathbb{P}^1;$$

as may be readily checked, this map is

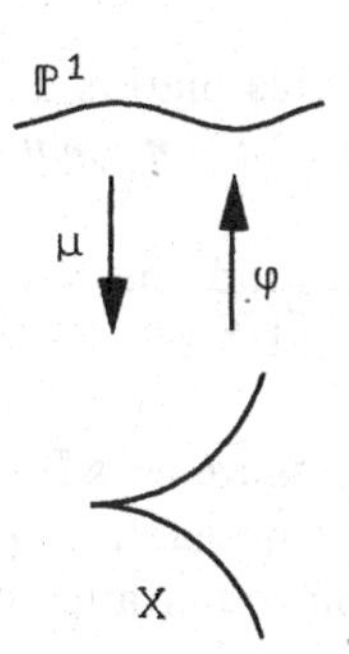

not regular, although its graph, being the same subset of $X \times \mathbb{P}^1$ as the graph of μ, is a subvariety of $X \times \mathbb{P}^1$.

Example 2.25. Fiber Products

The notion of categorical product has a direct generalization: in any category, given objects X, Y, and B and morphisms $\varphi: X \to B$ and $\eta: Y \to B$, the *fiber product* of X and Y over B, denoted $X \times_B Y$, is defined to be the object Z with morphisms $\alpha: Z \to X$ and $\beta: Z \to Y$ such that for any object W and maps $\alpha': W \to X$ and $\beta': W \to Y$ with $\varphi \circ \alpha' = \eta \circ \beta'$, there is a unique map $\gamma: W \to Z$ with $\alpha' = \alpha \circ \gamma$ and $\beta' = \beta \circ \gamma$. This property uniquely determines Z, if it exists. In the category of sets, fiber products exist; very simply, the fiber product is

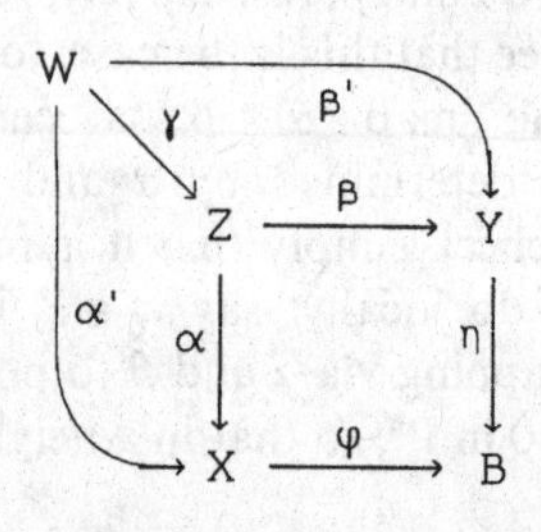

$$X \times_B Y = \{(x, y) \in X \times Y: \varphi(x) = \eta(y)\}.$$

What we may observe here is that the set-theoretic fiber product of two varieties may be given the structure of a variety in a reasonably natural way. To do this, let $\Gamma \subset X \times B$ be the graph of φ and $\Psi \subset Y \times B$ the graph of η. By Exercise 2.24, both $\Gamma \times Y$ and $\Psi \times X$ are subvarieties of the triple product $X \times Y \times B$, and hence so is their intersection, which is set-theoretically the fiber product. We will in the future refer to $X \times_B Y$ as the fiber product of X and Y over B.

Exercise 2.26. Show that the variety $X \times_B Y$ is indeed the fiber product of X and Y over B in the category of algebraic varieties.

Example 2.27. Combinations of Veronese and Segre Maps

We can combine the constructions of the Veronese and Segre maps to arrive at more varieties. To take the simplest case of this, let $v: \mathbb{P}^1 \to \mathbb{P}^2$ be the quadratic Veronese map, $\sigma: \mathbb{P}^1 \times \mathbb{P}^2 \to \mathbb{P}^5$ the Segre map, and consider the composition

$$\varphi: \mathbb{P}^1 \times \mathbb{P}^1 \xrightarrow{\mathrm{Id} \times v} \mathbb{P}^1 \times \mathbb{P}^2 \xrightarrow{\sigma} \mathbb{P}^5.$$

The image of this map is again an algebraic variety (in particular, it is another example of a *rational normal scroll* and is denoted $X_{2,2}$).

Exercise 2.28. Find the equations of the variety $X_{2,2}$. Show that the rational normal curve in $\mathbb{P}^4$ may be realized as a hyperplane section of $X_{2,2}$.

Exercise 2.29. Realize $\mathbb{P}^1 \times \mathbb{P}^1$ as the quadric surface Q in $\mathbb{P}^3$ given by the polynomial $Z_0Z_3 - Z_1Z_2$, and let $L \subset Q$ be the line $Z_0 = Z_1 = 0$. Show that the vector space of homogeneous quadratic polynomials in the Z_i vanishing on L has dimension 7, and show that the composite map φ may be given as

$$\varphi\colon [Z_0, Z_1, Z_2, Z_3] \mapsto [F_0(Z), \ldots, F_5(Z)],$$

where $\{Q, F_0(Z), \ldots, F_5(Z)\}$ is a basis for this vector space. Show that this map cannot be represented by a sixtuple of homogeneous polynomials of the same degree in $\mathbb{P}^3$ with no common zero locus on Q.

As with the Veronese map, there is a coordinate-free version of the Segre map: if we view the spaces $\mathbb{P}^n$ and $\mathbb{P}^m$ as the projective spaces associated to vector spaces V and W, respectively, then the target space $\mathbb{P}^{(n+1)(m+1)-1}$ may be naturally identified with the space $\mathbb{P}(V \otimes W)$, and the map

$$\sigma\colon \mathbb{P}V \times \mathbb{P}W \to \mathbb{P}(V \otimes W)$$

given as

$$\sigma\colon ([v], [w]) \mapsto [v \otimes w].$$

LECTURE 3

Cones, Projections, and More About Products

Example 3.1. Cones

We start here with a hyperplane $\mathbb{P}^{n-1} \subset \mathbb{P}^n$ and a point $p \in \mathbb{P}^n$ not lying on $\mathbb{P}^{n-1}$; if we like, we can take coordinates Z on $\mathbb{P}^n$ so that $\mathbb{P}^{n-1}$ is given by $Z_n = 0$ and the point $p = [0, \ldots, 0, 1]$. Let $X \subset \mathbb{P}^{n-1}$ be any variety. We then define the *cone* $\overline{X, p}$ over X with vertex p to be the union

$$\overline{X, p} = \bigcup_{q \in X} \overline{qp}$$

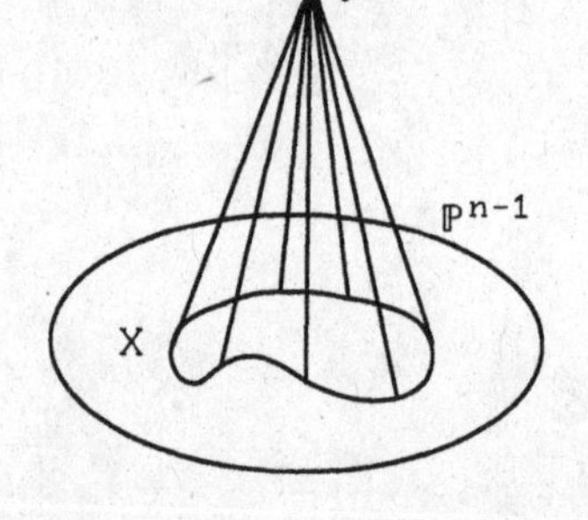

of the lines joining p to points of X. (If p lies on the hyperplane at infinity, $\overline{X, p}$ will look like a cylinder rather than a cone; in projective space these are the same thing.) $\overline{X, p}$ is easily seen to be a variety: if we choose coordinates as earlier and $X \subset \mathbb{P}^{n-1}$ is the locus of polynomials $F_\alpha = F_\alpha(Z_0, \ldots, Z_{n-1})$, the cone $\overline{X, p}$ will be the locus of the same polynomials F_α viewed as polynomials in $Z_0, \ldots, Z_n$.

As a slight generalization of the cone construction, let $\Lambda \cong \mathbb{P}^k \subset \mathbb{P}^n$ and $\Psi \cong \mathbb{P}^{n-k-1}$ be complementary linear subspaces (i.e., disjoint and spanning all of $\mathbb{P}^n$), and let $X \subset \Psi$ be any variety. We can then define the cone $\overline{X, \Lambda}$ over X with vertex Λ to be the union of the $(k + 1)$-planes $\overline{q, \Lambda}$ spanned by Λ together with points $q \in X$. Of course, this construction represents merely an iteration of the preceding one; we can also construct the cone $\overline{X, \Lambda}$ by taking the cone over X with vertex a point $k + 1$ times.

Exercise 3.2. Let Ψ and $\Lambda \subset \mathbb{P}^n$ be complementary linear subspaces as earlier, and $X \subset \Psi$ and $Y \subset \Lambda$ subvarieties. Show that the union of all lines joining points of X to points of Y is a variety.

In Lecture 8, we will see an analogous way of constructing a variety $\overline{X, Y}$ for any pair of varieties $X, Y \subset \mathbb{P}^n$.

Example 3.3. Quadrics

We can use the concept of cone to give a uniform description of quadric hypersurfaces, at least in case the characteristic of the field K is not 2. To begin with, a quadric hypersurface $Q \subset \mathbb{P}V = \mathbb{P}^n$ is given as the zero locus of a homogeneous quadratic polynomial Q: $V \to K$. Now assume that $\operatorname{char}(K) \neq 2$. The polynomial Q may be thought of as the quadratic form associated to a bilinear form Q_0 on V, that is, we may write

$$Q(v) = Q_0(v, v),$$

where Q_0: $V \times V \to K$ is defined by

$$Q_0(v, w) = \frac{Q(v + w) - Q(v) - Q(w)}{2}.$$

Note that Q_0 is both symmetric and bilinear. There is also associated to Q_0 the corresponding linear map

$$\tilde{Q}: V \to V^*$$

given by sending v to the linear form $Q(v, \cdot)$, i.e., by setting

$$\tilde{Q}(v)(w) = \tilde{Q}(w)(v) = Q_0(v, w).$$

Now, to classify quadrics, note that any quadric Q on a vector space V may be written, in terms of a suitably chosen basis, as

$$Q(X) = X_0^2 + X_1^2 + \cdots + X_k^2.$$

To see this, we choose the basis $e_0, \ldots, e_n$ for V as follows. First, we choose e_0 such that $Q(e_0) = 1$; then we choose $e_1 \in (Ke_0)^\perp$ (i.e., such that $Q_0(e_0, e_1) = 0$) such that $Q(e_1) = 1$, and so on, until Q vanishes identically on $(Ke_0 + \cdots + Ke_k)^\perp$. Finally, we may complete this to a basis with an arbitrary basis $e_{k+1}, \ldots, e_n$ for $(Ke_0 + \cdots + Ke_k)^\perp$. We say in this case that the quadric Q has *rank* $k + 1$; note that $k + 1$ is also the rank of the linear map $\tilde{Q}$. By this, a quadric is determined up to projective motion by its rank.

Note that as in Example 1.20, we are led to define a quadric hypersurface in general to be an equivalence class of nonzero homogeneous quadratic polynomials; two such polynomials are equivalent if they differ by multiplication by a scalar. The one additional object that this introduces into the class of quadrics is the *double plane*, that is, the quadric associated to the square $Q = L^2$ of a linear polynomial L.

If $\mathbb{P}^1$ there are two types of quadrics:

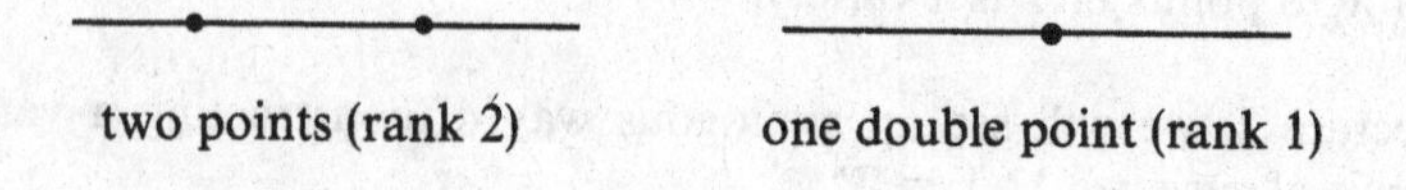

In $\mathbb{P}^2$, there are three:

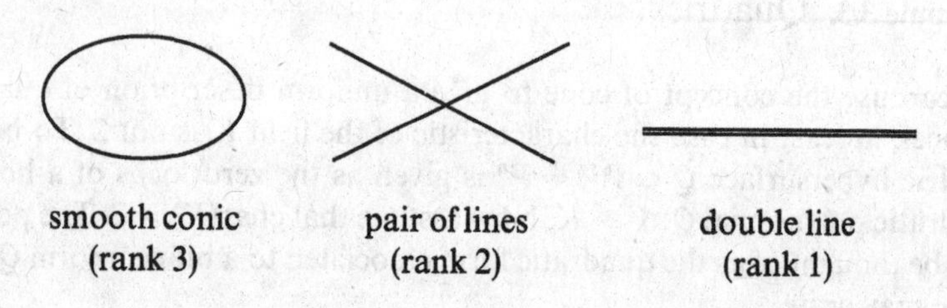

In $\mathbb{P}^3$ there are four:

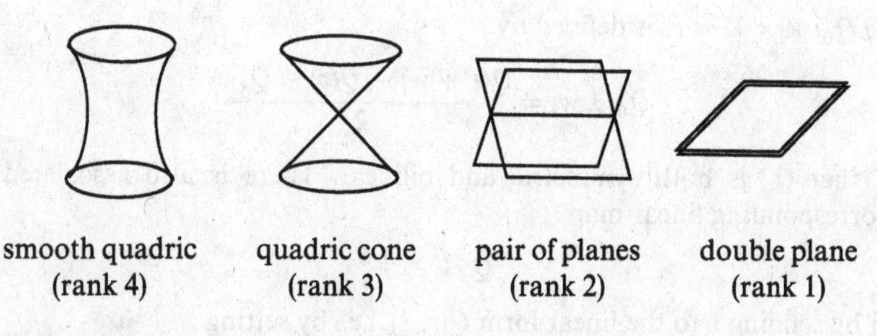

In general, we will call a quadric $Q \subset \mathbb{P}^n$ *smooth* if it has maximal rank $n + 1$, i.e., if the associated bilinear form Q_0 is nondegenerate (the reason for the term, if it is not already clear, will become so in Lecture 14). We have then the following geometric characterization: a quadric Q of rank $k \geq 2$ is the cone, with vertex $\Lambda \cong \mathbb{P}^{n-k}$, over a smooth quadric in $\overline{Q}$ in $\mathbb{P}^{k-1}$. To be specific, we can say that the vertex Λ is the subspace associated to the kernel of the map $\tilde{Q}$.

Example 3.4. Projections

We come now to a crucial example. Let the hyperplane $\mathbb{P}^{n-1} \subset \mathbb{P}^n$ and the point $p \in \mathbb{P}^n - \mathbb{P}^{n-1}$ be as in Example 3.1. We can then define a map

$$\pi_p \colon \mathbb{P}^n - \{p\} \to \mathbb{P}^{n-1}$$

by

$$\pi_p \colon q \mapsto \overline{qp} \cap \mathbb{P}^{n-1};$$

that is, sending a point $q \in \mathbb{P}^n$ other than p to the point of intersection of the line $\overline{pq}$ with the hyperplane $\mathbb{P}^{n-1}$. π_p is called *projection from the point p to the hyperplane* $\mathbb{P}^{n-1}$. In terms of coordinates Z used earlier, this is simple: we just set

$$\pi_p: [Z_0, \ldots, Z_n] \mapsto [Z_0, \ldots, Z_{n-1}].$$

Suppose now that X is any projective variety in $\mathbb{P}^n$ not containing the point p. We may then restrict the map π_p to the variety X to get a regular map $\pi_p: X \to \mathbb{P}^{n-1}$; the image $\overline{X} = \pi_p(X)$ of this map is called the *projection of* X *from* p *to* $\mathbb{P}^{n-1}$. We then have the following basic theorem.

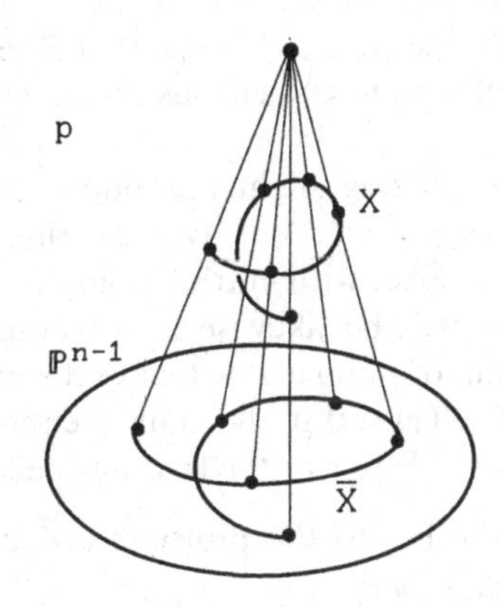

Theorem 3.5. *The projection* $\overline{X}$ *of* X *from* p *to* $\mathbb{P}^{n-1}$ *is a projective variety.*

Proof. The essential ingredient in this proof is elimination theory, which centers around the notion of the *resultant* of two polynomials in a variable z. We will take a moment out here to recall/describe this.

To begin with, suppose that $f(z)$ and $g(z)$ are two polynomials in a variable z with coefficients in a field K, of degrees m and n, respectively, and we ask whether they have a common factor. The answer is not hard: we observe simply that f and g will have a common factor if and only if there is a polynomial h of degree $m + n - 1$ divisible by both, i.e., if and only if the spaces of polynomials of degree $m + n - 1$ divisible by f and g individually meet nontrivially. This is equivalent to saying that the polynomials $f, z \cdot f, z^2 \cdot f, \ldots, z^{n-1} \cdot f, g, z \cdot g, \ldots, z^{m-1} \cdot g$ fail to be independent, or in other words that the determinant

$$\begin{vmatrix} a_0 & a_1 & . & . & a_m & 0 & 0 & . & . & . & 0 \\ 0 & a_0 & a_1 & . & . & a_m & 0 & . & . & . & 0 \\ . & & & & & & & & & & \\ . & & & & & & & & & & \\ 0 & 0 & . & . & a_0 & a_1 & . & . & . & . & a_m \\ b_0 & b_1 & . & . & . & b_n & 0 & . & . & . & 0 \\ 0 & b_0 & b_1 & . & . & . & b_n & 0 & . & . & 0 \\ \vdots & & & & & & & & & & \\ 0 & . & . & 0 & b_0 & b_1 & . & . & . & . & b_n \end{vmatrix}$$

of the $(m + n) \times (m + n)$ matrix of coefficients of these polynomials is zero. This determinant is called the resultant $R(f, g)$ of f and g with respect to z, and we may express our analysis as the following lemma.

Lemma 3.6. *Two polynomials f and g in one variable over a field K will have a common factor if and only if the resultant $R(f, g) = 0$.*

Note that if f and g are polynomials of degree strictly less than m and n, respectively—in other words, $a_m = b_n = 0$—then the determinant will also vanish. This corresponds to the fact that R really tests whether or not the homogenizations of f and g to homogeneous polynomials of degree m and n have a common zero in $\mathbb{P}^1$.

Generalizing slightly, suppose that f and g are polynomials in the variable z not over a field but over the ring $K[x_1, \ldots, x_n]$. We can still form the matrix of coefficients with entries a_i and b_j that are polynomials in $x_1, \ldots, x_n$; the determinant will be likewise a polynomial $R(f, g) \in K[x_1, \ldots, x_n]$, again called the resultant of f and g. It will have the property that for any n-tuple of elements $y_1, \ldots, y_n \in K$, $R(y) = 0$ if and only if either $f(y, z)$ and $g(y, z)$ have a common root as polynomials in z or the leading coefficient of both vanishes.

Returning to the projection $\overline{X}$ of X from p, suppose the projection map π_p is given as earlier by

$$\pi_p: [Z_0, \ldots, Z_n] \mapsto [Z_0, \ldots, Z_{n-1}].$$

The key point is that for any point $q = [Z_0, \ldots, Z_{n-1}] \in \mathbb{P}^{n-1}$, the line $l = \overline{pq}$ will meet X if and only if every pair of polynomials $F, G \in I(X)$ has a common zero on $l = \{[\alpha Z_0, \ldots, \alpha Z_{n-1}, \beta]\}$. To see this, observe that if l does not meet X, we can first find a polynomial F vanishing at finitely many points of l, and since not all $G \in I(X)$ vanish at any one of those points, we can find such a G nonzero at all of them. (Note that this statement would not be true if instead of taking all pairs $F, G \in I(X)$ we took only pairs of a set of generators.)

Now, for any pair F, G of homogeneous polynomials in $Z_0, \ldots, Z_n$, we may think of F and G as polynomials in Z_n with coefficients in $K[Z_0, \ldots, Z_{n-1}]$ and form the resultant with respect to Z_n accordingly (note that the degree of F and G in Z_n will be less than their homogeneous degree if they vanish at p). We denote this resultant by $R(F, G)$, noting that it is again homogeneous in $Z_0, \ldots, Z_{n-1}$. We have then for any point $q = [Z_0, \ldots, Z_{n-1}] \in \mathbb{P}^{n-1}$ the sequence of implications

the line $l = \overline{pq}$ meets X

$\Leftrightarrow$ every pair F, G of homogeneous polynomials in $I(X)$ has a common zero on l

$\Leftrightarrow$ the resultant $R(F, G)$ vanishes at q for all homogeneous pairs $F, G \in I(X)$.

In other words, the image $\overline{X}$ of the projection $\pi: X \to \mathbb{P}^{n-1}$ is the common zero locus of the polynomials $R(F, G)$, where F and G range over all pairs of homogeneous elements of $I(X)$. □

Exercise 3.7. Justify the first of the preceding implications, that is, show that if l is any line in $\mathbb{P}^n$ not meeting X we can find a pair of homogeneous polynomials $F, G \in I(X)$ with no common zeros on l.

Exercise 3.8. Find the equations of the projection of the twisted cubic curve from the point $[1, 0, 0, 1]$ and from $[0, 1, 0, 0]$. (Note that taking resultants may not be the most efficient way of doing this.) If you're feeling energetic, show that any projection of a twisted cubic from a point is projectively equivalent to one of these two.

The notion of projection may be generalized somewhat: if $\Lambda \cong \mathbb{P}^k$ is any subspace and $\mathbb{P}^{n-k-1}$ a complementary one, we can define a map

$$\pi_\Lambda\colon \mathbb{P}^n - \Lambda \to \mathbb{P}^{n-k-1}$$

by sending a point $q \in \mathbb{P}^n - \Lambda$ to the intersection of $\mathbb{P}^{n-k-1}$ with the $(k+1)$-plane $\overline{q, \Lambda}$. Again, for any $X \subset \mathbb{P}^n$ disjoint from Λ we may restrict to obtain a regular map π_Λ on X, whose image is called the *projection of X from Λ to $\mathbb{P}^{n-k-1}$*. Inasmuch as this map may also be realized as the composition of a sequence of projections from points $p_0, \ldots, p_k$ spanning Λ, Theorem 3.5 also implies that the projection $\pi_\Lambda(X) \subset \mathbb{P}^{n-k-1}$ is a variety.

Projection maps are readily expressed in intrinsic terms: if $\mathbb{P}^n$ is the projective space $\mathbb{P}V$ associated to an $(n+1)$-dimensional vector space V, $\Lambda = \mathbb{P}W$ corresponds to the $(k+1)$-plane $W \subset V$ and $\mathbb{P}^{n-k-1} = \mathbb{P}U$ for some $(n-k)$-plane $U \subset V$ complementary to W, then π_Λ is just the map associated to the projection $V = W \oplus U \to U$. In particular, note that if U' is another $(n-k)$-plane complementary to W, the projections to $\mathbb{P}U$ and to $\mathbb{P}U'$ will differ by composition with an isomorphism $\mathbb{P}U \cong \mathbb{P}U'$; thus the projection $\pi_\Lambda(X)$ of X from Λ to $\mathbb{P}^{n-k-1}$ will depend, up to projective equivalence, only on Λ and not on the subspace $\mathbb{P}^{n-k-1}$.

Exercise 3.9. Show that the curves $C_{\alpha,\beta}$ of Exercise 1.29 may be realized as projections of a rational normal curve in $\mathbb{P}^4$ from points $p_{\alpha,\beta} \in \mathbb{P}^4$, and use these to illustrate the point that projections $\pi_p(X)$, $\pi_{p'}(X)$ of a variety X from different points need not be projectively equivalent.

In fact, any regular map $\varphi\colon \mathbb{P}^n \to \mathbb{P}^m$ from one projective space to another may be realized, for some d, as the composition of the Veronese map $v_d\colon \mathbb{P}^n \to \mathbb{P}^N$ with a projection $\pi_\Lambda\colon \mathbb{P}^N \to \mathbb{P}^{m'}$ from a center Λ disjoint from the Veronese variety $v_d(\mathbb{P}^n) \subset \mathbb{P}^N$ and possibly an inclusion $\mathbb{P}^{m'} \subset \mathbb{P}^m$, though it would be difficult to prove that at this point.

Example 3.10. More Cones

We can also use projections to broaden our definition of cones, as follows. Suppose that $X \subset \mathbb{P}^n$ is any variety and $p \in \mathbb{P}^n$ any point not lying on X. Then the union

$$\overline{X, p} = \bigcup_{q \in X} \overline{qp},$$

which we will again call the *cone* over X with vertex p, is a variety; it is the cone, in the sense of Example 3.1, with vertex p over the projection $\overline{X} = \pi_p(X)$ of X from the point p to any hyperplane $\mathbb{P}^{n-1}$ not containing p. Similarly, for any k-plane $\Lambda \subset \mathbb{P}^n$ disjoint from X we can form the union of the $(k+1)$-planes

$\overline{\Lambda, q}$ spanned by Λ and points $q \in X$; this will be a variety called the cone over X with vertex Λ.

Example 3.11. More Projections

Consider for a moment a subvariety $X \subset Y \times \mathbb{P}^1$ where Y is an affine variety. Such a variety may be given as the zero locus of polynomials $F(W_0, W_1)$ homogeneous in the coordinates W_0, W_1 on $\mathbb{P}^1$, whose coefficients are regular functions on Y. The technique of elimination theory then tells us that the image $\overline{X}$ of X under the projection map $\pi_1: Y \times \mathbb{P}^1 \to Y$ is the zero locus of the resultants of all pairs of such polynomials; in particular, it is a closed subset of Y.

Next, suppose we have a subvariety X of $Y \times \mathbb{P}^2$ and would like to make the same statement. We can do this by choosing a point $p \in \mathbb{P}^2$ and first mapping $Y \times (\mathbb{P}^2 - \{p\})$ to $Y \times \mathbb{P}^1$, then projecting from $Y \times \mathbb{P}^1 \to Y$. This works except where X meets the locus $Y \times \{p\}$, i.e., if we let $V \subset Y$ be the closed subset $\{q \in Y: (q, p) \in X\}$ it shows that the image $\overline{X} = \pi_1(X) \subset Y$ intersects the open set $U = Y - V$ in a closed subset of U. But since $\overline{X}$ contains V, it follows that $\overline{X}$ is closed in Y. In general, this argument establishes the following theorem.

Theorem 3.12. *Let Y be any variety and $\pi: Y \times \mathbb{P}^n \to Y$ be the projection on the first factor. Then the image $\pi(X)$ of any closed subset $X \subset Y \times \mathbb{P}^n$ is a closed subset of Y.*

As an immediate consequence of this, we may combine it with Exercise 2.24 to deduce the following fundamental theorem.

Theorem 3.13. *If $X \subset \mathbb{P}^n$ is any projective variety and $\varphi: X \to \mathbb{P}^m$ any regular map, then the image of φ is a projective subvariety of $\mathbb{P}^m$.*

A regular function on a variety X may be thought of as a map $X \to \mathbb{A}^1 \subset \mathbb{P}^1$. Applying Theorem 3.13 to this map, we may deduce the following corollary.

Corollary 3.14. *If $X \subset \mathbb{P}^n$ is any connected variety and f any regular function on X, then f is constant.*

"Connected" here means not the disjoint union of two proper closed subsets. This in turn yields the following corollary.

Corollary 3.15. *If $X \subset \mathbb{P}^n$ is any connected variety other than a point and $Y \subset \mathbb{P}^n$ is any hypersurface then $X \cap Y \neq \varnothing$.*

PROOF. Let $F(X)$ be the homogeneous polynomial defining the hypersurface Y; say the degree of F is d. If $X \cap Y = \varnothing$, we can apply Corollary 3.14 to the regular functions G/F, where G ranges over homogeneous polynomials of degree d on $\mathbb{P}^n$, to deduce that X is a point. □

Constructible Sets

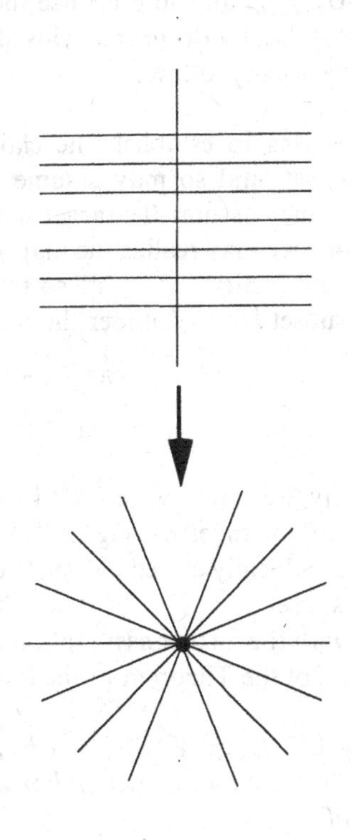

Naturally enough, Theorem 3.13 raises the question of what the image of an affine or quasi-projective variety X may be under a regular map $f: X \to \mathbb{P}^n$. The first thing to notice is that it does not have to be a quasi-projective variety. The primary example of this is the map $f: \mathbb{A}^2 \to \mathbb{A}^2$ given by

$$f(x, y) = (x, xy).$$

We note that under this map, horizontal lines are mapped into lines through the origin, with every line through the origin covered except for the vertical; vertical lines are mapped into themselves, except for the y-axis, which is collapsed to the point $(0, 0)$. The image is thus the union of the open subset $\{(z, w): z \neq 0\}$ with the origin, a set that is not locally closed at the origin.

Happily, this is about as bad as the situation gets; the images of quasi-projective varieties in general form a class of subsets of $\mathbb{P}^n$, called *constructible sets*, which look pretty much like what you'd expect on the basis of this example. A constructible set $Z \subset \mathbb{P}^n$ may be defined to be a finite disjoint union of locally closed subsets $U_i \subset \mathbb{P}^n$, that is, a set expressible as

$$Z = X_1 - (X_2 - (X_3 - \cdots - X_n))\ldots)$$

for $X_1 \supset X_2 \supset X_3 \supset \cdots \supset X_n$ a nested sequence of closed subsets of $\mathbb{P}^n$ (to see the correspondence between the definitions, take $U_1 = X_1 - X_2$, $U_2 = X_3 - X_4$ and so on). Equivalently, we may define the class of constructible subsets of $\mathbb{P}^n$ to be the smallest class including open subsets and closed under the operations of finite intersection and complementation.

The basic fact is that images of quasi-projective varieties under regular maps in general are constructible sets. In fact, it is not harder to prove that images of constructible sets are constructible.

Theorem 3.16. *Let $X \subset \mathbb{P}^m$ be a quasi-projective variety, $f: X \to \mathbb{P}^n$ a regular map, and $U \subset X$ any constructible set. Then $f(U)$ is a constructible subset of $\mathbb{P}^n$.*

PROOF. The key step in the proof is to establish an a priori weaker claim: that the image $f(U)$ contains a nonempty open subset $V \subset \overline{f(U)}$ of the closure of $f(U)$.

Given this, we set $U_1 = U \cap (X - f^{-1}(V))$ and observe that the theorem for U follows from the theorem for U_1. We then apply the claim to U_1 and define a closed subset $U_2 \subsetneqq U_1$ and so on; since the Zariski topology is Noetherian (see page 18), a chain of strictly decreasing closed subsets of a constructible set is finite, and the result eventually follows.

It remains to establish the claim. To begin with, we may replace U by an open subset, and so may assume that it is affine; restricting to a smaller affine open, we may assume the target space is also affine space. After replacing U by the graph of f we may realize the map f as the restriction to a closed subset $U \subset \mathbb{A}^n$ of a linear projection $\mathbb{A}^n \to \mathbb{A}^m$, so that it is enough to prove the claim for a locally closed subset $U \subset \mathbb{A}^n$ under the projection

$$\pi\colon \mathbb{A}^n \to \mathbb{A}^{n-1}$$
$$:(z_1, \ldots, z_n) \mapsto (z_1, \ldots, z_{n-1}).$$

Finally, we can replace $\mathbb{A}^{n-1}$ by the Zariski closure $Y = \overline{\pi(U)}$ of the image of U and $\mathbb{A}^n$ by the inverse image $\pi^{-1}(Y) = Y \times \mathbb{A}^1$. It will thus suffice to establish the claim for a locally closed subset U of a product $Y \times \mathbb{A}^1$ (or, equivalently, a locally closed subset $U \subset Y \times \mathbb{P}^1$) and the projection map $\pi\colon Y \times \mathbb{P}^1 \to Y$ on the first factor, with the further assumption that $\pi(U)$ is dense in Y. In sum, we have reduced the proof of the Theorem to the following Lemma.

Lemma 3.17. *Let $\pi\colon Y \times \mathbb{P}^1 \to Y$ be projection on the first factor and let $U \subset Y \times \mathbb{P}^1$ be any locally closed subset such that $\pi(U)$ is dense in Y. Then $\pi(U)$ contains an open subset of Y.*

PROOF. Let X be the closure of U in $Y \times \mathbb{P}^1$ and write

$$U = X \cap V$$

for some open $V \subset Y \times \mathbb{P}^1$; let T be the complement of V in $Y \times \mathbb{P}^1$. Note that if $X = Y \times \mathbb{P}^1$ we are done, since the locus of points $p \in Y$ such that T contains the fiber $\{p\} \times \mathbb{P}^1$ is a proper subvariety of Y; we assume accordingly that $X \subsetneqq Y \times \mathbb{P}^1$. By Theorem 3.12, $\pi(X)$ is closed, and so by our hypothesis $\pi(X) = Y$; thus we just have to show that the closed subvariety $\pi(X \cap T)$ does not equal Y.

Now, after restricting to an open subset of Y, the ideals of X and T will be generated by polynomials F of the form

$$F(Z, W) = a_0 Z^n + a_1 Z^{n-1} W + \cdots + a_n W^n.$$

What's more, not every pair of such polynomials $F \in I(X)$ and $G \in I(T)$ can have a common factor: if for example every $G \in I(T)$ had a factor H in common with a given $F \in I(X)$, it would follow that H was nowhere zero on U and hence that $F/H \in I(X)$ as well. But if $F \in I(X)$ and $G \in I(T)$ have no common factor, the image $\pi(X \cap T)$ will be contained in the proper subvariety of Y defined by their resultant $R(F, G)$. □

LECTURE 4

Families and Parameter Spaces

Example 4.1. Families of Varieties

Next, we will give a definition without much apparent content, but one that is fundamental in much of algebraic geometry. Basically, the situation is that, given a collection $\{V_b\}$ of projective varieties $V_b \subset \mathbb{P}^n$ indexed by the points b of a variety B, we want to say what it means for the collection $\{V_b\}$ to "vary algebraically with parameters." The answer is simple: for any variety B, we define a *family of projective varieties* in $\mathbb{P}^n$ with base B to be simply a closed subvariety $\mathscr{V}$ of the product $B \times \mathbb{P}^n$. The fibers $V_b = (\pi_1)^{-1}(b)$ of $\mathscr{V}$ over points of b are then referred to as the *members*, or *elements* of the family; the variety $\mathscr{V}$ is called the *total space*, and the family is said to be *parametrized* by B. The idea is that if $B \subset \mathbb{P}^m$ is projective, the family $\mathscr{V} \subset \mathbb{P}^m \times \mathbb{P}^n$ will be described by a collection of polynomials $F_\alpha(Z, W)$ bihomogeneous in the coordinates Z on $\mathbb{P}^m$ and W on $\mathbb{P}^n$, which we may then think of as a collection of polynomials in W whose coefficients are polynomials on B; similarly, if B is affine we may describe $\mathscr{V}$ by a collection of polynomials $F_\alpha(z, W)$, which we may think of as homogeneous polynomials in the variables W whose coefficients are regular functions on B.

There are many further conditions we can impose on families to insure that they do indeed vary continuously in various senses; we will discuss some of these further in Lecture 21.

We should remark here that as a general rule a geometric condition on the members of a family of varieties $V_b \subset \mathbb{P}^n$ will determine a constructible, and often an open or a closed subset of the parameter space B; for example, we use Theorem 3.13 to show that for any point $p \in \mathbb{P}^n$ the set of $b \in B$ such that $p \in V_b$ will be a closed subvariety of B. More generally, we have the following.

Exercise 4.2. Let $X \subset \mathbb{P}^n$ be any projective variety and $\{V_b\}$ any family of projective varieties in $\mathbb{P}^n$ with base B. Show that the set

$$\{b \in B: X \cap V_b \neq \emptyset\}$$

is closed in B. More generally, if $\{W_b\}$ is another family of projective varieties in $\mathbb{P}^n$ with base B, show that the set

$$\{b: V_b \cap W_b \neq \emptyset\}$$

is a closed subvariety of B.

Exercise 4.3. More generally still, if $\{W_c\}$ is another family of projective varieties in $\mathbb{P}^n$ with base C, show that the set

$$\{(b, c): V_b \cap W_c \neq \emptyset\}$$

is a closed subvariety of the product $B \times C$. Show that this implies the preceding exercise.

In a similar vein, for any $X \subset \mathbb{P}^n$ and family $\{V_b\}$ we can consider the subset

$$\{b \in B: V_b \subset X\}.$$

This is always constructible, though we cannot prove that here. It is not, however, in general closed; for example, take $\mathscr{V} \subset \mathbb{A}^2 \times \mathbb{P}^1$ given in terms of Euclidean coordinates z on $\mathbb{A}^2$ and homogeneous coordinates W on $\mathbb{P}^1$ by $z_1 W_2 = z_2 W_1$ and $X \subset \mathbb{P}^1$ any point. By contrast, we have the following.

Exercise 4.4. For any $X \subset \mathbb{P}^n$ and any family $\{V_b\}$, show that the subset

$$\{b \in B: X \subset V_b\}$$

is closed in B.

Example 4.5. The Universal Hyperplane

If we think of the projective space $\mathbb{P}^{n*}$ as the set of hyperplanes $H \subset \mathbb{P}^n$, we may define a subset of the product $\mathbb{P}^{n*} \times \mathbb{P}^n$ simply as

$$\Gamma = \{(H, p): p \in H\}.$$

This is a subvariety of $\mathbb{P}^{n*} \times \mathbb{P}^n$: in terms of coordinates Z on $\mathbb{P}^n$ and W on $\mathbb{P}^{n*}$ corresponding to dual bases for K^{n+1} and K^{n+1*}, it is given by the single bilinear polynomial

$$\sum W_i \cdot Z_i = 0.$$

In particular, it may be realized as a hyperplane section of the Segre variety $\mathbb{P}^{n*} \times \mathbb{P}^n \subset \mathbb{P}^{n^2+2n}$.

Γ is the simplest example of a family; inasmuch as the fibers of Γ over the first factor $B = \mathbb{P}^{n*}$ are all the hyperplanes in $\mathbb{P}^n$, we think of Γ as the family of all hyperplanes in $\mathbb{P}^n$, parameterized by $\mathbb{P}^{n*}$. (Needless to say, the situation is symmetric; via projection on the second factor, we may view Γ as the family of all hyperplanes in $\mathbb{P}^{n*}$, parameterized by $\mathbb{P}^n$.)

The reason for the adjective "universal" is the following property of Γ: if $\mathscr{V} \subset B \times \mathbb{P}^n$ is any family of hyperplanes satisfying a technical condition called *flatness* (see page 267) then there is a (unique) regular map $B \to \mathbb{P}^{n*}$ such that $\mathscr{V}$ is the fiber product $B \times_{\mathbb{P}^{n*}} \Gamma$—in other words,

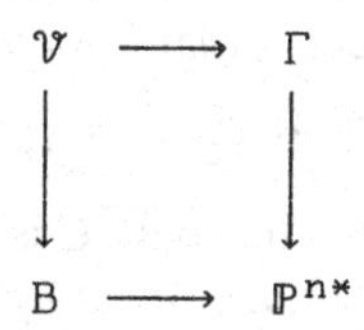

the map $\varphi: B \to \mathbb{P}^{n*}$ sending a point $b \in B$ to the hyperplane $V_b \in \mathbb{P}^{n*}$ is regular. It is unfortunate that we have to throw in the presently undefined condition "flat" here (it will be defined in Lecture 21), but the statement is false without it, as the example following Exercise 2.24 shows. This is one of the many places where the language of schemes would be useful.

Example 4.6. The Universal Hyperplane Section

We can use the preceding construction to see that the set of hyperplane sections of a given variety $X \subset \mathbb{P}^n$ forms a family. Simply, with $\Gamma \subset \mathbb{P}^{n*} \times \mathbb{P}^n$ as in Example 4.5, set

$$\begin{aligned}\Omega_X &= \{(H, p): p \in H \cap X\} \\ &= (\pi_2)^{-1}(X),\end{aligned}$$

where $\pi_2: \Gamma \to \mathbb{P}^n$ is projection on the second factor. From the second description, we see that Ω_X is a subvariety of $\mathbb{P}^{n*} \times X$, which we may view as the family of hyperplane sections of X.

One question we may ask about any family of varieties $\mathscr{V} \subset B \times \mathbb{P}^n$ is whether it admits a *section*, that is, a map $\sigma: B \to \mathscr{V}$ such that $\pi_1 \circ \sigma$ is the identity on B. Similarly, we define a *rational section* to be a section σ defined on some nonempty open subset $U \subset B$ (the reason for the term "rational" will be made clearer in Lecture 7). The problem of determining whether a given family admits a section can be subtle; for example, even in the relatively simple case of the universal hyperplane section of a variety $X \subset \mathbb{P}^n$, it is not known in general under what conditions Ω_X admits a rational section (even in case X is a surface). The following exercise will be simple enough to do after we have introduced some further machinery (specifically, it will be an immediate consequence of Theorem 11.14), but it may be instructive to try it now.

Exercise 4.7. Show that the universal hyperplane section Ω_X of X does not admit a rational section in case (i) $X \subset \mathbb{P}^2$ is a smooth plane conic and (ii) $X \subset \mathbb{P}^3$ is a twisted cubic.

Example 4.8. Parameter Spaces of Hypersurfaces

The parametrization of the family of hyperplanes in $\mathbb{P}^n$ by $\mathbb{P}^{n*}$ is the first example of a general construction, which we will now discuss. We start with a family already discussed in Example 1.20: the set Σ of all conic curves $C \subset \mathbb{P}^2$. A conic $C \subset \mathbb{P}^2$ may be given, in homogeneous coordinates X_i on $\mathbb{P}^2$, as the locus of a polynomial

$$F(X) = a \cdot X_0^2 + b \cdot X_1^2 + c \cdot X_2^2 + d \cdot X_0 X_1 + e \cdot X_0 X_2 + f \cdot X_1 X_2$$

with not all the coefficients zero. The conic C is determined by the 6-tuple (a, b, c, d, e, f) up to scalars, that is, $(\lambda a, \lambda b, \lambda c, \lambda d, \lambda e, \lambda f)$ and (a, b, c, d, e, f) determine the same conic for any $\lambda \in K^*$. Thus, we see that the set Σ may be identified with a projective space $\mathbb{P}^5$. (Without coordinates, if $\mathbb{P}^2 = \mathbb{P}V$, then the homogeneous polynomials of degree 2 on V form the vector space $W = \mathrm{Sym}^2(V^*)$, and we have an identification $\Sigma = \mathbb{P}W$ obtained by sending the zero locus of F to the point $[F] \in \mathbb{P}W$.)

Of course, in order for this to be completely accurate, we have to more or less define a plane conic to correspond to such a polynomial; thus, for example, we have to define loci such as pairs of lines—the zero locus of $X_0 X_1$—and double lines—the zero locus of X_0^2—to be in the set Σ. (We will see in Lecture 22 another approach to defining the set of conics if we are too fastidious to include line pairs and double lines as conics.)

The variety $\Sigma = \mathbb{P}^5$ parametrizing plane conics is an example of a *parameter space*, a basic construction in algebraic geometry. It is peculiar to algebraic geometry in that in most geometric categories it is relatively rare to find the set of geometric objects of a given type naturally endowed with the structure of a geometric object of the same type; for example, the family of submanifolds of a given manifold is not even locally a manifold in the usual sense. Within algebraic geometry, though, this construction is ubiquitous; virtually every object introduced in the subject varies with parameters in the sense that the set of all such objects is naturally endowed with the structure of an algebraic variety. This is true not only for subvarieties of projective space, but for subvarieties of a given projective variety $X \subset \mathbb{P}^n$; and, by applying this notion to their graphs, to maps between two given projective varieties.

We lack at this point a number of the basic notions necessary to describe the general construction of these parameter spaces, and so will have to defer this discussion to Lecture 21. We can say, though, that the construction of the parameter space $\mathbb{P}^5$ for the set of plane conics generalizes immediately to the set of hypersurfaces in $\mathbb{P}^n$ of a given degree d: such a hypersurface X is given by a homogeneous polynomial

$$F(Z_0, \ldots, Z_n) = \sum a_{i_0, \ldots, i_n} \cdot Z_0^{i_0} \cdot \ldots \cdot Z_n^{i_n}$$

so that the set of hypersurfaces is parametrized by the points of a projective space $\mathbb{P}^N$ with homogeneous coordinates $a_{i_0, \ldots, i_n}$.

Example 4.9. Universal Families of Hypersurfaces

In the terminology we have introduced, to say that the set of hypersurfaces of degree d in $\mathbb{P}^n$ is parametrized by a projective space $\mathbb{P}^N$ suggests the existence of a family of hypersurfaces with base $B = \mathbb{P}^N$. Such a family does indeed exist; for example, consider once again the equation of the general conic in $\mathbb{P}^2$

$$a \cdot X_0^2 + b \cdot X_1^2 + c \cdot X_2^2 + d \cdot X_0 X_1 + e \cdot X_0 X_2 + f \cdot X_1 X_2.$$

If we think of $[X_0, X_1, X_2]$ as coordinates on $\mathbb{P}^2$ and $[a, b, c, d, e, f]$ as coordinates on $\mathbb{P}^5$, then we may view this equation as defining a hypersurface $\mathfrak{X} \subset \mathbb{P}^5 \times \mathbb{P}^2$, with the property that the fiber of $\mathfrak{X}$ over any point $C \in \mathbb{P}^5$ is the conic curve $C \subset \mathbb{P}^2 = \mathbb{P}^2 \times \{C\}$ corresponding to the point C. $\mathfrak{X}$ is called the *universal family* of conics; as in the case of the universal hyperplane, it is called universal because any flat family of smooth conics $\mathscr{V} \subset B \times \mathbb{P}^2$ may be realized as the fiber product

$$\mathscr{V} = B \times_{\mathbb{P}^5} \mathfrak{X}$$

for a unique regular map $B \to \mathbb{P}^5$. (The specification "smooth conics" in the last sentence may be broadened to include line pairs, but things get trickier when we include double lines; see Lecture 21 for a discussion of these issues.)

More generally, universal families exist for the parameter spaces of hypersurfaces of any degree d in projective space $\mathbb{P}^n$ of any dimension. Note that if we had taken $d = 1$, the parameter space would be simply the dual projective space $\mathbb{P}^{n*}$, and the universal family just the universal hyperplane $\Gamma \subset \mathbb{P}^{n*} \times \mathbb{P}^n$ described in Example 4.5.

The basic observation made earlier for families, that as a general rule geometric conditions on the members of a family determine constructible subsets of the base, applies to parameter spaces. For example, let $\Sigma \cong \mathbb{P}^5$ be the parameter space for plane conics, and consider the subset $\Psi \subset \mathbb{P}^5$ corresponding to double lines. This may be realized as the image of the space $\mathbb{P}^{2*}$ of lines in $\mathbb{P}^2$, under the map sending a line l to the "conic" l^2; that is, the map taking a linear form

$$l(X) = a \cdot X_0 + b \cdot X_1 + c \cdot X_2$$

and sending it to the quadratic polynomial

$$l(X)^2 = a^2 \cdot X_0^2 + b^2 \cdot X_1^2 + c^2 \cdot X_2^2 + 2ab \cdot X_0 X_1 + 2ac \cdot X_0 X_2 + 2bc \cdot X_1 X_2$$

(each defined only up to scalars, naturally). In coordinates, then, this is the map

$$v_2 \colon \mathbb{P}^{2*} \to \mathbb{P}^5$$

given by

$$v_2 \colon [a, b, c] \mapsto [a^2, b^2, c^2, 2ab, 2ac, 2bc],$$

which we may recognize (if the characteristic of K is not 2) as the quadratic Veronese map. We see thus that the variety Ψ is the Veronese surface in $\mathbb{P}^5$.

We can similarly characterize the subset $\Delta \subset \mathbb{P}^5$ of conics that consist of a union of lines (i.e., corresponding to quadratic polynomials that factor as a product of linear forms) as the image of the map $\mathbb{P}^{2*} \times \mathbb{P}^{2*} \to \mathbb{P}^5$ sending a pair of linear forms $([l], [m])$ to their product $[l \cdot m]$. We may recognize this map as a composition of the Segre map $\mathbb{P}^{2*} \times \mathbb{P}^{2*} \to \mathbb{P}^8$ followed by a projection (in intrinsic terms, this is the map $\mathbb{P}V \times \mathbb{P}V \to \mathbb{P}(V \otimes V) \to \mathbb{P}(\mathrm{Sym}^2 V)$); in any event, it is clearly a regular map and so its image is, as claimed, a subvariety of $\mathbb{P}^5$. As we will see in Lecture 8, we can also realize Δ as the chordal variety of the Veronese surface.

Exercise 4.10. Let $\mathbb{P}^N$ be as earlier the parameter space of hypersurfaces of degree d in $\mathbb{P}^n$. Show that the subset Σ of $\mathbb{P}^N$ corresponding to nonprime polynomials $F(X_0, \ldots, X_n)$ is a projective subvariety of $\mathbb{P}^N$. (Σ corresponds to hypersurfaces X that contain a hypersurface Y of degree strictly less than d, though we will need Theorem 5.1 to establish this.)

Exercise 4.11. Let $X \subset \mathbb{P}^n$ be any hypersurface of degree d given by a homogeneous polynomial $F(Z_0, \ldots, Z_n)$. Show that the subset of hyperplanes $H \subset \mathbb{P}^n$ such that the restriction of F to H factors (i.e., such that $H \cap X$ contains a hypersurface of degree $< d$ in H) is a subvariety of $\mathbb{P}^{n*}$.

The question of the existence of sections may be raised in the case of the universal hypersurfaces; in this case, the answer is known, as indicated later.

Exercise 4.12. (a) For any n, find a rational section of the universal hyperplane $\Gamma \subset \mathbb{P}^{n*} \times \mathbb{P}^n$. (b) For n odd, find a section of Γ. (c) For n even, show that there does not exist a section of Γ. (For part (c), you may want to use the fact, stated on page 37 following Exercise 3.9, that any regular map from $\mathbb{P}^n$ to $\mathbb{P}^m$ is given by an $(m + 1)$-tuple of homogeneous polynomials.) (*)

Exercise 4.13. Show that the universal plane conic $\mathfrak{X} \subset \mathbb{P}^5 \times \mathbb{P}^2$ does not admit even a rational section. (*)

It is in general true that the universal family of hypersurfaces of any degree $d > 1$ admits no rational sections[1].

Exercise 4.14. (a) Let $\mathfrak{X} \subset B \times \mathbb{P}^3$ be any family of twisted cubics. Show that $\mathfrak{X}$ admits a rational section. (b) By contrast, exhibit a family of rational normal curves of degree 4 that does not have a rational section. (*) (The general pattern is that a family $\mathfrak{X} \subset B \times \mathbb{P}^{2n+1}$ of rational normal curves of odd degree will always

[1] I don't know of a reference for this; a proof can be given by applying the Lefschetz hyperplane theorem to the universal family $\mathfrak{X} \subset \mathbb{P}^N \times \mathbb{P}^n$.

admit a rational section, but this is not the case for families of rational normal curves of even degree.)

Example 4.15. A Family of Lines

Our final examples of families are of linear spaces in $\mathbb{P}^n$. To start with the simplest case, we let $U \subset \mathbb{P}^n \times \mathbb{P}^n$ be the complement of the diagonal and consider the subset

$$\Omega = \{(p, q; r): r \in \overline{pq}\} \subset U \times \mathbb{P}^n.$$

We first observe that Ω is indeed a subvariety of $U \times \mathbb{P}^n$; it is the family whose fiber over a point $(p, q) \in U$ is the line spanned by p and q. Since every line $l \subset \mathbb{P}^n$ occurs as a fiber of $\pi_1: \Omega \to U$, we may think of this as a parameter space for lines; it is not optimal, however, because every line occurs many times, rather than just once. We will see how to fix this when we discuss Grassmannians in Lecture 6.

Exercise 4.16. Show that Ω is indeed a subvariety of $U \times \mathbb{P}^n$. More generally, show that for any k the subset $U \subset (\mathbb{P}^n)^k$ of k-tuples $(p_1, \ldots, p_k)$ such that $p_1, \ldots, p_k$ are linearly independent is open, and that the locus

$$\Omega = \{((p_1, \ldots, p_k); r): r \in \overline{p_1, \ldots, p_k}\} \subset U \times \mathbb{P}^n$$

is again a subvariety. What is the closure of Ω in $(\mathbb{P}^n)^k \times \mathbb{P}^n$?

Exercise 4.17. For another family of linear spaces, let $V \subset (\mathbb{P}^{n*})^{n-k}$ be the open subset of $(n-k)$-tuples of linearly independent hyperplanes, and set

$$\Xi = \{((H_1, \ldots, H_{n-k}); r): r \in H_1 \cap \ldots \cap H_{n-k}\} \subset V \times \mathbb{P}^n.$$

Show that Ξ is a family of k-planes. In the case of lines in $\mathbb{P}^3$, compare the family constructed in this exercise to that constructed earlier: are they isomorphic?

LECTURE 5

Ideals of Varieties, Irreducible Decomposition, and the Nullstellensatz

Generating Ideals

The time has come to talk about the various senses in which a variety may be defined by a set of equations. There are three different meanings of the statement that a collection of polynomials $\{F_\alpha(Z)\}$ "cut out" a variety $X \subset \mathbb{P}^n$, and several different terms are used to convey each of these meanings.

Let's start with the affine case, where there are only two possibilities. Let $X \subset \mathbb{A}^n$ be a variety and $\{f_\alpha(z_1, \ldots, z_n)\}_{\alpha=1,\ldots,m}$ a collection of polynomials. When we say that the polynomials f_α determine X, we could a priori mean one of two things: either

(i) the common zero locus $V(f_1, \ldots, f_m)$ of the polynomials f_α is X or
(ii) the polynomials f_α generate the ideal $I(X)$.

Clearly, the second is stronger. For example, the zero locus of the polynomial $x^2 \in K[x]$ is the origin $0 \in \mathbb{A}^1$, but the ideal of functions vanishing at 0 is (x), not (x^2). In general, the ideal of functions vanishing on a variety has the property that, for any polynomial $f \in K[z_1, \ldots, z_n]$, if a power $f^k \in I$ then $f \in I$. We formalize this by observing that for any ideal I in a ring R, the set of all elements $f \in R$ such that $f^k \in I$ for some $k > 0$ is again an ideal, called the *radical* of I and denoted $\mathfrak{r}(I)$. We call an ideal I *radical* if it is equal to $\mathfrak{r}(I)$; as we have just observed, an ideal without this property cannot be of the form $I(X)$.

To put it another way, we have a two-way correspondence

$$\left\{\begin{matrix}\text{subvarieties}\\ \text{of } \mathbb{A}^n\end{matrix}\right\} \underset{V}{\overset{I}{\rightleftarrows}} \left\{\begin{matrix}\text{ideals}\\ I \subset K[z_1, \ldots, z_n]\end{matrix}\right\}$$

but this is not by any means bijective: in one direction, the composition of the two is the identity—the definition of a variety $X \subset \mathbb{A}^n$ amounts to the statement that $V(I(X)) = X$—but going the other way the composition is neither injective nor surjective. We can fix this by simply restricting our attention to the image of the map I, and happily there is a nice characterization of this image (and indeed of the composition $I \circ V$). This is the famous Nullstellensatz:

Theorem 5.1. *For any ideal $I \subset K[z_1, \ldots, z_n]$, the ideal of functions vanishing on the common zero locus of I is the radical of I, i.e.,*

$$I(V(I)) = \mathfrak{r}(I)$$

Thus, there is a bijective correspondence between subvarieties $X \subset \mathbb{A}^n$ and radical ideals $I \subset K[z_1, \ldots, z_n]$.

We will defer both the proof of the Nullstellensatz and some of its corollaries to later in this lecture and will proceed with our discussion now.

Note that, as one consequence of the Nullstellensatz, we can say that a K-algebra A occurs as the coordinate ring of an affine variety if and only if A is finitely generated and has no nilpotents. Clearly these two conditions are necessary; if they are satisfied, we can write

$$A = K[x_1, \ldots, x_n]/(f_1, \ldots, f_m)$$

so that we will have $A = A(X)$, where $X \subset \mathbb{A}^n$ is the zero locus of the polynomials f_α.

At this point we can take a minute out and mention one of the fundamental notions of scheme theory. Basically, if one is going to fix the correspondence on page 48 so as to make it bijective, there are naively two ways of going about it. We can either restrict the class of objects on the right or enlarge the class of objects on the left. In classical algebraic geometry, as we have just said, we do the former; in scheme theory, we do the latter. Thus, we define an affine scheme $X \subset \mathbb{A}^n$ to be an object associated to an arbitrary ideal $I \subset K[z_1, \ldots, z_n]$.

What sense can this possibly make? This is not the place to go into it in any detail, but we may remark that, in fact, most of the notions that we actually deal with in algebraic geometry are defined in terms of rings and ideals as well as in terms of subsets of affine or projective spaces. For example, if $X \subset \mathbb{A}^n$ is a variety with ideal $I = I(X)$, we define a function on X to be an element of the ring $A(X) = K[z_1, \ldots, z_n]/I$; the intersection of two such varieties $X, Y \subset \mathbb{A}^n$ is given by the join of their ideals; the data of a map between two such varieties X and Y are equivalent to the data of a map $A(Y) \to A(X)$, and so on. The point is that all these things make sense whether or not I is a radical ideal. The scheme associated to an arbitrary ideal $I \subset K[z_1, \ldots, z_n]$ may not seem like a geometric object, especially in case I is not radical, but it does behave formally like one and it encodes extra information that is of geometric interest.

Before going on, we will introduce some terminology. We say that a collection $\{f_\alpha\}$ of polynomials cut out a variety $X \subset \mathbb{A}^n$ *set-theoretically* to mean just that their common zero locus $V(\{f_\alpha\}) = X$; we say that they cut out X *scheme-theoretically*, or *ideal-theoretically*, if in fact they generate the ideal $I(X)$.

Ideals of Projective Varieties

The case of projective space is in one respect like that of affine space: we have a correspondence between projective varieties $X \subset \mathbb{P}^n$ and homogeneous ideals $I \subset K[Z_0, \ldots, Z_n]$ that becomes almost a bijection when we restrict ourselves to radical ideals (the almost is because we have to exclude the radical ideal $\mathfrak{m} = (Z_0, \ldots, Z_n)$). There is, however, one other sense in which a collection of polynomials can cut out a variety.

For example, suppose $X \subset \mathbb{P}^n$ is any variety and $I \subset K[Z_0, \ldots, Z_n]$ its ideal. Consider the ideal I' formed by simply intersecting I with the ideal $(Z_0, \ldots, Z_n)^k$, that is, if we write I as the direct sum of its homogeneous pieces

$$I = \bigoplus_{m \in \mathbb{Z}} I_m,$$

then

$$I' = \bigoplus_{m \geq k} I_m.$$

Certainly the radical of I' is I, since any element of I, raised to a sufficiently high power, will lie in I'. But the relationship between I and I' is closer than that: for any polynomial $F \in K[Z_0, \ldots, Z_n]$, F will lie in I if and only if the product of F with any homogeneous polynomial of sufficiently high degree lies in I'. What this means is that if we restrict to the affine open subset $(Z_i \neq 0) \cong \mathbb{A}^n \subset \mathbb{P}^n$—that is to say, we consider the 0th graded piece of the localization $(I' \cdot K[Z_0, \ldots, Z_n, Z_i^{-1}])_0$ in the coordinate ring $(K[Z_0, \ldots, Z_n, Z_i^{-1}])_0 \cong K[z_1, \ldots, z_n]$ of $\mathbb{A}^n$—we get exactly the ideal of the affine open subset $X \cap \mathbb{A}^n \subset \mathbb{A}^n$. We may say, in this case, that the ideal I' cuts out the variety X *locally*, even though it does not equal I.

To formalize this, we introduce the notion of the *saturation* $\bar{I}$ of an ideal $I \subset K[Z_0, \ldots, Z_n]$. This is given by

$$\bar{I} = \{F \in K[Z_0, \ldots, Z_n]: (Z_0, \ldots, Z_n)^k \cdot F \subset I \text{ for some } k\}.$$

Note that since $K[Z_0, \ldots, Z_n]$ is Noetherian, $\bar{I}/I$ is finitely generated, so that $\bar{I}$ will agree with I in large enough degree. Indeed, we have the following.

Exercise 5.2. Show that the following conditions on a pair of homogeneous ideals I and $J \subset K[Z_0, \ldots, Z_n]$ are equivalent:

(i) I and J have the same saturation.
(ii) $I_m = J_m$ for all $m \gg 0$.
(iii) I and J agree locally, that is, they generate the same ideal in each localization $K[Z_0, \ldots, Z_n, Z_i^{-1}]$ of $K[Z_0, \ldots, Z_n]$.

In the language of schemes, all three conditions of Exercise 5.2 amount to saying that I and J define the same subscheme of $\mathbb{P}^n$; we often say that a collection of functions cuts out a variety $X \subset \mathbb{P}^n$ *scheme-theoretically* if the saturation of the ideal they generate in $K[Z_0, \ldots, Z_n]$ is the homogeneous ideal $I(X)$. In sum, then, the three statements we can make about a collection of polynomials F_α in relation to a variety X are, in order of increasing strength, that they

(i) cut out the variety X set-theoretically, if their common zero locus in $\mathbb{P}^n$ is X;
(ii) cut out the variety X scheme-theoretically, if the saturation of the ideal they generate is $I(X)$; and
(iii) generate the homogeneous ideal $I(X)$ of X.

By way of example, if $I = I(X)$ is the ideal of the line $(X = 0)$ in the plane $\mathbb{P}^2$ with homogeneous coordinates X, Y, and Z, the ideal $I' = (X^2, XY, XZ)$ generates I locally, though it does not equal I; while the ideal (X^2) does not even generate I locally, though its zero locus is the same.

Exercise 5.3. Consider once again the rational normal curve $C \subset \mathbb{P}^d$ given in Example 1.14. Show that the homogeneous quadratic polynomials

$$F_{i,i}(Z) = Z_i^2 - Z_{i-1}Z_{i+1}$$

and

$$F_{i,i+1}(Z) = Z_i Z_{i+1} - Z_{i-1}Z_{i+2}$$

generate the ideal of the rational normal curve locally but do not generate the homogeneous ideal $I(C)$ for $d \geq 4$.

Exercise 5.4. Show that the polynomials $F_{i,j}(Z) = Z_i Z_j - Z_{i-1}Z_{j+1}$ for $1 \leq i \leq j \leq d-1$ do generate the homogeneous ideal of the rational normal curve $C \subset \mathbb{P}^d$. Similarly, check that the equations given earlier for the Veronese and Segre varieties in general do generate their homogeneous ideals.

As a final example, note that if $X \subset \mathbb{P}^n$ is a variety and $p \in \mathbb{P}^n - X$ a point, the equations we have exhibited that cut out the projection $\overline{X} = \pi_p(X)$ of a variety X from p—the pairwise resultants of the polynomials F, $G \in I(X)$—do not in general generate its homogeneous ideal (see, for example, Exercise 3.8). In fact, they do generate the ideal of $\overline{X}$ locally, though our proof does not show this.

Lastly, we should remark that there is a projective version of the Nullstellensatz: the ideal of polynomials on $\mathbb{P}^n$ vanishing on the common zero locus of a collection $\{F_\alpha\}$ of homogeneous polynomials is the radical of the ideal they generate (or the unit ideal, if the F_α have no common zeros). As a consequence, we deduce that any finitely generated graded algebra

$$A = \bigoplus A_i$$

over K is the homogeneous coordinate ring of a projective variety if it has no nilpotent elements and is generated by its first graded piece A_1.

Irreducible Varieties and Irreducible Decomposition

Definition. We say that a variety is *irreducible* if for any pair of closed subvarieties $Y, Z \subset X$ such that $Y \cup Z = X$, either $Y = X$ or $Z = X$.

Observe that an affine variety $X \subset \mathbb{A}^n$ is irreducible if and only if its ideal $I(X) \subset K[x_1, \ldots, x_n]$ is prime. To see this, note first that if Y and Z are proper closed subvarieties of X, there exist $f \in I(Y)$ not in $I(X)$ and $g \in I(Z)$ not in $I(X)$; if $X = Y \cup Z$ it follows that $f \cdot g \in I(X)$. Conversely, if $f, g \in K[x_1, \ldots, x_n]$ with $f \cdot g \in I(X)$, then the subvarieties of X defined by f and g—that is, the subvarieties $Y = V(I(X), f)$ and $Z = V(I(X), g)$—have union X. More generally, we say that an ideal $I \subset K[x_1, \ldots, x_n]$ is *primary* if $\forall f, g \in K[x_1, \ldots, x_n]$, $f \cdot g \in I \Rightarrow f^m \in I$ for some m or $g \in I$; this implies that the radical of I is prime. The same argument then shows that for any I, the variety $V(I)$ will be irreducible if I is primary; moreover, if we use the Nullstellensatz, we can deduce that in fact $V(I)$ will be irreducible if and only if the radical of I is prime.

The analogous statements apply to projective varieties and their homogeneous ideals: a variety $X \subset \mathbb{P}^n$ is irreducible if and only if its homogeneous ideal $I(X)$ is prime, and the zero locus $V(I)$ of a homogeneous ideal $I \subset K[X_0, \ldots, X_n]$ is irreducible if I is primary. Note that if a projective variety $X \subset \mathbb{P}^n$ is irreducible then so is any nonempty affine open subset $U = X \cap \mathbb{A}^n$, though the converse is true only in the sense that if the affine open subset $X \cap \mathbb{A}^n$ of X is irreducible for every hyperplane complement $\mathbb{A}^n \subset \mathbb{P}^n$ (not just the standard $U_0, \ldots, U_n$) then X must be irreducible.

Exercise 5.5. Show that a variety X is irreducible if and only if every Zariski open subset of X is dense, i.e., every two Zariski open subsets of X meet.

A basic theorem of commutative algebra is the following proposition.

Proposition 5.6. *Any radical ideal $I \subset K[x_1, \ldots, x_n]$ is uniquely expressible as a finite intersection of prime ideals $\mathfrak{p}_i$ with $\mathfrak{p}_i \not\subset \mathfrak{p}_j$ for $j \neq i$.*

Given this equivalence, this implies the following.

Theorem 5.7. *Any variety X may be uniquely expressed as a finite union of irreducible subvarieties X_i with $X_i \not\subset X_j$ for $i \neq j$.*

The varieties X_i appearing in the expression of X as a finite union of irreducible varieties are called the *irreducible components* of X.

A few notes: first, those familiar with commutative algebra will recognize Proposition 5.6 as a very weak form of the general theorem on primary decomposition of ideals in Noetherian rings. We will prove this weak form in the section at the end of this lecture; for a proof of the full statement see [AM] or [E]. Second, observe that the uniqueness of the expression of a radical ideal as a finite intersection of prime ideals is formal; if we had $I = \bigcap \mathfrak{p}_i = \bigcap \mathfrak{q}_j$ then for each i we would have $\mathfrak{p}_i \supset \bigcap \mathfrak{q}_j \Rightarrow \mathfrak{p}_i \supset \mathfrak{q}_k$ for some k and vice versa. The same argument (with inclusions reversed and intersections and unions exchanged) shows that the expression of an arbitrary variety as a union of irreducible components is likewise unique.

It is worthwhile to go through some of the varieties introduced earlier and verify that they are irreducible (one useful tool in this regard is the observation that the image of an irreducible projective variety under a regular map is an irreducible variety). We will not do much of this explicitly here, since it will become much easier once we have introduced the notion of dimension, and in particular Theorem 11.14. One example that is worth doing, and that will be useful shortly, is the following.

Theorem 5.8. *Let $X \subset \mathbb{P}^n$ be an irreducible variety, and let $\Omega_X \subset \mathbb{P}^{n*} \times X$ be its universal hyperplane section, as in Example 4.6. Then Ω_X is irreducible.*

PROOF. For each point $p \in X$, let $\Gamma_p = (\pi_2)^{-1}(p)$ be the fiber of Ω_X over p. We claim that for any irreducible component Ψ of Ω_X, the locus

$$\varphi(\Psi) = \{p \in X : \Gamma_p \subset \Psi\}$$

is a closed subset of X. The theorem follows from this claim: since Γ_p is isomorphic to $\mathbb{P}^{n-1}$, which is irreducible, for any irreducible decomposition $\Omega_X = \Psi_1 \cup \cdots \cup \Psi_k$ of Ω_X we must have

$$X = \varphi(\Psi_1) \cup \cdots \cup \varphi(\Psi_k).$$

It will follow that $X = \varphi(\Psi_i)$ for some i and hence that $\Omega_X = \Psi_i$.

To establish the claim, we may work locally, say over the open subset $U \subset X$ given by $Z_0 \neq 0$. Now, for each $\alpha = (\alpha_1, \ldots, \alpha_n) \in \mathbb{A}^n$, let $\Phi_\alpha \subset \mathbb{P}^{n*} \times U$ be the locus given in terms of homogeneous coordinates Z on $\mathbb{P}^n$ and dual coordinates W on $\mathbb{P}^{n*}$ by the equations

$$W_0 = -\alpha_1 Z_1 - \cdots - \alpha_n Z_n \quad \text{and}$$
$$W_i = \alpha_i Z_0, \quad i = 1, \ldots, n.$$

Φ_α is then a closed subvariety of $\mathbb{P}^{n*} \times U$, meeting each fiber Γ_p in exactly one point; moreover, the union of the Φ_α is the inverse image of U in Ω_X. It follows that $\varphi(\Psi) \cap U$ may be written as the intersection

$$\begin{aligned} \varphi(\Psi) \cap U &= \bigcap \{p \in U : \Phi_\alpha \cap \Gamma_p \in \Psi\} \\ &= \bigcap \pi_2(\Psi \cap \Phi_\alpha); \end{aligned}$$

by Theorem 3.12, this is closed. □

Exercise 5.9. By a similar argument, show that the product of two irreducible varieties is irreducible.

General Objects

Having introduced the notion of irreducible variety, we can also mention a fairly ubiquitous piece of terminology: the notion of a *general* object. Basically, when a family of objects $\{X_p\}_{p \in \Sigma}$—varieties, maps, or whatever—is parametrized by the points of an irreducible algebraic variety Σ, the statement that "the general object X has property P" is taken to mean that "the subset of points $p \in \Sigma$ such that the corresponding object X_p has property P contains a Zariski open dense subset of Σ." Thus, for example, given a point $p_0 \in \mathbb{P}^2$, we say that "a general line $L \subset \mathbb{P}^2$ does not contain p_0" to refer to the fact that the set of lines containing p_0 is contained in a proper subvariety of the dual plane $\mathbb{P}^{2*}$. As another example, we say that "the general conic has rank 3 (i.e., is projectively equivalent to the image of the Veronese map $v_2: \mathbb{P}^1 \to \mathbb{P}^2$)." This refers to the fact that, as observed in the discussion preceding Exercise 1.21, the set of conics in $\mathbb{P}^2$ can be parametrized by the points of $\mathbb{P}^5$, and asserts that in this $\mathbb{P}^5$ the subset of those not projectively equivalent to $v_2(\mathbb{P}^1)$ is contained in a proper subvariety.

We should note that the word "generic" is sometimes used in place of "general." "General" is preferable, since in some sources the word "generic" is given a technical meaning (it is also sometimes the practice to use the phrase "the generic object X has property P" to mean that the set of $p \in \Sigma$ such that X_p does not have this property is contained in a countable union of proper subvarieties of Σ). Nonetheless, using "generic" and "general" interchangably is one of the more venial sins associated to the use of the word(s).

Another remark to be made here is that there is also an adverbial usage of the term; for example, if X is an irreducible variety, we say that a map $f: X \to \mathbb{P}^n$ is *generically finite* to mean that for a general point $p \in X$ the inverse image $f^{-1}(f(p))$ is finite.

The example of "the general conic" points up one possibly troublesome issue: every time we use this terminology we will be implicitly invoking the existence of a parameter space. In general, we will not refer explicitly to the construction of this parameter space; in some cases it may seem ambiguous. In fact, there are standard constructions of parameter spaces in algebraic geometry, which we will discuss in Lecture 21; it is these to which we implicitly refer. In practice, however, we can approach the matter on an ad hoc basis. We give some examples of this usage, starting with an exercise.

Exercise 5.10. Consider the parameter space for lines in $\mathbb{P}^n$ introduced in Example 4.15. Show that, given any linear space $\Gamma \subset \mathbb{P}^n$ of dimension $n - 2$ or less the general line in $\mathbb{P}^n$ does not meet Γ. Given a twisted cubic curve $C \subset \mathbb{P}^3$, show that the general line in $\mathbb{P}^3$ does not meet C. (As we noted at the time, the parameter space for lines introduced in Example 4.15 is not the standard one; but we will see when we do introduce the Grassmannian that the same statements apply.)

Example 5.11. General Projections

We have seen that projections of the twisted cubic curve $C \subset \mathbb{P}^3$ are projectively equivalent to one of two curves, the nodal cubic $Z_0 Z_2^2 = Z_1^3 + Z_0 Z_1^2$ and the cuspidal cubic $Z_0 Z_2^2 = Z_1^3$. We can further make the statement "the general projection of a twisted cubic curve to $\mathbb{P}^2$ is projectively equivalent to the nodal cubic." Implicit in this statement is the idea that the set of projections of a twisted cubic

$C \subset \mathbb{P}^3$ is parametrized by the set of points in the complement $\mathbb{P}^3 - C$ of C in $\mathbb{P}^3$; the content of the statement is that for all points p in an open subset $U \subset \mathbb{P}^3 - C$, the projection $\pi_p(C)$ is projectively equivalent to the nodal cubic. We will see more statements about the general projections of varieties in Lecture 15.

Example 5.12. General Twisted Cubics

As another example, consider again twisted cubics. Any twisted cubic C can be written as the image of a map of the form

$$t \mapsto [a_{0,3}t^3 + a_{0,2}t^2 + a_{0,1}t + a_{0,0}, \ldots, a_{3,3}t^3 + a_{3,2}t^2 + a_{3,1}t + a_{3,0}],$$

where the determinant of the matrix $(a_{i,j})$ of coefficients is nonzero, so that we can describe C by specifying a nonsingular 4×4 matrix. Of course, this does not give a bijection between the variety $U \subset K^{16}$ of invertible 4×4 matrices and the set of twisted cubics, since the curve C does not determine this expression. Alternately, recall from Theorem 1.18 that any twisted cubic is determined by six points in general position in $\mathbb{P}^3$. Thus we get a twisted cubic for any point in the open subset $V \subset (\mathbb{P}^3)^6$ corresponding to configurations in general position, though as in the previous case, this is not a bijection.

The point is, we can take the statement that "the general twisted cubic has property X" to mean either that the set of $C \in U$ with property X contains an open dense subset, or the same for the analogous set of $C \in V$. This apparent ambiguity will be dealt with in Lecture 21, where we will see that a universal parameter space $\mathscr{H}$ for twisted cubic curves in $\mathbb{P}^3$ does exist and that the corresponding maps $U \to \mathscr{H}$ and $V \to \mathscr{H}$ are regular. It will follow that to say that a property holds for an open dense subset of $C \in U$ is equivalent to saying that it holds for an open dense subset of $C \in \mathscr{H}$, which is in turn equivalent to saying that it holds for an open dense subset of V.

Two further variations on the theme of "general" objects: first, it is clear that if a family of objects is parametrized by a variety Σ, the family of (ordered) pairs (or n-tuples) of these objects is parametrized by the product Σ^2 (or Σ^n); when we talk about a property of a "general pair" of these objects, we mean a property enjoyed by the pairs corresponding to an open dense subset of this product. Thus, for example, "a general triple of points in $\mathbb{P}^2$ does not lie on a line." Sometimes the usage dictates using the fiber product rather than the regular product; for example, a "general pair of points on a general line in $\mathbb{P}^2$" would refer to a point in an open subset of the (irreducible) variety $\Gamma \times_{\mathbb{P}^{2*}} \Gamma$, where $\Gamma \subset \mathbb{P}^{2*} \times \mathbb{P}^2$ is the universal hyperplane (line) in $\mathbb{P}^2$, as in Example 4.5. Also, we say that an object $X \in \Sigma$ arising in some construction is "general" if, given no further specification, X could be the object corresponding to any point in an open dense subset of Σ. Thus, "a general point p on a general line $l \subset \mathbb{P}^2$ is a general point of the plane," and "a general pair of points on a general line in the plane is a general pair of points in the plane," but "a general triple of points on a general line is not a general triple of points in the plane." This terminology may seem opaque at first but it is extremely useful; it becomes, if anything, too easy to use with a little practice.

Exercise 5.13. Show that for any d and $n \leq (d+1)(d+2)/2$, a general set of n points in $\mathbb{P}^2$ imposes independent conditions on curves of degree d, in the sense that the space of polynomials vanishing at the points has codimension n in the space of all homogeneous polynomials of degree d on $\mathbb{P}^2$. In case $n \leq 2d+1$, exactly what open subset of $(\mathbb{P}^2)^n$ is implicitly referred to?

Exercise 5.14. Let $C \subset \mathbb{P}^2$ be a curve of degree d, that is, the zero locus of a homogeneous polynomial $F(Z_0, Z_1, Z_2)$ of degree d without repeated factors. Show that a general line $L \subset \mathbb{P}^2$ will intersect C in d points.

Finally, here is an example of the usage "general" that we will need in the following example (and that is worthwhile in its own right).

Proposition 5.15. *Let $\pi: X \to Y$ be any regular map with Y irreducible, and let $Z \subset X$ be any locally closed subset. Then for a general point $p \in Y$ the closure of the fiber $Z_p = Z \cap \pi^{-1}(p)$ is the intersection of the closure $\bar{Z}$ of Z with the fiber $X_p = \pi^{-1}(p)$.*

We will defer the proof of this proposition until after the proof of Theorem 11.12.

Example 5.16. Double Point Loci

One classic example of the decomposition of a variety into irreducibles is the definition of the double point locus associated to a generically finite map. We suppose that X is an irreducible projective variety and $\varphi: X \to \mathbb{P}^n$ a map; we also assume that φ is generically finite, that is, for a general point $p \in \varphi(X)$ the fiber $\varphi^{-1}(p)$ is finite. Recall from Example 2.25 that the set Z of pairs of points $p, q \in X$ that map to the same point of $\mathbb{P}^n$, that is, the set-theoretic fiber product $X \times_{\mathbb{P}^n} X$, is a subvariety of $X \times X$. Now, to say that φ is generically finite implies in particular that the diagonal $\Delta \subset Z \subset X \times X$ is an irreducible component of Z (this follows from applying Proposition 5.15 to the subvariety $W = Z - \Delta \subset Z \to X$). In this case, we define the *double point locus* of the map φ to be the union of the remaining components of Z, or its image in X under projection.

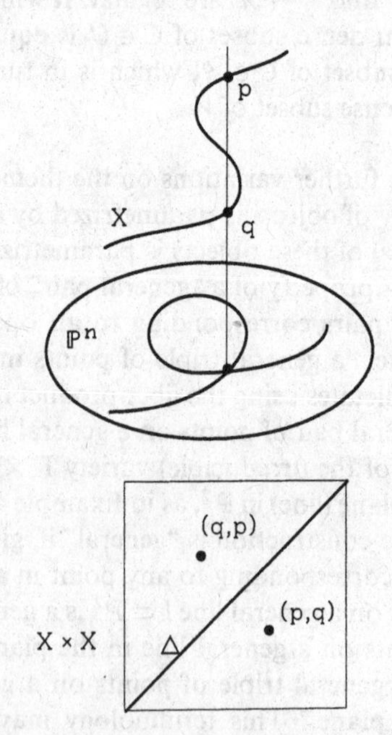

We should give one warning here: all we are really doing in this example is saying that the set of distinct pairs of points of X mapping to the same point of $\mathbb{P}^n$ is a quasi-projective variety—more precisely, a closed subset of the complement of the diagonal in the product $X \times X$. To actually prove meaningful theorems about the double point locus in general requires a much more sensitive definition. To see why this is so, consider the simplest example, of a map φ of a curve to the plane $\mathbb{P}^2$ that is generically one to one but may be two to one over a finite collection of points—for example, the projection of a space curve $C \subset \mathbb{P}^3$ from a point r in $\mathbb{P}^3$ described in Exercises 1.27 and 3.8. Clearly every chord $\overline{pq}$ of the curve C containing the point r contributes a pair of points (p, q) and (q, p) to the double point locus Z of φ. But what happens if, as we vary the point r of projection, one of the chords specializes to a tangent line? The points (p, q) and (q, p) both approach the point (p, p) on the diagonal in $X \times X$, and we'd like to say that this point lies in the double point locus; but the preceding definition does not see it. In fact, to define the double point locus correctly we have to be both more imaginative and working in the category of schemes rather than varieties. Good references for this are [F1], [K].

A Little Algebra

In this section we will give proofs of several of the algebraic lemmas: the Nullstellensatz (Theorem 5.1); one of its corollaries (Lemma 2.1, which states that the ring of regular functions on an affine variety X is simply its coordinate ring $A(X)$); and Proposition 5.6, which asserts that every radical ideal is an intersection of primes. We will try to prove these statements with a minimum of algebraic machinery (in the case of the Nullstellensatz, we give two proofs, the now-classic proof of Artin and Tate and a "quick and dirty" alternative); for a more thorough treatment of these areas, the reader can read the standard sources [AM] and [E] on commutative algebra. One notion from commutative algebra that cannot be readily dispensed with, however, is that of a Noetherian ring; we will assume that the reader knows what this means and that the polynomial ring $K[x_1, \ldots, x_n]$ is one.

Proof of the Nullstellensatz. We start with an arbitrary ideal in the ring $K[x_1, \ldots, x_n]$; we let X be its zero locus $V(I)$. We have a trivial inclusion

$$\mathfrak{r}(I) \subset I(V(I)),$$

and we have to establish the opposite inclusion. We will do this in two stages. We will first prove the result in the special case $V(I) = \emptyset$, that is, we will prove the following.

Theorem 5.17. (Weak Nullstellensatz). *Any ideal $I \subset K[x_1, \ldots, x_n]$ with no common zeros is the unit ideal.*

We will then see that this implies the apparently stronger form.

PROOF OF THEOREM 5.17. We have to show that any ideal I properly contained in the ring $K[x_1, \ldots, x_n]$ must have nonempty zero locus. Since we know that such an ideal I must be contained in some maximal ideal, this will follow from the following proposition.

Proposition 5.18. *Any maximal ideal* $\mathfrak{m}$ *in the ring* $K[x_1, \ldots, x_n]$ *is of the form* $(x_1 - a_1, \ldots, x_n - a_n)$ *for some* $a_1, \ldots, a_n \in K$.

PROOF. The statement that $\mathfrak{m}$ is of the form $(x_1 - a_1, \ldots, x_n - a_n)$ is equivalent to the statement that the quotient $L = K[x_1, \ldots, x_n]/\mathfrak{m}$ is K itself; since K is algebraically closed, this in turn is equivalent to saying just that L is algebraic over K. The key step in showing this is the following lemma.

Lemma 5.19. *Let* R *be a Noetherian ring and* $S \supset R$ *any subring of the polynomial ring* $R[x_1, \ldots, x_n]$. *If* $R[x_1, \ldots, x_n]$ *is finitely generated as an* S*-module, then* S *itself is finitely generated as an* R*-algebra.*

PROOF OF LEMMA 5.19. Let $y_1, \ldots, y_m \in R[x_1, \ldots, x_n]$ be generators of $R[x_1, \ldots, x_n]$ as an S-module; we can write

$$x_i = \sum a_{i,j} \cdot y_j$$

and likewise

$$y_i \cdot y_j = \sum b_{i,j,k} \cdot y_k$$

with $a_{i,j}$, $b_{i,j,k} \in S$. Let $S_0 \subset S$ be the subring generated over R by the coefficients $a_{i,j}$ and $b_{i,j,k}$; being finitely generated over R, S_0 is again Noetherian. By virtue of these relations, the elements $y_1, \ldots, y_m$ generate $R[x_1, \ldots, x_n]$ as an S_0-module. But a submodule of a finitely generated module over a Noetherian ring is again finitely generated; thus S is a finitely generated S_0-module and hence a finitely generated R-algebra. □

PROOF OF PROPOSITION 5.18. Consider again our extension field $L = K[x_1, \ldots, x_n]/\mathfrak{m}$ of K. We can, after reordering the x_i, assume that $x_1, \ldots, x_k \in L$ are algebraically independent over K, with $x_{k+1}, \ldots, x_n$ algebraic over the subfield $K(x_1, \ldots, x_k) \subset L$. Since L is thus a finitely generated $K(x_1, \ldots, x_k)$-module, we can apply Lemma 5.19 to deduce that the purely transcendental extension $K(x_1, \ldots, x_k)$ is a finitely generated K-algebra.

This is where we finally run into a contradiction. Let $z_1, \ldots, z_l \in L$ be a collection of generators of $K(x_1, \ldots, x_k)$ as a K-algebra; write

$$z_i = \frac{P_i(x_1, \ldots, x_k)}{Q_i(x_1, \ldots, x_k)}$$

for some collection of polynomials P_i, Q_i. Now let $f \in K[x_1, \ldots, x_k]$ be any irreducible polynomial. By hypothesis, we can write $1/f$ as a polynomial in the rational functions z_i; clearing denominators, we deduce that f must divide at least one of the polynomials Q_i. This implies in particular that there can be only finitely many irreducible polynomials in $K[x_1, \dots, x_k]$. But for $k \geq 1$ the ring $K[x_1, \dots, x_k]$ contains infinitely many irreducible polynomials (this is true for any field K; in our case, since K is necessarily infinite, we can just exhibit the polynomials $\{x - a\}_{a \in K}$). We may thus deduce that $k = 0$, i.e., L is algebraic over K and hence equal to K. □

Lastly, we want to deduce the Nullstellensatz from the a priori weaker Theorem 5.17. Once more, suppose that $I \subset K[x_1, \dots, x_n]$ is any ideal, and suppose that $f \in K[x_1, \dots, x_n]$ is any polynomial vanishing on the common zeros of I—that is, $f \in I(V(I))$. We want to show that $f^m \in I$ for some $m > 0$.

To do this, we use what is classically called the *trick of Rabinowitsch*. This amounts to realizing the complement $U_f = \{(x_1, \dots, x_n): f(x_1, \dots, x_n) \neq 0\} \subset \mathbb{A}^n$ as an affine variety in its own right—specifically, as the variety

$$\Sigma = \{(x_1, \dots, x_{n+1}): x_{n+1} \cdot f(x_1, \dots, x_n) = 1\} \subset \mathbb{A}^{n+1}$$

and applying the Weak Nullstellensatz there. In other words, we simply observe that the ideal $J \subset K[x_1, \dots, x_{n+1}]$ generated by I and the polynomial $x_{n+1} \cdot f(x_1, \dots, x_n) - 1$ has no common zero locus, and so must be the unit ideal. Equivalently, if

$$A = K[x_1, \dots, x_n][f^{-1}] = K[x_1, \dots, x_{n+1}]/(x_{n+1} \cdot f - 1)$$

is the coordinate ring of U_f, we must have $I \cdot A = (1)$. We can thus write

$$1 = \sum g_i \cdot a_i$$

with $g_i \in I$ and $a_i \in A$; collecting terms involving x_{n+1} we can express this as

$$1 = h_0 + h_1 \cdot x_{n+1} + \cdots + h_m \cdot (x_{n+1})^m$$

with $h_i \in I$. Finally, multiplying through by f^m we have

$$f^m = f^m \cdot h_0 + \cdots + h_m;$$

in particular, $f^m \in I$. □

Alternative Proof of the Nullstellensatz. As promised, we give here a shorter proof of a marginally weaker statement (we have to assume that our ground field K is of infinite transcendence degree over the prime field $\mathbb{Q}$ or $\mathbb{F}_p$).

To begin with, we may replace the ideal I in the statement of the Nullstellensatz by its radical. This is then expressible as an intersection of prime ideals

$$I = \mathfrak{p}_1 \cap \mathfrak{p}_2 \cap \cdots \cap \mathfrak{p}_k.$$

On the other hand, we have $X = \bigcup X_i$ where $X_i = V(\mathfrak{p}_i)$, so

$$I(V(I)) = I(V(\mathfrak{p}_1)) \cap \cdots \cap I(V(\mathfrak{p}_k)).$$

Thus, it suffices to establish the Nullstellensatz for a prime ideal $\mathfrak{p}$, i.e., to establish the following.

Lemma 5.20. *Let $\mathfrak{p} \subset K[x_1, \ldots, x_n]$ be a prime ideal. If $f \in K[x_1, \ldots, x_n]$ is any polynomial not contained in $\mathfrak{p}$, then there exists $a_1, \ldots, a_n \in K$ such that*

$$\mathfrak{p} \subset (x_1 - a_1, \ldots, x_n - a_n),$$

but $f(a_1, \ldots, a_n) \neq 0$, i.e., $f \notin (x_1 - a_1, \ldots, x_n - a_n)$.

PROOF. We will prove this only under one additional hypothesis: that K is of infinite transcendence degree over the prime field $k = \mathbb{Q}$ or $\mathbb{F}_p$. Given this, we write f in the form

$$f = \sum c_I \cdot x^I;$$

we suppose that $\mathfrak{p}$ is generated by polynomials g_α, which we write as

$$g_\alpha = \sum c_{I,\alpha} \cdot x^I$$

and let

$$L = k(\ldots, c_I, \ldots, c_{I,\alpha}, \ldots) \subset K$$

be the field generated over k by the coefficients of f and the g_α. Now set

$$\mathfrak{p}_0 = \mathfrak{p} \cap L[x_1, \ldots, x_n] \subset K[x_1, \ldots, x_n].$$

Note that since all the generators g_α of $\mathfrak{p}$ lie in the subring $L[x_1, \ldots, x_n] \subset K[x_1, \ldots, x_n]$, we have $\mathfrak{p}_0 \cdot K[x_1, \ldots, x_n] = \mathfrak{p}$; also, $\mathfrak{p}_0$ is again prime in $L[x_1, \ldots, x_n]$, so that the quotient $L[x_1, \ldots, x_n]/\mathfrak{p}_0$ is an integral domain. Let M be its quotient field; since M is finitely generated over k, there exists an embedding

$$\iota: M \hookrightarrow K.$$

Let $a_i \in K$ be the image under this map of the element $x_i \in L[x_1, \ldots, x_n]/\mathfrak{p}_0 \subset M$.

It is not hard to see now that $a_1, \ldots, a_n$ fulfill the conditions of the lemma: by construction, the ideal $\mathfrak{p}_0$ is contained in the ideal $(x_1 - a_1, \ldots, x_n - a_n) \subset L[x_1, \ldots, x_n]$, and so $\mathfrak{p} = \mathfrak{p}_0 \cdot K[x_1, \ldots, x_n]$ is contained in the ideal $(x_1 - a_1, \ldots, x_n - a_n) \subset K[x_1, \ldots, x_n]$. On the other hand, $f \in L[x_1, \ldots, x_n]$, but $f \notin \mathfrak{p}_0$, so $\iota(f) = f(a_1, \ldots, a_n) \neq 0$, i.e., $f \notin (x_1 - a_1, \ldots, x_n - a_n)$. □

Restatements and Corollaries

We can reexpress the Nullstellensatz in the often useful form of the following theorem.

Theorem 5.21. *Every prime ideal in $K[x_1, \ldots, x_n]$ is the intersection of the ideals of the form $(x_1 - a_1, \ldots, x_n - a_n)$ containing it.*

Note that the same statement for the ring $K[x_1, \ldots, x_n]/I$ (i.e., that every prime ideal is the intersection of the ideals of the form $(x_1 - a_1, \ldots, x_n - a_n)$ containing it) follows immediately by applying Theorem 5.21 to the prime ideal $(f_1, \ldots, f_k, g_1, \ldots, g_l)$ where $I = (g_1, \ldots, g_l)$. In particular, note that if a polynomial f does not vanish anywhere in $V(I)$, it must be a unit in $K[x_1, \ldots, x_n]/I$.

As one corollary of the Nullstellensatz, we can now give a proof of Lemma 2.1. The circumstances are these: $X \subset \mathbb{A}^n$ is any affine variety, $f \in K[x_1, \ldots, x_n]$ any polynomial and $U_f = \{p \in X : f(p) \neq 0\} \subset X$ the corresponding distinguished open subset. By definition, a *regular* function on U_f is a function g such that for any point $p \in U_f$ we can write

$$g = h/k$$

in some neighborhood of p with $k(p) \neq 0$. We claim that the ring of such functions is just the localization $A(X)[1/f]$ of $A(X)$.

To prove this, note first that by the Noetherian property of the Zariski topology (page 18), if g is any regular function on U_f then we can find a finite open cover $\{U_\alpha\}$ of U_f such that in each U_α we can write

$$g = h_\alpha/k_\alpha$$

with k_α nowhere zero on U_α; we can further take the open sets U_α in this cover to be distinguished, i.e., we can assume $U_\alpha = U_f \cap U_{f_\alpha}$ for some collection f_α. Now, since the open sets U_{f_α} cover U_f and k_α is nowhere zero on $U_f \cap U_{f_\alpha}$, the common zero locus of the k_α must be contained in the zero locus of f; by the Nullstellensatz we must have $f^m \in (\ldots, k_\alpha, \ldots)$ for some m, or in other words, we can write

$$f^m = \sum l_\alpha \cdot k_\alpha.$$

But now

$$f^m \cdot g = \sum (l_\alpha \cdot k_\alpha) \cdot (h_\alpha/k_\alpha) = \sum l_\alpha h_\alpha;$$

that is,

$$g = \frac{\sum l_\alpha h_\alpha}{f^m} \in A(X)[1/f]. \qquad \square$$

We note one immediate corollary of this: that any regular function on $\mathbb{A}^n$ itself must be a polynomial. It follows in particular that any regular function on $\mathbb{P}^n$ must be a constant; this is a special case of Corollary 3.14.

Proof of Proposition 5.6. This proposition asserts that every radical ideal $I \subset K[x_1, \ldots, x_n]$ is a finite intersection of prime ideals.

We use here the property of Noetherian rings that every collection $\{\mathfrak{a}_i\}$ of ideals contains maximal elements, that is, ideals $\mathfrak{a}_i$ not contained in any other ideal of the collection. We apply this to the collection of radical ideals $I \subset K[x_1, \ldots, x_n]$ such that I is not a finite intersection of prime ideals; we let I_0 be a maximal such ideal. By construction, I_0 is not itself prime; let a and $b \in K[x_1, \ldots, x_n]$ be polynomials not in I_0 such that $ab \in I_0$, and let

$$I_1 = \mathfrak{r}(I_0, a) \quad \text{and} \quad I_2 = \mathfrak{r}(I_0, b)$$

be the radicals of the ideals generated by I_0 together with a and b. Since I_1 and I_2 are radical and strictly contain I_0, by hypothesis each will be a finite intersection of prime ideals; Proposition 5.6 will thus follow once we establish that

$$I_0 = I_1 \cap I_2.$$

To show this, suppose that $f \in I_1 \cap I_2$. By definition, we will have $f^m \in (I_0, a)$ and $f^n \in (I_0, b)$ for some m and n, i.e., we can write

$$f^m = g_1 + h_1 \cdot a \quad \text{and} \quad f^n = g_2 + h_2 \cdot b$$

with $g_1, g_2 \in I_0$. But then

$$f^{m+n} = g_1 g_2 + g_1 h_2 b + g_2 h_1 a + h_1 h_2 \cdot ab \in I_0$$

and since I_0 is radical it follows that $f \in I_0$.

LECTURE 6

Grassmannians and Related Varieties

Example 6.1. Grassmannians

Grassmannians are fundamental objects in algebraic geometry: they are simultaneously objects of interest in their own right and basic tools in the construction and study of other varieties. We will be dealing with Grassmannians constantly in the course of this book; here we introduce them and mention a few of their basic properties.

By way of notation, we let $G(k, n)$ denote the set of k-dimensional linear subspaces of the vector space K^n; if we want to talk about the set of k-planes in an abstract vector space V without making a choice of basis for V we also write $G(k, V)$. Of course, a k-dimensional subspace of a vector space K^n is the same thing as a $(k-1)$-plane in the corresponding projective space $\mathbb{P}^{n-1}$, so that we can think of $G(k, n)$ as the set of such $(k-1)$-planes; when we want to think of the Grassmannian this way we will write it $\mathbb{G}(k-1, n-1)$ or $\mathbb{G}(k-1, \mathbb{P}V)$.

In most contexts, Grassmannians are defined initially via coordinate patches or as a quotient of groups; it is then observed that they may be embedded in a projective space. Since our main objects of interest here are projective varieties, we will do it differently, describing the Grassmannian first as a subset of projective space. This is straightforward: if $W \subset V$ is the k-dimensional linear subspace spanned by vectors $v_1, \ldots, v_k$, we can associate to W the multivector

$$\lambda = v_1 \wedge \cdots \wedge v_k \in \bigwedge^k(V).$$

λ is determined up to scalars by W: if we chose a different basis, the corresponding vector λ would simply be multiplied by the determinant of the change of basis matrix. We thus have a well-defined map of sets

$$\psi: G(k, V) \to \mathbb{P}(\wedge^k V).$$

In fact, this is an inclusion: for any $[\omega] = \psi(W)$ in the image, we can recover the corresponding subspace W as the space of vectors $v \in V$ such that $v \wedge \omega = 0 \in \wedge^{k+1} V$. This inclusion is called the *Plücker embedding* of $G(k, V)$.

The homogeneous coordinates on $\mathbb{P}^N = \mathbb{P}(\wedge^k V)$ are called *Plücker coordinates* on $G(k, V)$. Explicitly, if we choose an identification $V \cong K^n$ we can represent the plane W by the $k \times n$ matrix M_W whose rows are the vectors v_i; the matrix M_W is determined up to multiplication on the left by an invertible $k \times k$ matrix. The Plücker coordinates are then just the maximal minors of the matrix M_W.

We have described the Grassmannian $G(k, V)$ as a subset of $\mathbb{P}(\wedge^k V)$; we should now check that it is indeed a subvariety. This amounts to characterizing the subset of *totally decomposable* vectors $\omega \in \wedge^k V$, that is, products $\omega = v_1 \wedge \cdots \wedge v_k$ of linear factors. We begin with a basic observation: given a multivector $\omega \in \wedge^k V$ and a vector $v \in V$, the vector v will divide ω—that is, ω will be expressible as $v \wedge \varphi$ for some $\varphi \in \wedge^{k-1} V$—if and only if the wedge product $\omega \wedge v = 0$. Moreover, a multivector ω will be totally decomposable if and only if the space of vectors v dividing it is k-dimensional. Thus, $[\omega]$ will lie in the Grassmannian if and only if the rank of the map

$$\begin{aligned} \varphi(\omega): V &\to \wedge^{k+1} V \\ : v &\mapsto \omega \wedge v \end{aligned}$$

is $n - k$. Since the rank of $\varphi(\omega)$ is never strictly less than $n - k$, we can say

$$[\omega] \in G(k, V) \Leftrightarrow \operatorname{rank}(\varphi(\omega)) \le n - k.$$

Now, the map $\wedge^k V \to \operatorname{Hom}(V, \wedge^{k+1} V)$ sending ω to $\varphi(\omega)$ is linear, that is, the entries of the matrix $\varphi(\omega) \in \operatorname{Hom}(V, \wedge^{k+1} V)$ are homogeneous coordinates on $\mathbb{P}(\wedge^k V)$; we can say that $G(k, V) \subset \mathbb{P}(\wedge^k V)$ is the subvariety defined by the vanishing of the $(n - k + 1) \times (n - k + 1)$ minors of this matrix.

This is the simplest way to see that $G(k, V)$ is a subvariety of $\mathbb{P}(\wedge^k V)$, but the polynomials we get in this way are far from the simplest possible; in particular, they do not generate the homogeneous ideal of $G(k, V)$. To find the actual generators of the ideal, we need to invoke also the natural identification of $\wedge^k V$ with the exterior power $\wedge^{n-k} V^*$ of the dual space V^* (this is natural only up to scalars, but that's okay for our purposes). In particular, an element $\omega \in \wedge^k V$ corresponding to $\omega^* \in \wedge^{n-k} V^*$ gives rise in this way to a map

$$\begin{aligned} \psi(\omega): V^* &\to \wedge^{n-k+1} V^* \\ : v^* &\mapsto v^* \wedge \omega^*; \end{aligned}$$

by the same argument ω will be totally decomposable if and only if the map $\psi(\omega)$ has rank at most k. What's more, in case ω is totally decomposable, the kernel of the map $\varphi(\omega)$—the subspace W itself—will be exactly the annihilator of the kernel of $\psi(\omega)$; equivalently, the images of the transpose maps

$${}^t\varphi(\omega): \wedge^{k+1} V^* \to V^*$$

and

$$^t\psi(\omega)\colon \wedge^{n-k+1} V \to V$$

annihilate each other. In sum, then, we see that $[\omega] \in G(k, V)$ if and only if for every pair $\alpha \in \wedge^{k+1} V^*$ and $\beta \in \wedge^{n-k+1} V$, the contraction

$$\Xi_{\alpha,\beta}(\omega) = \langle {}^t\varphi(\omega)(\alpha), {}^t\psi(\omega)(\beta) \rangle = 0.$$

The $\Xi_{\alpha,\beta}$ are thus quadratic polynomials whose common zero locus is the Grassmannian $G(k, V)$. They are called the *Plücker relations*, and they do in fact generate the homogeneous ideal of $G(k, V)$, though we will not prove that here.

Exercise 6.2. In the special case $k = 2$, assuming $\text{char}(K) \neq 2$ show directly that a vector $\omega \in \wedge^2 V$ is decomposable if and only if $\omega \wedge \omega = 0$ and hence that the Grassmannian $G(2, V) \subset \mathbb{P}(\wedge^2 V)$ is a variety cut out by quadrics. (In fact, the equation $\omega \wedge \omega = 0$ represents $\binom{n}{4}$ independent quadratic relations, which are exactly the span of the Plücker relations.)

Observe in particular that the first nontrivial Grassmannian—the first one that is not a projective space—is $G(2, 4)$, and this sits as a quadric hypersurface in $\mathbb{P}(\wedge^2 K^4) \cong \mathbb{P}^5$.

We can get another picture of the Grassmannian by looking at certain special affine open subsets. To describe these first intrinsically, let $\Gamma \subset V$ be a subspace of dimension $n - k$, corresponding to a multivector $\omega \in \wedge^{n-k} V = \wedge^k V^*$. We can think of ω as a homogeneous linear form on $\mathbb{P}(\wedge^k V)$; let $U \subset \mathbb{P}(\wedge^k V)$ be the affine open subset where $\omega \neq 0$. Then the intersection of $G(k, V)$ with U is just the set of k-dimensional subspaces $\Lambda \subset V$ complementary to Γ. Any such subspace can be viewed as the graph of a map from V/Γ to Γ and vice versa, so that we have an identification

$$G(k, V) \cap U \cong \text{Hom}(V/\Gamma, \Gamma) \cong K^{k(n-k)}.$$

To see this in coordinates, identify V with K^n and say the subspace Γ is spanned by the last $n - k$ basis vectors $e_{k+1}, \ldots, e_n \in K^n$. Then $U \cap G(k, n)$ is the subset of spaces Λ such that the $k \times n$ matrix M_Λ whose first $k \times k$ minor is nonzero. It follows that any $\Lambda \in G(k, V) \cap U$ is represented as the row space of a unique matrix of the form

$$\begin{bmatrix} 1 & 0 & 0 & . & . & 0 & a_{1,1} & a_{1,2} & . & . & . & a_{1,n-k} \\ 0 & 1 & 0 & . & . & 0 & a_{2,1} & a_{2,2} & . & . & . & a_{2,n-k} \\ \vdots & & & & & & & & & & & \\ 0 & 0 & . & . & 0 & 1 & a_{k,1} & a_{k,2} & . & . & . & a_{k,n-k} \end{bmatrix}$$

and vice versa. The entries $a_{i,j}$ of this matrix then give the bijection of $U \cap G(k, V)$ with $K^{k(n-k)}$.

Note that the affine coordinates on the affine open subset of $G(k, V)$ are just the $k \times k$ minors of this matrix, which is to say the minors of all sizes of the $k \times (n-k)$ matrix $(a_{i,j})$. In particular, expansion of any of these determinants along any row or column yields a quadratic relation among these minors; thus, for example,

$$a_{1,1} \cdot a_{2,2} - a_{1,2} \cdot a_{2,1} = \begin{vmatrix} a_{1,1} & a_{1,2} \\ a_{2,1} & a_{2,2} \end{vmatrix}$$

is a relation among the affine coordinates on $\mathbb{P}(\wedge^k K^n)$ restricted to $G(k, n)$. In this way, we can write down all the Plücker relations explicitly in coordinates.

There is, finally, another way to describe the affine coordinates on the open subset $U \cap G(k, n)$ of k-planes Λ complementary to a given $(n-k)$-plane Γ: we take vectors $v_1, \ldots, v_k \in K^n$ that, together with Γ, span all of K^n, and set

$$v_i(\Lambda) = \Lambda \cap (\Gamma + v_i).$$

The vectors $v_i(\Lambda)$ then give a basis for Λ, for all $\Lambda \in U$; and the k-tuple of vectors $v_i(\Lambda) - v_i \in \Gamma$ gives an identification of $U \cap G(k, n)$ with Γ^k.

Subvarieties of Grassmannians

To begin with, an inclusion of vector spaces $W \hookrightarrow V$ induces an inclusion of Grassmannians $G(k, W) \hookrightarrow G(k, V)$; likewise, a quotient map $V \to V/U$ to the quotient of V by an l-dimensional subspace U induces an inclusion $G(k-l, V/U) \hookrightarrow G(k, V)$. More generally, if $U \subset W \subset V$, we have an inclusion $G(k-l, W/U) \hookrightarrow G(k, V)$. The images of such maps are called sub-Grassmannians and are subvarieties of $G(k, V)$ (in terms of the Plücker embedding $G(k, V) \hookrightarrow \mathbb{P}(\wedge^k V)$, they are just the intersection of $G(k, V)$ with linear subspaces in $\mathbb{P}(\wedge^k V)$, as we will see in the following paragraph).

If we view the Grassmannian as the set of linear subspaces in a projective space $\mathbb{P}V$, the sub-Grassmannians are just the subsets of planes contained in a fixed subspace and/or containing a fixed subspace. We can also consider the subset $\Sigma(\Lambda) \subset \mathbb{G}(k, \mathbb{P}V)$ of k-planes that meet a given m-dimensional linear subspace $\Lambda \subset \mathbb{P}V$, or more generally the subset $\Sigma_l(\Lambda)$ of k-planes that meet a given Λ in a subspace of dimension of at least l. These are again subvarieties of the Grassmannian; $\Sigma_l(\Lambda)$ may be described as the locus

$$\Sigma_l(\Lambda) = \{[\omega] \colon \omega \wedge v_1 \wedge \cdots \wedge v_{m-l+1} = 0 \quad \forall v_1, \ldots, v_{m-l+1} \in \Lambda\}$$

from which we see in particular that it, like the sub-Grassmannians, is the intersection of the Grassmannian with a linear subspace of $\mathbb{P}(\wedge^k V)$. These are in turn special cases of a class of subvarieties of $\mathbb{G}(k, \mathbb{P}V)$ called *Schubert cycles*, about which we will write more later.

There are also analogs for Grassmannians of projection maps on projective space. Specifically, suppose $W \subset V$ is a subspace of codimension l in the n-dimensional vector space V. For $k \leq l$, we have a map $\pi \colon U \to G(k, V/W)$ defined on the open set $U \subset G(k, V)$ of k-planes meeting W only in (0) simply by taking the image; for $k \geq l$ we have a map $\eta\colon U' \to G(k-l, W)$ defined on the open subset $U' \subset G(k, V)$ of planes transverse to W by taking the intersection. Note that both these maps may be realized, via the Plücker embeddings of both target and domain, by a linear projection on the ambient projective space $\mathbb{P}(\wedge^k V)$—for example, the map π is the restriction to $G(k, V)$ of the linear map $\mathbb{P}(\wedge^k V) \to \mathbb{P}(\wedge^k(V/W))$ induced by the projection $V \to V/W$.

Example 6.3. The Grassmannian $\mathbb{G}(1, 3)$

The next few exercises deal specifically with the geometry of the Grassmannian $\mathbb{G} = \mathbb{G}(1, 3)$ parametrizing lines in $\mathbb{P}^3$, which as we have seen may be realized (via the Plücker embedding) as a quadric hypersurface in $\mathbb{P}^5$.

Exercise 6.4. For any point $p \in \mathbb{P}^3$ and plane $H \subset \mathbb{P}^3$ containing p, let $\Sigma_{p,H} \subset \mathbb{G}$ be the locus of lines in $\mathbb{P}^3$ passing through p and lying in H. Show that under the Plücker embedding $\mathbb{G} \to \mathbb{P}^5$, $\Sigma_{p,H}$ is carried to a line, and that conversely every line in $\mathbb{P}^5$ lying on $\mathbb{G}$ is of the form $\Sigma_{p,H}$ for some p and H.

Exercise 6.5. For any point $p \in \mathbb{P}^3$, let $\Sigma_p \subset \mathbb{G}$ be the locus of lines in $\mathbb{P}^3$ passing through p; for any plane $H \subset \mathbb{P}^3$, let $\Sigma_H \subset \mathbb{G}$ be the locus of lines in $\mathbb{P}^3$ lying in H. Show that under the Plücker embedding, both Σ_p and Σ_H are carried into two-planes in $\mathbb{P}^5$, and the conversely any two-plane $\Lambda \cong \mathbb{P}^2 \subset \mathbb{G} \subset \mathbb{P}^5$ is either equal to Σ_p for some p or to Σ_H for some H.

Exercise 6.6. Let $l_1, l_2 \subset \mathbb{P}^3$ be skew lines. Show that the set $Q \subset \mathbb{G}$ of lines in $\mathbb{P}^3$ meeting both is the intersection of $\mathbb{G}$ with a three-plane $\mathbb{P}^3 \subset \mathbb{P}^5$, and so is a quadric surface. Deduce yet again that $Q \cong \mathbb{P}^1 \times \mathbb{P}^1$. What happens if l_1 and l_2 meet?

Exercise 6.7. Now let $Q \subset \mathbb{P}^3$ be a smooth quadric surface. Show that the two families of lines on Q correspond to plane conic curves on $\mathbb{G}$ lying in complementary two-planes $\Lambda_1, \Lambda_2 \subset \mathbb{P}^5$. Show that, conversely, the lines in $\mathbb{P}^3$ corresponding to a plane conic curve $C \subset \mathbb{G}$ sweep out a smooth quadric surface if and only if the plane Λ spanned by C is not contained in $\mathbb{G}$. What happens to this correspondence if either the quadric becomes a cone or the plane Λ lies in G?

The next exercise is a direct generalization of the preceding one; it deals with Segre varieties other than $\mathbb{P}^1 \times \mathbb{P}^1$.

Exercise 6.8. Let $\Sigma_{1,k} \cong \mathbb{P}^1 \times \mathbb{P}^k \subset \mathbb{P}^{2k+1}$ be the Segre variety, and for each $p \in \mathbb{P}^1$ let Λ_p be the fiber of $\Sigma_{1,k}$ over p. We have seen that Λ_p is a k-plane in $\mathbb{P}^{2k+1}$; show that the assignment $p \mapsto \Lambda_p$ defines a regular map of $\mathbb{P}^1$ to the Grassmannian $\mathbb{G}(k, 2k+1)$ whose image is a rational normal curve lying in a $(k+1)$-plane in $\mathbb{P}(\wedge^{k+1} K^{2k+2})$.

Before proceeding, we should mention here a generalization of Exercise 6.4, which will be crucial in the proof of Theorem 10.19. (The reader is encouraged to skip ahead and read this theorem, which does not require much more than what we have introduced already.)

Exercise 6.9. Let $G = G(k, V) \subset \mathbb{P}^N = \mathbb{P}(\wedge^k V)$. (i) Show that for any pair of points $\Lambda, \Lambda' \in G$ the line $\overline{\Lambda, \Lambda'}$ they span in $\mathbb{P}^N$ lies in G if and only if the corresponding k-planes intersect in a $(k-1)$-plane (equivalently, lie in a $(k+1)$-plane). Thus, any line $L \subset G \subset \mathbb{P}^N$ consists of the set of k-planes in V containing a fixed $(k-1)$-plane $\Gamma \subset V$ and contained in a fixed $(k+1)$-plane $\Omega \subset V$.

(ii) Use part (i) to show that any maximal linear subspace $\Phi \subset G \subset \mathbb{P}^N$ is either the set of k-planes containing a fixed linear subspace of V or the set of k-planes contained in a fixed linear subspace of V.

Example 6.10. An Analog of the Veronese Map

There is a somewhat esoteric analog of the Veronese map for Grassmannians. Let $S = K[Z_0, \ldots, Z_n]$ be the homogeneous coordinate ring of projective space $\mathbb{P}^n$, and denote by S_d the dth graded piece of S, that is, the vector space of homogeneous polynomials of degree d in $Z_0, \ldots, Z_n$. Now, for any k-plane $\Lambda \subset \mathbb{P}^n$, let $I(\Lambda)$ be its homogeneous ideal, and let $I(\Lambda)_d \subset S_d$ be its dth graded piece. Then $I(\Lambda)_d$ is a subspace of codimension $\binom{k+d}{d}$ in S_d, and so we get a regular map

$$v_d^*: \mathbb{G}(k, n) \to G\left(\binom{n+d}{d} - \binom{k+d}{d}, \binom{n+d}{d}\right)$$

or, dually, a map

$$v_d: \mathbb{G}(k, n) \to G\left(\binom{k+d}{d}, \binom{n+d}{d}\right).$$

Exercise 6.11. Verify the preceding statements about the codimension of $I(\Lambda)_d$ in S_d and that the map v_d is a regular map.

It is perhaps easier (at least in characteristic 0) to express this map in intrinsic terms: if we view $\mathbb{P}^n$ as the projective space $\mathbb{P}V$ associated to a vector space V, and $\mathbb{G}(k, n) = G(k+1, V)$ as the Grassmannian of $(k+1)$-dimensional subspaces of V, it is just the map sending a subspace $\Lambda \subset V$ to the subspace $\mathrm{Sym}^d(\Lambda) \subset \mathrm{Sym}^d(V)$. (In particular, in the case $k = 0$ we have the usual Veronese map.)

Example 6.12. Incidence Correspondences

Let $\mathbb{G} = \mathbb{G}(k, n)$ be the Grassmannian of k-planes in $\mathbb{P}^n$. We may then define a subvariety $\Sigma \subset \mathbb{G} \times \mathbb{P}^n$ by setting

$$\Sigma = \{(\Lambda, x): x \in \Lambda\}.$$

Σ is simply the subvariety of the product whose fiber over a given point $\Lambda \in \mathbb{G}$ is the k-plane $\Lambda \subset \mathbb{P}^n$ itself; in the language of Lecture 4, it is the "universal family" of k-planes. The simplest example of this is the universal hyperplane, the variety $\Sigma \subset \mathbb{P}V^* \times \mathbb{P}V$ whose fiber over a point $H \in \mathbb{P}V^*$ is just the hyperplane $H \subset \mathbb{P}V$, discussed earlier in Example 4.5. In general, this is the universal family referred to in Example 4.15 when we indicated that the family constructed there was not optimal.

It's not hard to see that Σ is a projective variety; in fact, we may write

$$\Sigma = \{([v_1 \wedge \cdots \wedge v_k], [w]): v_1 \wedge \cdots \wedge v_k \wedge w = 0\};$$

or in the case of the universal hyperplane,

$$\Sigma = \{([v^*], [w]): \langle v^*, w \rangle = 0\} \subset \mathbb{P}V^* \times \mathbb{P}V.$$

The construction of Σ is just the paradigm for a general construction that will arise over and over in elementary algebraic geometry. One example of its usefulness is the following proposition.

Proposition 6.13. *Let $\Phi \subset \mathbb{G}(k, n)$ be any subvariety. Then the union*

$$\Psi = \bigcup_{\Lambda \in \Phi} \Lambda \subset \mathbb{P}^n$$

is also a variety.

PROOF. Let π_1, π_2 be the projection maps from the incidence correspondence Σ to $\mathbb{G}(k, n)$ and to $\mathbb{P}^n$. We can write

$$\Psi = \pi_2(\pi_1^{-1}(\Phi))$$

from which it follows that Ψ is a subvariety of $\mathbb{P}^n$. □

Example 6.14. Varieties of Incident Planes

Let $X \subset \mathbb{P}^n$ be a projective variety. We claim that the locus $\mathscr{C}_k(X)$ of k-planes meeting X is a subvariety of the Grassmannian $\mathbb{G}(k, n)$. To see this, we may use the incidence correspondence $\Sigma \subset \mathbb{G} \times \mathbb{P}^n$ introduced in Example 6.12: we write

$$\mathscr{C}_k(X) = \pi_1(\pi_2^{-1}(X)) \subset \mathbb{G}(k, n)$$

where Σ is the incidence correspondence and $\pi_1: \Sigma \to \mathbb{G}(k, n)$, $\pi_2: \Sigma \to \mathbb{P}^n$ are the projection maps. This variety, called the *variety of incident planes*, will be useful in a number of contexts, most notably the construction of the *Chow variety*. Note that we have already seen that $\mathscr{C}_k(X) \subset \mathbb{G}(k, n)$ is a subvariety in the special case of X a linear subspace of $\mathbb{P}^n$.

Exercise 6.15. Now let $X \subset \mathbb{P}^n$ be a locally closed subset. Show that the closure in $\mathbb{G}(k, n)$ of the locus of k-planes meeting X is the variety of k-planes meeting the closure $\overline{X}$ of X.

Exercise 6.16. (i) Let $C \subset \mathbb{P}^2 \subset \mathbb{P}^3$ be the plane conic curve given by $Z_3 = Z_0Z_2 - Z_1^2 = 0$. Find the equations of the variety of incident lines $\mathscr{C}_1(C) \subset \mathbb{G}(1, 3)$. (ii) Do the same for the twisted cubic given parametrically by $t \mapsto [1, t, t^2, t^3]$. (*)

Example 6.17. The Join of Two Varieties

Let $X, Y \subset \mathbb{P}^n$ be any two disjoint projective varieties. We can combine Proposition 6.13 and Example 6.14 to deduce that the union $J(X, Y) \subset \mathbb{P}^n$ of the lines joining X to Y is again a projective variety. First, by Example 6.14, the set $\mathscr{J}(X, Y)$ of lines joining X and Y is a subvariety of the Grassmannian, since it is expressible as the intersection $\mathscr{C}_1(X) \cap \mathscr{C}_1(Y)$; then by Proposition 6.13 the union of these lines is a subvariety of $\mathbb{P}^n$. We call the variety $J(X, Y)$ the *join* of X and Y.

This construction generalizes that of the cone. We will give another proof that $J(X, Y)$ is indeed a subvariety of $\mathbb{P}^n$ in Example 8.1; this alternate construction will in particular allow us to generalize the definition of $J(X, Y)$ to the case where X and Y do meet (or even are equal).

Exercise 6.18. Give another proof that $J(X, Y)$ is a subvariety of $\mathbb{P}^n$, as follows. First, show that in case X and Y are contained in complementary linear subspaces $\Gamma, \Lambda \subset \mathbb{P}^n$ the join $J(X, Y)$ is simply the intersection of the cones $\overline{\Gamma, Y}$ and $\overline{\Lambda, X}$. Second, reduce to this case by arguing that any pair of disjoint varieties $X, Y \subset \mathbb{P}^n$ can be realized as the images of varieties $\tilde{X}, \tilde{Y}$ contained in complementary linear spaces of a larger projective space $\mathbb{P}^N$ under a linear projection from $\mathbb{P}^N$ to $\mathbb{P}^n$, and that $J(X, Y)$ is simply the image of $J(\tilde{X}, \tilde{Y})$ under π. Does this approach allow you to extend the definition to the case where X and Y meet?

Example 6.19. Fano Varieties

A fundamental type of subvariety of the Grassmannian $\mathbb{G}(k, n)$ is the *Fano variety* associated to a variety $X \subset \mathbb{P}^n$. This is just the variety of k-planes contained in X, that is,

$$F_k(X) = \{\Lambda : \Lambda \subset X\} \subset \mathbb{G}(k, n).$$

To see that $F_k(X)$ is indeed a variety, observe first that it is enough to do this in case X is the hypersurface given by a polynomial $G(Z)$: in general, the Fano variety $F_k(X)$ will be the intersection in $\mathbb{G}(k, n)$ of the Fano varieties associated to the hypersurfaces containing it. To show it in this case, we work locally; we restrict our attention to the affine open subset $U \subset G(k + 1, n + 1)$ of $(k + 1)$-planes $\Lambda \subset K^{n+1}$ complementary to a given $(n - k)$-plane Λ_0 and exhibit explicitly equations for $F_k(X) \cap U \subset U \cong K^{(k+1)(n-k)}$. We start by choosing a basis $v_0(\Lambda), \ldots, v_k(\Lambda)$ for each $\Lambda \in U$ by taking vectors $v_0, \ldots, v_k \in V$ that, together with Λ_0, span all of V, and setting

$$v_i(\Lambda) = \Lambda \cap (\Lambda_0 + v_i).$$

As we saw in the discussion of Grassmannians, the coordinates of these vectors are regular functions on U. Now, we can view the homogeneous polynomial G as an element of $\mathrm{Sym}^d(K^{n+1})^* \subset ((K^{n+1})^{\otimes d})^*$, and set, for each multi-index $I = \{i_1, \ldots, i_d\}$,

$$a_I(\Lambda) = G(v_{i_1}(\Lambda), \ldots, v_{i_d}(\Lambda))$$

(to put it another way, the a_I are the coefficients of the restriction of G to Λ, written in terms of the basis for Λ dual to the basis $\{v_0(\Lambda), \ldots, v_k(\Lambda)\}$). The $a_I(\Lambda)$ then provide a system of polynomials cutting out $F_k(X)$ in U.

For a more intrinsic version of this argument, recall from Example 6.10 that the map sending a k-plane $\Lambda \subset \mathbb{P}^n$ to the dth graded piece $I(\Lambda)_d$ of its ideal, viewed as a subspace of the space S_d of homogeneous polynomials of degree d, is a regular map

$$v_d^*\colon \mathbb{G}(k, n) \to G(l, N)$$

where $N = \binom{n+d}{d}$, $l = \binom{n+d}{d} - \binom{k+d}{d}$.

Now, the subset $\Phi \subset G(l, N)$ of l-planes in S_d containing the homogeneous polynomial $G \in S_d$ is a subvariety, and we can write

$$F_k(X) = (v_d^*)^{-1}(\Phi)$$

so $F_k(X)$ is indeed a subvariety of $\mathbb{G}(k, n)$.

Exercise 6.20. Carry this out in the case of the quadric surface $Q \subset \mathbb{P}^3$ given by the polynomial $Z_0Z_3 - Z_1Z_2$, and show that the Fano variety $F_1(Q)$ is a union of two conic curves. Compare this with the parametric description of $F_1(Q)$ given in the discussion of the Segre map.

LECTURE 7

Rational Functions and Rational Maps

Rational Functions

Let $X \subset \mathbb{A}^n$ be an irreducible affine variety. Since its coordinate ring $A(X)$ is an integral domain, we can form its quotient field; this is called the *rational function field* of X and is usually denoted $K(X)$; its elements are called *rational functions* on X. Note that if $Y \subset X$ is an open subset that is an affine variety in its own right (as in the discussion on page 19), the function field of Y will be the same as that of X.

One warning: a rational function $h \in K(X)$ is written as a quotient f/g, where f and $g \in A(X)$ are regular functions on X; but despite the name, h itself is not a function on X; even if we allow ∞ as a value at points where $g = 0$, we cannot in general make sense of h at points where both f and g vanish. We will see shortly in what sense we can deal with these objects as maps.

Next, consider an irreducible projective variety $X \subset \mathbb{P}^n$. We can define its function field in two ways. We can either take the rational function field $K(U)$ of any nonempty affine open $U = X \cap \mathbb{A}^n$ or we can form the field of fractions of the homogeneous coordinate ring $S(X)$ and take the 0th graded piece of that field, that is, take expressions of the form $h(Z) = F(Z)/G(Z)$ where F and G are homogeneous of the same degree.

Exercise 7.1. Show that this all makes sense, i.e., that the first definition of $K(X)$ is independent of the choice of affine open U and that the second agrees with the first.

We can extend this discussion to reducible varieties; for example, we can form the quotient ring of the coordinate ring of any variety by inverting all non-zero divisors (though it won't in general be a field; indeed, by Theorem 5.7 it will be a direct sum of the fields $K(X_i)$). Likewise, a rational function on a reducible variety X will be given by specifying (arbitrarily) a rational function on each irreducible component of X, equivalently, by specifying a regular function on some dense open subset of X.

Rational Maps

Having introduced the notion of rational function, we can now discuss rational maps. We will proceed in four steps. We will give a provisional definition, explain why the definition needs to be fixed up (this is essentially just a matter of abuse of terminology), indicate how we may fix it up, and then proceed with our discussion.

Provisional Definition 7.2. Let X be an irreducible variety. A *rational map* φ from X to $\mathbb{A}^n$ is given by an n-tuple of rational functions, i.e.,

$$\varphi(x) = (h_1(x), \ldots, h_n(x))$$

where $h_i \in K(X)$. Likewise, a rational map of X to $\mathbb{P}^n$ is given by an $(n + 1)$-tuple of rational functions

$$\varphi(x) = [h_0(x), \ldots, h_n(x)].$$

A rational map is usually represented by a dotted arrow:

$$\varphi: X \dashrightarrow \mathbb{P}^n.$$

Note that if $X \subset \mathbb{P}^m$ is projective, there is another way to represent a rational map to $\mathbb{P}^n$: we may write the rational functions $h_i(x)$ in the form $F_i(x)/G_i(x)$, where F_i and G_i are homogeneous polynomials of the same degree d_i and then multiply the vector $[h_0(x), \ldots, h_n(x)]$ by the product of the G_i to arrive at an expression

$$\varphi(x) = [H_0(x), \ldots, H_n(x)]$$

where the H_i are homogeneous polynomials of the same degree. The difference between this and the corresponding description of a regular map to $\mathbb{P}^n$, of course, is that we do not require that the H_i do not all vanish simultaneously at some points of X.

Lastly, note that, since the field of rational functions on a variety X is the same as the function field of an affine open $U \subset X$, there is essentially no difference between giving a rational map on X and giving one on U. In particular, every rational map on X is regular on some open subset of X, and conversely any regular map on an affine open subset of X extends to a rational one on all of X.

What's wrong? Essentially the same objection as raised in connection with rational functions: a rational map, despite the name, is not a map, since it may not be defined at some points of X. But if a rational map is not a map, what sort of object is it? Definition 7.2, which says that a rational map is "given by" a collection of rational functions, but does not say what it actually is, is clearly unsatisfactory from a formal standpoint. For example, since a rational map $\varphi: X \dashrightarrow Y$ cannot be described simply as an assignment to each point $p \in X$ of a point of Y, it may not a priori be clear when two rational maps are to be considered the same.

The solution to this problem lies in the preceding observations that a rational map $\varphi: X \dashrightarrow Y$ should be defined on a Zariski open subset $U \subset X$; that conversely every regular map on an open $U \subset X$ should extend to a rational one on X; and that a rational map should be determined by its values on any open subset $U \subset X$ where it is defined. Combining these observations we are led to the following definition.

Definition 7.3. Let X be an irreducible variety and Y any variety. A *rational map*

$$\varphi: X \dashrightarrow Y$$

is defined to be an equivalence class of pairs (U, γ) with $U \subset X$ a dense Zariski open subset and $\gamma: U \to Y$ a regular map, where two such pairs (U, γ) and (V, η) are said to be equivalent if $\gamma|_{U \cap V} = \eta|_{U \cap V}$.

Note that if $Z \subset X$ is any open set, there is a natural bijection between the sets of rational maps from X to Y and of rational maps from Z to Y. Similarly, if $W \subset Y$ is any open subset, we have an inclusion of the set of rational maps $X \dashrightarrow W$ into the set of rational maps $X \dashrightarrow Y$; the image of this inclusion is just the set of those rational maps $[(U, \gamma)]$ such that $\gamma(U) \not\subset Y - W$.

Let $\varphi: X \dashrightarrow Y$ and $\eta: Y \dashrightarrow Z$ be a pair of rational maps. In case there exist pairs (U, f) representing φ and (V, g) representing η such that $f^{-1}(V) \neq \emptyset$, we define the *composition* $\eta \circ \varphi$ to be the equivalence class of $(f^{-1}(V), g \circ f)$. If φ is the inclusion of a subvariety $X \subset Y$, we also call this the *restriction* of η to X. Note that neither is defined in general: it may well be that for any (U, f) and (V, g) the image of f lies entirely outside V.

If Definition 7.3 is the correct one, why start with a provisional definition? The answer is that Definition 7.2 is in fact much closer to the way we actually think about rational maps in practice, that is, as maps given by rational functions. For the most part, this is how we will work with them, being always careful not to assume formal properties they don't possess (such as restriction or composition). In any event, Definition 7.3 is what we may consider our *first point of view on rational maps*: a rational map $\varphi: X \dashrightarrow Y$ is a regular map on an open dense subset of X.

As in the case of rational functions, we can extend the preceding discussion to the case of reducible varieties, for example, by applying Definition 7.3. In particular, giving a rational map on a variety X with irreducible components $X_1, \ldots, X_k$ will be the same as giving a rational map on each component separately, without any conditions coming from their intersections. For this reason, nothing is lost if in discussing rational maps we restrict ourselves to the case where the domain is irreducible. By way of terminology, then, we will adopt this convention; for example, if we write "let $\varphi: X \dashrightarrow \mathbb{P}^n$ be a rational map" we will be implicitly assuming that X is irreducible.

Example 7.4.

The simplest example of a rational map that is not regular is the map

$$\varphi: \mathbb{A}^2 \dashrightarrow \mathbb{P}^1$$

given by

$$\varphi(x, y) = [x, y].$$

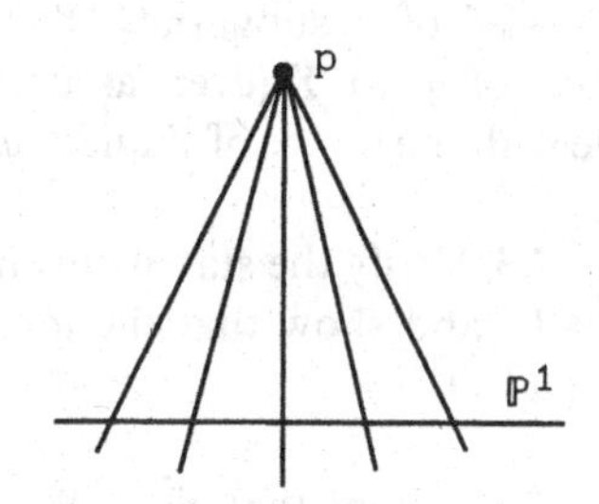

Note that this map is defined exactly on $\mathbb{A}^2 - \{(0, 0)\}$, and sends lines through the origin in $\mathbb{A}^2$ to the points of $\mathbb{P}^1$ they represent; in particular, there is no way it can be continuously extended to all of $\mathbb{A}^2$. Completing, we may think of φ as a rational map from $\mathbb{P}^2$ to $\mathbb{P}^1$, which may then be described geometrically: it is just the projection from the point $p = [0, 0, 1] \in \mathbb{P}^2$ to a line.

This has an obvious generalization: for any plane $\Lambda \cong \mathbb{P}^k$ in $\mathbb{P}^n$, the projection map π_Λ from Λ to a subspace $\mathbb{P}^{n-k-1} \subset \mathbb{P}^n$, regular on the complement $\mathbb{P}^n - \Lambda$, may be thought of as a rational map from $\mathbb{P}^n$ to $\mathbb{P}^{n-k-1}$; and similarly for any $X \subset \mathbb{P}^n$ not contained in Λ the restriction of π_Λ to X gives a rational map $X \dashrightarrow \mathbb{P}^{n-k-1}$, whether or not X meets Λ.

Graphs of Rational Maps

Let X be a projective variety and $\varphi: X \dashrightarrow \mathbb{P}^n$ a rational map; let $U \subset X$ be an open subset where φ is defined. The graph of $\varphi|_U$ is as we have seen a closed subvariety of $U \times \mathbb{P}^n$; by the *graph* of the rational map φ we will mean the closure Γ_φ in $X \times \mathbb{P}^n$ of the graph of $\varphi|_U$. Note that this is independent of the choice of open subset $U \subset X$ (in particular, if φ is regular, this is just the ordinary graph).

Since the graph Γ_φ of the rational map φ is a closed subvariety of $X \times \mathbb{P}^n$, it is also a projective variety. It follows that the projection $\pi_2(\Gamma_\varphi)$ of the graph Γ_φ to $\mathbb{P}^n$ is again a projective variety; we will define the *image* of the rational map φ to be the image of its graph Γ_φ.

We can also use the graph of φ to define the notion of image and inverse image for rational maps. With φ as earlier, for any closed subvariety $Z \subset \mathbb{P}^n$ we take the *inverse image* $\varphi^{-1}(Z)$ of Z in X to be the image

$$\varphi^{-1}(Z) = \pi_1(\pi_2^{-1}(Z)),$$

where $\pi_1: \Gamma_\varphi \to X$ and $\pi_2: \Gamma_\varphi \to \mathbb{P}^n$ are the projections. Likewise, for any closed $Y \subset X$, we define the *image*, or *total transform* of Y under φ to be the image

$$\varphi(Y) = \pi_2(\pi_1^{-1}(Y)).$$

One warning: as in the case of the terminology "rational map" itself, the terms "image" and "inverse image" may be misleading; for example, it is not the case that for any point $p \in \mathbb{P}^n$ in the image of φ there exists a point $q \in X$ with $\varphi(q) = p$. It gets especially dangerous when we talk about the image under a rational map $\varphi: X \dashrightarrow \mathbb{P}^n$ of a subvariety $Y \subset X$; as Exercise 7.6 shows, the image of the restriction of φ to Y (even assuming this restriction exists) will not in general coincide with the image of Y under φ (the former is called the *proper transform* of Y).

Exercise 7.5. Verify the statement that the graph Γ_φ is independent of the choice of open set U, and show that the image of the rational map φ is the closure of the image of $\varphi(U)$.

Exercise 7.6. Show that if $\varphi: X \dashrightarrow \mathbb{P}^n$ is a rational map and $Z \subset X$ a subvariety such that the restriction η of φ to Z is defined, the graph $\Gamma_\eta \subset Z \times \mathbb{P}^n$ will be contained in the inverse image $(\pi_1)^{-1}(Z) = \Gamma_\varphi \cap (Z \times \mathbb{P}^n) \subset X \times \mathbb{P}^n$, but may not be equal to it.

Exercise 7.7. Show that the image of a rational normal curve $C \subset \mathbb{P}^n$ under projection from a point $p \in C$ is a rational normal curve $C' \subset \mathbb{P}^{n-1}$.

By way of notation, for any projective variety $Y \subset \mathbb{P}^n$, we will now define a rational map $\varphi: X \dashrightarrow Y$ to be a rational map from X to $\mathbb{P}^n$ whose image is contained in Y; by the preceding this is the same as requiring $\varphi(U) \subset Y$ for any $U \subset X$ on which φ is regular.

We observed, following Exercise 2.24, that it is not the case that a map φ: $X \to \mathbb{P}^m$ on a projective variety $X \subset \mathbb{P}^n$ is regular if and only if its graph Γ_φ is a subvariety of the product $\mathbb{P}^n \times \mathbb{P}^m$. It is true, however, if we replace the word "regular" with "rational."

Exercise 7.8. Let $X \subset \mathbb{P}^m$ be an irreducible quasiprojective variety. Assuming that the characteristic of the field K is 0, show that a set map $\varphi: X \to \mathbb{P}^n$ is rational if and only if its graph is closed in $X \times \mathbb{P}^n$—in other words, a subvariety $\Gamma \subset X \times \mathbb{P}^n$ that meets the general fiber $\{p\} \times \mathbb{P}^n$ in one point determines a rational map from X to $\mathbb{P}^n$. (*)

Exercise 7.8 is difficult, as is suggested by the fact that it is false in positive characteristic. For example, suppose $\text{char}(K) = p$ and consider the graph $\Gamma_\alpha \subset \mathbb{A}^1 \times \mathbb{A}^1$ of the map $\alpha: \mathbb{A}^1 \to \mathbb{A}^1$ given by $x \mapsto x^p$. Γ_α meets the general (indeed, every) fiber $\mathbb{A}^1 \times \{p\}$ in a single point, so that α has an inverse φ as a set map and the graph of φ is closed in $\mathbb{A}^1 \times \mathbb{A}^1$; but φ is clearly not rational. In any event, Exercise 7.8 gives us (at least in characteristic 0) our *second point of view on rational maps*: rational maps $\varphi: X \to \mathbb{P}^n$ correspond one to one to irreducible closed subvarieties $\Gamma \subset X \times \mathbb{P}^n$ such that for general $p \in X$ the intersection $\Gamma \cap \{p\} \times \mathbb{P}^n$ is a single point.

We can use this to extend Chow's theorem to rational functions. Let $X \subset \mathbb{P}^n$ be any complex submanifold of $\mathbb{P}^n$, and f any meromorphic function on X. By Theorem 1.9, X is a subvariety of $\mathbb{P}^n$; we can use Theorem 1.9 to further deduce that f is rational.

Exercise 7.9. Let $\varphi: X \dashrightarrow \mathbb{P}^n$ be a rational map, Γ_φ its graph. Show that there is a unique maximal open subset $U \subset X$ such that the projection π_1 on the first factor induces an isomorphism between $(\pi_1)^{-1}(U) \subset \Gamma_\varphi$ and U, i.e. (harking back to Definition 7.3), there is a pair (U, f) in the equivalence class φ such that for all $(V, g) \in \varphi$, $V \subset U$.

The maximal open set $U \subset X$ described in this exercise is called the *domain of regularity* of the rational map, and its complement $X - U$ is called the *indeterminacy locus* of φ.

Birational Isomorphism

As we have noted, among the ways in which rational maps fail to behave like ordinary maps, the composition of rational maps is not always defined. Given $\varphi: X \dashrightarrow Y$ and $\gamma: Y \dashrightarrow Z$, it may happen that the image of φ lies outside any open subset of Y on which γ is defined, making the composition undefined even as a rational map. One circumstance in which this will not happen, though, is if the map φ is *dominant*—that is, if the image of φ is all of Y. (We do not use the word "surjective" because, of course, it is not necessarily true in case $\mathrm{Im}(\varphi) = Y$ that $\forall p \in Y\ \exists q \in X: \varphi(q) = p$.)

It is clear from the definitions that given a pair of rational maps $\varphi: X \dashrightarrow Y$ and $\gamma: Y \dashrightarrow Z$ with φ dominant, the composition $\gamma \circ \varphi$ is a well-defined rational map. In particular, if $f \in K(Y)$ is any rational function, $\varphi^* f$ is a well-defined rational function on X. Thus, a dominant rational map $\varphi: X \dashrightarrow Y$ induces an inclusion of function fields $\varphi^*: K(Y) \hookrightarrow K(X)$ and is in turn determined by this inclusion. This in turn suggests the definition of an equivalence relation among varieties weaker than ordinary isomorphism.

Definition. We say that a rational map $\varphi: X \dashrightarrow Y$ is *birational* if there exists a rational map $\gamma: Y \dashrightarrow X$ such that $\varphi \circ \gamma$ and $\gamma \circ \varphi$ are both defined and equal to the identity. We say that two irreducible varieties X and Y are *birationally isomorphic*, or just *birational*, if there exists a birational map between them.

By the characterization of rational maps in Exercise 7.8, in characteristic 0 a birational map is simply a rational map $\varphi: X \dashrightarrow Y$ that is *generically one to one*, i.e., such that for general $q \in Y$ the graph Γ_φ intersects the fiber $X \times \{q\}$ in exactly one point, or equivalently the inverse image $\varphi^{-1}(q)$ consists of one point. We also have the following.

Exercise 7.10. Show that two varieties X and Y are birational if and only if either of the following equivalent conditions hold:

(i) $K(X) \cong K(Y)$

(ii) there exist nonempty open subsets $U \subset X$ and $V \subset Y$ isomorphic to one another.

Observe that if $\varphi: X \dashrightarrow \mathbb{P}^n$ is any rational map, the projection π_1 of the graph Γ_φ of φ to X is a birational map. (In fact, every regular birational map is tautologously of this form.) This, in turn, we may take as our *third point of view on rational maps*: a rational map $\varphi: X \dashrightarrow \mathbb{P}^n$ is a regular map on a variety X' birational to X.

A special case of the definition of a birational map is the case $Y = \mathbb{P}^n$; we define the following.

Definition. We say that a variety is *rational* if, equivalently,

(i) X is birational to $\mathbb{P}^n$;

(ii) $K(X) \cong K(x_1, \ldots, x_n)$; or

(iii) X possesses an open subset U isomorphic to an open subset of $\mathbb{A}^n$.

If these conditions do not hold, the variety is called *irrational.*

Example 7.11. The Quadric Surface

The simplest (nontrivial) example of a birational isomorphism is that between a quadric surface in $\mathbb{P}^3$ and the plane $\mathbb{P}^2$ given by projection. Specifically, let $Q \subset \mathbb{P}^3$ be a quadric surface (e.g., the surface $Z_0Z_3 - Z_1Z_2 = 0$) and $p \in Q$ any point (e.g., $[0, 0, 0, 1]$). Let $\pi_p: Q - \{p\} \to \mathbb{P}^2$ be the projection map; with the preceding choices, this is just the map

$$\pi_p: [Z_0, Z_1, Z_2, Z_3] \mapsto [Z_0, Z_1, Z_2].$$

This then defines a rational map $\pi: Q \dashrightarrow \mathbb{P}^2$. Since a general line through p will meet Q in exactly one other point (you can check that this is true of any line not lying in the plane $Z_0 = 0$, which intersects Q in the union of the two lines $Z_0 = Z_1 = 0$ and $Z_0 = Z_2 = 0$), the map is generically one to one, and so has a rational inverse; to be explicit, this is the map

$$\pi^{-1}: \mathbb{P}^2 \dashrightarrow Q$$

$$: [Z_0, Z_1, Z_2] \mapsto [Z_0^2, Z_0Z_1, Z_0Z_2, Z_1Z_2].$$

Thus the quadric surface Q is birational to $\mathbb{P}^2$. Needless to say, we could also see from the isomorphism $Q \cong \mathbb{P}^1 \times \mathbb{P}^1$ that the function field of Q is $K(x, y)$ (or that Q contains $\mathbb{A}^2$ as a Zariski open subset).

We will discuss this example further in Example 7.22. In the meantime, the following exercises give more examples of rational varieties.

Exercise 7.12. Consider the cubic curve $C \subset \mathbb{P}^2$ given by the equation $ZY^2 = X^3 + X^2Z$. Show that the projection map π_p from the point $p = [0, 0, 1]$ gives a birational isomorphism of C with $\mathbb{P}^1$.

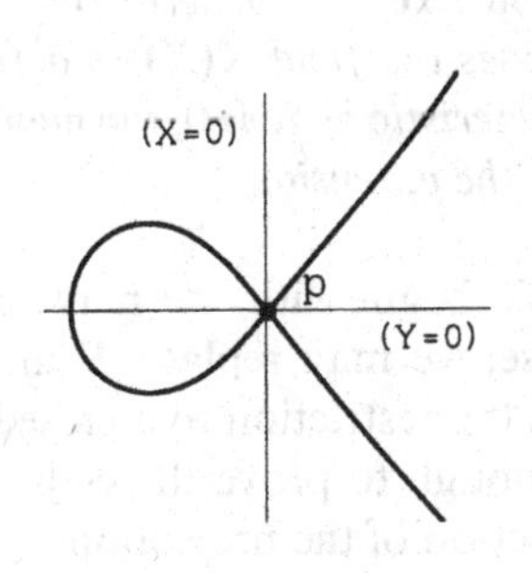

Exercise 7.13. Show that for any m and n the product $\mathbb{P}^m \times \mathbb{P}^n$ is rational. Give an explicit birational isomorphism of $\mathbb{P}^m \times \mathbb{P}^n$ with $\mathbb{P}^{m+n}$.

Exercise 7.14. Let $Q \subset \mathbb{P}^n$ be a quadric hypersurface of rank $r \geq 2$ and $p \in Q$ any point not lying on the vertex of Q. Show that the projection map $\pi_p: Q \dashrightarrow \mathbb{P}^{n-1}$ is a birational isomorphism.

Example 7.15. Hypersurfaces

We should mention at this juncture a famous remark that every irreducible variety X is birational to a hypersurface. There are two ways to see this. The first relies on the statement that if X is not a hypersurface, a general projection $\pi_p: X \to \mathbb{P}^{n-1}$ gives a birational isomorphism of X with its image $\overline{X}$. (This follows in characteristic 0 from the characterization of birational maps as generically one to one, given that a general line meeting X meets it in only one point; but unfortunately, a proof of this seemingly obvious fact will have to be deferred to Lecture 11 (Exercise 11.23).) Iterating this projection, we arrive in the end at a birational isomorphism of X with a hypersurface (or with $\mathbb{P}^k$, if X is a linear subspace of $\mathbb{P}^n$).

Alternatively, we can simply invoke the primitive element theorem to say that if $x_1, \ldots, x_k$ is a transcendence base for the function field of X, then $K(X)$ is generated over the field $K(x_1, \ldots, x_k)$ by a single element x_{k+1}, satisfying an irreducible polynomial relation

$$F(x_{k+1}) = a_d(x_1, \ldots, x_k) \cdot x_{k+1}^d + \cdots + a_0(x_1, \ldots, x_k)$$

with coefficients in $K(x_1, \ldots, x_k)$. Clearing denominators, we may take F to be an irreducible polynomial in all $k + 1$ variables x_i; it follows that X is birational to the hypersurface in $\mathbb{A}^{k+1}$ given by this polynomial.

Degree of a Rational Map

Let $f: X \dashrightarrow Y$ be a dominant rational map, corresponding to an inclusion $f^*: K(Y) \to K(X)$ of function fields. We can extend the preceding characterization of birational maps as generically one to one as follows.

Proposition 7.16. *The general fiber of the map f is finite if and only if the inclusion f^* expresses the field $K(X)$ as a finite extension of the field $K(Y)$. In this case, if the characteristic of K is 0, the number of points in a general fiber of f is equal to the degree of the extension.*

PROOF. To begin with, we may reduce (as in the proof of Theorem 3.16) to the affine case; we may replace X and Y by affine open subsets, and so realize the map f as the restriction to a closed subset of $\mathbb{A}^n$ of a linear projection $\mathbb{A}^n \to \mathbb{A}^m$. It is thus enough to prove the claim for a map $f: X \to Y$ of affine varieties given as the restriction of the projection

$$\pi: \mathbb{A}^n \to \mathbb{A}^{n-1}$$

$$: (z_1, \ldots, z_n) \mapsto (z_1, \ldots, z_{n-1}).$$

In this case the function field $K(X)$ is generated over $K(Y)$ by the element z_n. Suppose first that z_n is algebraic over $K(Y)$, and let

$$G(z_1, \ldots, z_n) = a_0(z_1, \ldots, z_{n-1}) \cdot z_n^d$$

$$+ a_1(z_1, \ldots, z_{n-1}) \cdot z_n^{d-1} + \cdots$$

(with $a_i \in K(Y)$) be the minimal polynomial satisfied by z_n. After clearing denominators, we may take a_i to be regular functions on Y, i.e., polynomials in $z_1, \ldots, z_{n-1}$. Let $\Delta(z_1, \ldots, z_{n-1})$ be the discriminant of G as a polynomial in z_n; since G is irreducible in $K(Y)[z_n]$ and $\operatorname{char}(K) = 0$, Δ cannot vanish identically on Y. It follows that the loci $(a_0 = 0)$ and $(\Delta = 0)$ are proper subvarieties of Y, and on the complement of their union the fibers of f consist of exactly d points.

Conversely, if z_n is transcendental over $K(Y)$, then for any polynomial $G(z_1, \ldots, z_n) \in I(X)$, written as

$$G(z_1, \ldots, z_n) = a_0(z_1, \ldots, z_{n-1}) \cdot z_n^d$$

$$+ a_1(z_1, \ldots, z_{n-1}) \cdot z_n^{d-1} + \cdots$$

the coefficient functions a_i must all vanish identically on Y. It follows that X contains the entire fiber of $\pi: \mathbb{A}^n \to \mathbb{A}^{n-1}$ over any point $p \in Y$, i.e., f is not generically finite. □

A rational map satisfying the conditions of Proposition 7.16 may be called either *generically finite* or *of finite degree*; the cardinality of the general fiber is called the *degree* of the map.

Blow-Ups

We come now to what will probably be a fairly difficult topic: the construction of the blow-up of a variety X along a subvariety Y. This is a regular birational map $\pi: \tilde{X} \to X$ associated to a subvariety $Y \subset X$ that is an isomorphism away from

Y but that may have nontrivial fibers over Y. It is a fundamental construction, and one that is, as we will see, easy to define. The difficulty is simply that we will not be able to give a very good description of what a blow-up looks like in general.

Example 7.17. Blowing Up Points

The simplest example of a blow-up is the graph Γ_φ of the rational map $\varphi: \mathbb{A}^2 \dashrightarrow \mathbb{P}^1$ introduced in Example 7.4. This graph, denoted $\tilde{\mathbb{A}}^2$, together with the projection map $\pi: \tilde{\mathbb{A}}^2 \to \mathbb{A}^2$, is called the *blow-up* of $\mathbb{A}^2$ at the point $(0, 0)$. It's not hard to draw a picture of this map; it looks like a spiral staircase (with the stairs extending in both directions). In particular, observe that the map $\tilde{\mathbb{A}}^2 \to \mathbb{A}^2$ is an isomorphism away from the origin $(0, 0) \in \mathbb{A}^2$, and that the fiber over that point is a copy of $\mathbb{P}^1$ corresponding to the lines through that point.

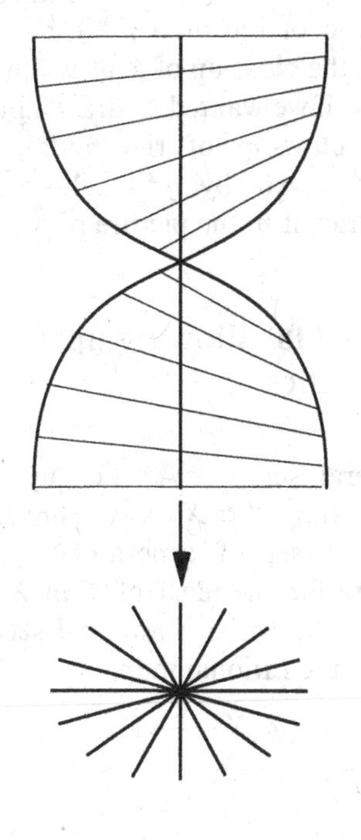

To see this (slightly) differently, note that if $[W_0, W_1]$ are homogeneous coordinates on $\mathbb{P}^1$, the open subset $W_0 \neq 0$ in $\tilde{\mathbb{A}}^2 \subset \mathbb{A}^2 \times \mathbb{P}^1$ is isomorphic to $\mathbb{A}^2$ and that the restriction of π to this open subset is just the map $f: \mathbb{A}^2 \to \mathbb{A}^2$ given by

$$f(x, y) = (x, xy),$$

that is, the basic example in our discussion of constructible sets in Lecture 3, pictured again later. Recall that under this map horizontal lines are mapped into lines through the origin, with every line through the origin covered except for the vertical; vertical lines are mapped into themselves, except for the y-axis, which is collapsed to the point $(0, 0)$.

More generally, let $\varphi: \mathbb{P}^n \dashrightarrow \mathbb{P}^{n-1}$ be the rational map given by projection from a point $p \in \mathbb{P}^n$ and $\tilde{\mathbb{P}}^n = \Gamma_\varphi \subset \mathbb{P}^n \times \mathbb{P}^{n-1}$ its graph. The map $\pi: \tilde{\mathbb{P}}^n \to \mathbb{P}^n$ is called the *blow-up* of $\mathbb{P}^n$ at the point p. As in the case of $\mathbb{A}^2$, the map π projects $\tilde{\mathbb{P}}^n$ isomorphically to $\mathbb{P}^n$ away from p, while over p the fiber is isomorphic to $\mathbb{P}^{n-1}$. (The variety $\tilde{\mathbb{P}}^n$ is also sometimes referred to as the blow-up.)

Still more generally, let $X \subset \mathbb{P}^n$ be a quasi-projective variety and $p \in X$ any point; let $\tilde{X} = \Gamma_\varphi \subset X \times \mathbb{P}^{n-1}$ be the graph of the projection map of X to $\mathbb{P}^{n-1}$ from p. The map $\pi: \tilde{X} \to X$ is then called the blow-up of X at p. The inverse image $E = \pi^{-1}(p) \subset \tilde{X}$ of the point p itself is called the *exceptional divisor* of the blow-up.

Another way to realize this is as a subvariety of the blow-up $\widetilde{\mathbb{P}}^n$ of $\mathbb{P}^n$ at p. If $\pi: \widetilde{\mathbb{P}}^n \to \mathbb{P}^n$ is the blow-up of $\mathbb{P}^n$ at p, then for $X \subset \mathbb{P}^n$ closed we can define the *proper transform* $\tilde{X}$ of X in $\widetilde{\mathbb{P}}^n$ to be the closure in $\widetilde{\mathbb{P}}^n$ of the inverse image $\pi^{-1}(X - \{p\})$ of the complement of p in X. (If X is only locally closed, we take the closure in the inverse image $(\pi_1)^{-1}(X) \subset \widetilde{\mathbb{P}}^n$.) This—or rather the restriction of the map π to it—is the same as the blow-up of X at p. Thus, for example, if we wanted to draw a picture of the blow-up of the nodal cubic $X \subset \mathbb{A}^2$ given by $y^2 = x^3 + x^2$, we could draw it in the picture of $\widetilde{\mathbb{A}}^2$.

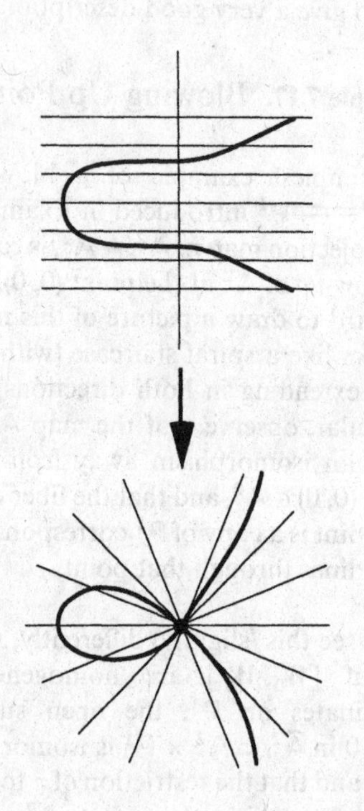

Example 7.18. Blowing up Subvarieties

In general, let $X \subset \mathbb{A}^m$ be any affine variety and $Y \subset X$ any subvariety. Choose a set of generators $f_0, \ldots, f_n \in A(X)$ for the ideal of Y in X (this doesn't have to be a minimal set) and consider the rational map

$$\varphi: X \dashrightarrow \mathbb{P}^n$$

given by

$$\varphi(x) = [f_0, \ldots, f_n].$$

Clearly, φ is regular on the complement $X - Y$, and in general won't be on Y; thus the graph Γ_φ will map isomorphically to X away from Y, but not in general over Y. The graph Γ_φ, together with the projection $\pi: \Gamma_\varphi \to X$, is called the *blow-up of* X *along* Y and sometimes denoted $\mathrm{Bl}_Y(X)$ or simply $\tilde{X}$. As before, the inverse image $E = \pi^{-1}(Y) \subset \tilde{X}$ is called the *exceptional divisor.*

Taking X affine is essentially irrelevant here: if $X \subset \mathbb{P}^m$ is a projective variety and $Y \subset X$ a subvariety, we can similarly define the blow-up of X along Y by taking a collection $F_0, \ldots, F_n$ of homogeneous polynomials of the same degree generating an ideal with saturation $I(Y)$ and letting $\mathrm{Bl}_Y(X)$ be the graph of the rational map $\varphi: X \dashrightarrow \mathbb{P}^n$ given by $[F_0, \ldots, F_n]$.

Exercise 7.19. (a) Show that the construction of $\mathrm{Bl}_Y(X)$ is local, i.e., that for X any variety, Y a subvariety and $U \subset X$ an affine open subset, the inverse image of U in

$\mathrm{Bl}_Y(X)$ is $\mathrm{Bl}_{Y \cap U}(U)$. (b) Show that the construction does not depend on the choice of generators of the ideal of Y in X.

An alternate construction of the blow-up of a variety X along a subvariety Y in the category of complex manifolds is a priori local. To start with, we can define the blow-up of a polydisc $\Delta \subset \mathbb{C}^n$ at the origin to be the inverse image of Δ in Γ_φ, where $\varphi: \mathbb{C}^n \dashrightarrow \mathbb{P}^{n-1}$ is the rational map sending a point z to $[z]$, that is, the manifold

$$\tilde{\Delta} = \{((z_1, \ldots, z_n), [w_1, \ldots, w_n]): z_i w_j = z_j w_i \ \forall i, j\} \subset \Delta \times \mathbb{P}^{n-1}.$$

Then if M is a complex manifold and $p \in M$ any point, we may take U a neighborhood of p isomorphic to the polydisc $\Delta \subset \mathbb{C}^n$, and define

$$\mathrm{Bl}_p(M) = (M - \{p\}) \textstyle\bigcup_{\Delta^*} \tilde{\Delta}$$

where Δ^* is the punctured polydisc $\Delta - \{(0, 0)\}$, which is isomorphic to its inverse image in $\tilde{\Delta}$.

More generally, we can define the blow-up of the polydisc Δ along the coordinate plane Γ given by the equations $z_1 = z_2 = \cdots = z_k = 0$ to be the manifold

$$\tilde{\Delta} = \{((z_1, \ldots, z_n), [w_1, \ldots, w_k]): z_i w_j = z_j w_i \ \forall i, j\} \subset \Delta \times \mathbb{P}^{k-1}.$$

Like the blow-up at the point, this maps isomorphically to the complement of Γ in Δ, and over a point $p \in \Gamma$ has as fiber the projective space $\mathbb{P}^{k-1}$ of normal directions to Γ at p. Moreover, it satisfies the basic naturality property that if $f: \Delta \to \Delta$ is any biholomorphic map carrying Γ to itself, f lifts (uniquely) to a biholomorphic map $\tilde{f}: \tilde{\Delta} \to \tilde{\Delta}$.

This last property allows us to globalize the blow-up in this setting. If M is a complex manifold of dimension n, $N \subset M$ a submanifold of dimension $n - k$, then we can find open subsets $U_i \subset M$ covering N such that each U_i is isomorphic to a polydisc via an isomorphism carrying $U_i \cap N$ to the coordinate plane Γ. We may then define the blow-up of M along N as the union

$$\mathrm{Bl}_N(M) = (M - N) \cup (\textstyle\bigcup \tilde{\Delta}_i),$$

where $\tilde{\Delta}_i$ is the blow-up of U_i along $U_i \cap N$, $\tilde{\Delta}_i$ is glued to $M - N$ along the common open subset $U_i - (U_i \cap N)$, and $\tilde{\Delta}_i$ is glued to $\tilde{\Delta}_j$ along the inverse image of $U_i \cap U_j$, via the induced map on blow-ups.

Exercise 7.20. (a) Verify the naturality property of blow-ups described earlier and use it to show that the construction of $\mathrm{Bl}_N(M)$ makes sense. (b) Show that if the complex manifold M is a subvariety of $\mathbb{C}^n = \mathbb{A}^n_{\mathbb{C}}$, the two notions of the blow-up $\mathrm{Bl}_N(M)$ of M along N coincide.

Modulo the fact that we have not defined tangent spaces to varieties yet, this construction gives us a good way to describe the blow-up of a variety X along a subvariety Y, at least in case both the varieties in question are smooth: the blow-up map $\mathrm{Bl}_Y(X) \to X$ is an isomorphism away from Y, and the fiber over a point $p \in Y$ is the projectivization of the normal space $N_p = T_p(X)/T_p(Y)$ to Y in X at p.

Blow-ups are just a special class of birational maps, but they play an important role in the study of rational maps in general. To say what this is, observe first that by their very definition, for any blow-up $\tilde{X} = \mathrm{Bl}_Y(X) \to X$ of a variety X along a subvariety Y there is a rational map $\varphi: X \dashrightarrow \mathbb{P}^n$ that may not be regular, but that extends to a regular map on all of $\tilde{X}$. In general, we have the following fundamental theorem, which we state here without proof.

Theorem 7.21. *Let X be any variety and $\varphi: X \dashrightarrow \mathbb{P}^n$ any rational map. Then φ can be resolved by a sequence of blow-ups, that is, there is a sequence of varieties $X = X_1, X_2, \ldots, X_k$, subvarieties $Y_i \subset X_i$ and maps $\pi_i: X_{i+1} \to X_i$ such that*

(i) *$\pi_i: X_{i+1} \to X_i$ is the blow-up of X_i along Y_i and*

(ii) *the map φ factors into a composition $\tilde{\varphi} \circ \pi_k^{-1} \circ \cdots \circ \pi_1^{-1}$ with $\tilde{\varphi}: X_{k+1} \to \mathbb{P}^n$ regular.*

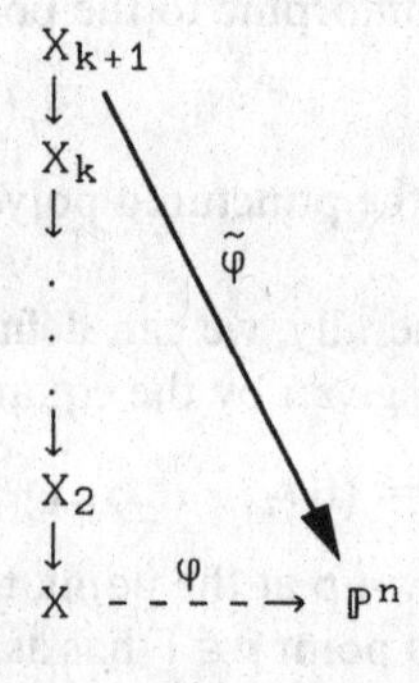

In other words, after we blow X up a finite number of times, we arrive at a variety $\tilde{X}$ and a birational isomorphism of $\tilde{X}$ with X such that the induced rational map $\tilde{\varphi}: \tilde{X} \to \mathbb{P}^n$ is in fact regular. In this sense, we can say that, in order to understand rational maps, we have only to understand regular ones and blow-ups. In particular, we may take this as our *fourth point of view on rational maps*: a rational map $\varphi: X \to \mathbb{P}^n$ is a regular map on a blow-up of X.

One last bit of terminology: if $\pi: Z = \mathrm{Bl}_p(X) \to X$ is the blow-up of a variety X at a point p and $E = \pi^{-1}(p) \subset Z$ the exceptional divisor, the process of passing from Z to X is called *blowing down*; thus, for example, we will say that the map π blows down E to a point, or that X is obtained from Z by blowing down E.

Example 7.22. The Quadric Surface Again

Consider again the map discussed in Example 7.11

$$\pi: Q \dashrightarrow \mathbb{P}^2$$

obtained by projecting the quadric surface $Q \subset \mathbb{P}^3$ given as the locus $Z_0Z_3 - Z_1Z_2 = 0$ from the point $p = [0, 0, 0, 1] \in Q$ to the plane $(Z_3 = 0)$. We may describe this map in terms of blow-ups, as follows. First, let $\Gamma \subset Q \times \mathbb{P}^2$ be the graph of π. Then since the homogeneous polynomials Z_0, Z_1, Z_2 giving the map π generate the ideal of the point p, we see that $\pi_1: \Gamma \to Q$ is the blow-up of Q at the point p. Note that the exceptional divisor $E \subset \Gamma$ of the blow-up maps isomorphically via the projection π_2 to the line given in homogeneous coordinates $[W_0, W_1, W_2]$ on $\mathbb{P}^2$ by $W_0 = 0$.

On the other hand, the map $\Gamma \to \mathbb{P}^2$ is one to one except over the two points

$q = [0, 0, 1]$ and $r = [0, 1, 0]$ corresponding to the two lines on Q through the point p; over these two points the fiber of the projection $\pi_2 \colon \Gamma \to \mathbb{P}^2$ is a line. Indeed, the inverse map π^{-1} is given by

$$\pi^{-1} \colon [W_0, W_1, W_2] \mapsto [W_0^2, W_0 W_1, W_0 W_2, W_1 W_2];$$

since the polynomials W_0^2, $W_0 W_1$, $W_0 W_2$, and $W_1 W_2$ generate an ideal whose saturation is the homogeneous ideal of the set $\{q, r\} \subset \mathbb{P}^2$, we may conclude that $\pi_2 \colon \Gamma \to \mathbb{P}^2$ is the blow-up of $\mathbb{P}^2$ at the points q and r. Finally, observe that the images in Q of the two exceptional divisors E_1 and E_2 are the lines on Q passing through the point p.

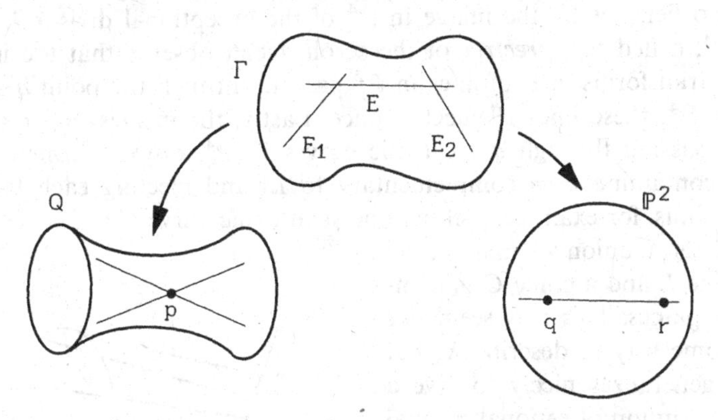

In sum, then, we can describe the map π as a map that blows up the point $p \in Q$ and blows down the lines on Q through p; we can describe the map π^{-1} as a map that blows up two points $q, r \in \mathbb{P}^2$ and blows down the line joining them.

Exercise 7.23. Show that the proper transforms in Γ of the two families of lines on Q (with the exception of the lines through p) are the proper transforms of the families of lines in $\mathbb{P}^2$ through q and r (with the exception of the line $\overline{qr}$).

Example 7.24. The Cubic Scroll in $\mathbb{P}^4$

Consider the rational map from $\mathbb{P}^2$ to $\mathbb{P}^4$ given by

$$\varphi([Z_0, Z_1, Z_2]) = [Z_0^2, Z_1^2, Z_0 Z_1, Z_0 Z_2, Z_1 Z_2].$$

The image X of this map is called the *cubic scroll* in $\mathbb{P}^4$; it is an example of a class of varieties called *rational normal scrolls*, which we will describe in Example 8.17.

There are several ways to describe the cubic scroll. First, it is the image of the Veronese surface $S \subset \mathbb{P}^5$ under the projection $\pi_p \colon S \to \mathbb{P}^4$ from a point $p \in S$; specifically, if the Veronese surface is given as the image of the map

$$\psi([Z_0, Z_1, Z_2]) = [Z_0^2, Z_1^2, Z_2^2, Z_0 Z_1, Z_0 Z_2, Z_1 Z_2]$$

then φ is just the composition of ψ with the projection map

$$\pi_p: [W_0, \ldots, W_5] \mapsto [W_0, W_1, W_3, W_4, W_5]$$

from the point $p = [0, 0, 1, 0, 0, 0]$.

Second, X is isomorphic to the blow-up of the plane $\mathbb{P}^2$ at the point $q = [0, 0, 1] \in \mathbb{P}^2$ where φ is not defined. In fact, if $\Gamma \subset \mathbb{P}^2 \times \mathbb{P}^1$ is the graph of the projection from q, then X is simply the image of Γ under the Segre embedding $\mathbb{P}^2 \times \mathbb{P}^1 \to \mathbb{P}^5$ (since $\Gamma \subset \mathbb{P}^2 \times \mathbb{P}^1$ is defined by a bilinear form, its image under the Segre map will lie in a hyperplane $\mathbb{P}^4 \subset \mathbb{P}^5$). To put it another way, φ is simply the composition of the inverse of the projection $\pi_1: \Gamma \to \mathbb{P}^2$ with the Segre map.

We can also describe the surface X in terms of the images of curves in $\mathbb{P}^2$ under φ. To begin with, the image in $\mathbb{P}^4$ of the exceptional divisor $E \subset \Gamma$ is a line $L \subset \mathbb{P}^4$, called the *directrix* of the scroll. Next, observe that the images of the proper transforms in Γ of lines in $\mathbb{P}^2$ passing through the point q are again lines M_λ in $\mathbb{P}^4$; these lines all meet L once. Lastly, the images under φ of lines in $\mathbb{P}^2$ not passing through q are conic curves in $\mathbb{P}^4$, disjoint from L (in fact, the plane containing C is complementary to L) and meeting each line M_λ in one point. Thus, for example, picking one such conic curve $C \subset X \subset \mathbb{P}^4$, we can describe X as a union of lines in $\mathbb{P}^4$ joining a line L and a conic C in complementary planes. This may seem like a cumbersome way to describe X, but in fact, it generalizes nicely to give a uniform description of rational normal scrolls in general.

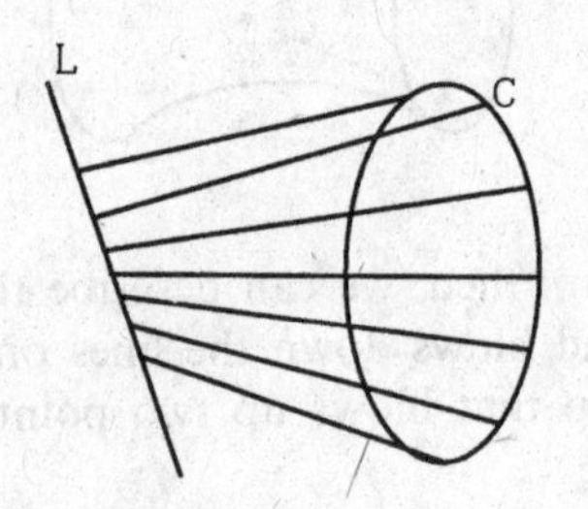

Exercise 7.25. Verify the statements made in the preceding paragraph about the images in X of curves on Γ.

Exercise 7.26. Let $r \in X$ be any point, $\pi_r: X \dashrightarrow \mathbb{P}^3$ the projection from r, and $Q \subset \mathbb{P}^3$ the image of π_r. Show that if r does not lie on the directrix of X then Q is a quadric of rank 4, while if r does lie on the directrix then the rank of Q is 3.

Here are some more exercises involving blowing up and down.

Exercise 7.27. As in Exercise 7.14, consider the projection of a quadric hypersurface $Q \subset \mathbb{P}^n$ from a point $p \in Q$; assume now that Q is of rank $n + 1$ (see Example 3.3). Show that this is a birational isomorphism of Q with $\mathbb{P}^{n-1}$ and describe this map in terms of blowing up and blowing down.

Exercise 7.28. As another generalization of Example 7.11, show that for any n the varieties $\mathbb{P}^m \times \mathbb{P}^n$ and $\mathbb{P}^{m+n}$ are birationally isomorphic via the map

$$\varphi: \mathbb{P}^m \times \mathbb{P}^n \dashrightarrow \mathbb{P}^{m+n}$$

given by

$$\varphi([Z_0, \ldots, Z_m], [W_0, \ldots, W_n])$$
$$= [Z_0 W_0, Z_1 W_0, \ldots, Z_m W_0, Z_0 W_1, \ldots, Z_0 W_n].$$

Describe the graph of this map and describe the map in terms of blowing up and down.

Exercise 7.29. Consider the group law

$$A: K^n \times K^n \to K^n$$

on the vector space K^n. This gives a rational map

$$A: \mathbb{P}^n \times \mathbb{P}^n \dashrightarrow \mathbb{P}^n;$$

for $n = 1$ and 2 describe the graph of this map and describe the map in terms of blowing up and down.

Unirationality

In closing, we should mention one other related notion. We say that a variety X is *unirational* if, equivalently, there exists a dominant rational map

$$\varphi: \mathbb{P}^n \dashrightarrow X$$

for some n or if the function field $K(X)$ can be embedded in a purely transcendental extension $K(z_1, \ldots, z_n)$ of K.

It was a classical theorem of Luroth that a curve is rational if and only if it is unirational; Castelnuovo and Enriques proved the same for surfaces. For over half a century after that, it was an open question whether or not the two notions coincided in general; then, in the 1970s, Clemens and Griffiths [CG] showed that most cubic threefolds $X \subset \mathbb{P}^4$ are unirational but not rational and Iskovskih and Manin similarly analyzed quartic threefolds [IM]. (We will see in Example 18.19 that a cubic threefold is unirational; the fact that it is not rational, however, is far beyond the scope of this book.)

In fact, the notion of unirationality may be the better generalization. For one thing, the notion of rationality is in general a fairly quirky one. For example, while we have seen that all quadric hypersurfaces are rational, the general cubic hypersurface $S \subset \mathbb{P}^n$ will be irrational for $n = 2$, rational for $n = 3$, irrational again for $n = 4$, and unknown in general (at least some smooth cubics of every even dimension are rational; it is suspected that not all are). By contrast, it is pretty elementary to see (as in Exercise 18.22) that all cubic hypersurfaces $X \subset \mathbb{P}^n$ are unirational for $n \geq 3$.

LECTURE 8

More Examples

Now that we have developed a body of fundamental notions, we are able to make a number of standard constructions.

Example 8.1. The Join of Two Varieties

Let $X, Y \subset \mathbb{P}^n$ be any two irreducible projective varieties. In Example 6.17 we saw that if X and Y are disjoint, then the locus $\mathscr{J}(X, Y) \subset \mathbb{G}(1, n)$ of lines meeting both is a subvariety of the Grassmannian, and hence the union $J(X, Y)$ of these lines is a subvariety of $\mathbb{P}^n$. We will now give another proof of this that will allow us to generalize the construction to the case where X and Y do meet (or even are equal). To do this, we first observe that if X and Y are disjoint, the map

$$j: X \times Y \to \mathbb{G}(1, n)$$

defined by sending the pair (p, q) to the line $\overline{pq}$, that is, by

$$j: ([v], [w]) \mapsto [v \wedge w],$$

is a regular map. More generally, even if X and Y meet, we get in this way a rational map $j: X \times Y \dashrightarrow \mathbb{G}(1, n)$; the image of this map will be just the closure in $\mathbb{G}(1, n)$ of the locus of lines $\overline{xy}$ with $x \in X$, $y \in Y$, and $x \neq y$. We call this the *variety of lines joining* X and Y and denote it $\mathscr{J}(X, Y)$. In case X and Y meet at a point p, it is not at all clear which of the lines through p will lie in $\mathscr{J}(X, Y)$; we will see the answer to this in at least some cases in Example 15.14.

Now, by Proposition 6.13, the union of all the lines $L \in \mathscr{J}(X, Y)$ is again a variety; we call this variety the *join* of X and Y and denote it $J(X, Y)$. If X and Y are disjoint, this will be exactly the union of the lines meeting both X and Y; if X meets Y, as the following exercise shows, we may get something less.

Exercise 8.2. To take the simplest nontrivial case, let Λ_1 and Λ_2 be two-planes in $\mathbb{P}^4$ meeting in one point p, and let $C_i \subset \Lambda_i$ be conic curves. First of all, take the case where C_1 and C_2 are disjoint; for example,

$$\Lambda_1 : W_0 = W_1 = 0 \qquad \Lambda_2 : W_3 = W_4 = 0$$

$$C_1 : W_2^2 = W_3 W_4 \qquad C_2 : W_0 W_1 = W_2^2.$$

Show in this case that the join $J(C_1, C_2)$ is a quartic hypersurface. To see what happens if one or both of the C_i pass through p, keep Λ_1 and Λ_2 the same but let C_1 and C_2 be given by

$$C_1 : W_4^2 = W_2 W_3 \qquad C_2 : W_1 W_2 = W_0^2.$$

Show that in this case $J(C_1, C_2)$ is a cubic hypersurface. Note in particular that not every line passing through p lies in $\mathcal{J}(C_1, C_2) \subset \mathbb{G}(1, 4)$. (*)

In general, it is a natural question to ask: if X and Y are varieties meeting at a point p, what lines through p lie in the image of $j: X \times Y \dashrightarrow \mathbb{G}(1, n)$? We do not have the language, let alone the techniques, to answer that question here, but for those willing to take both on faith, here is a partial answer: if X and Y are smooth and meet transversely at p, lines through p lying in the span of the projective tangent planes to X and Y at p will be in $\mathcal{J}(X, Y)$.

Note that for any subvariety $Z \subset X \times Y$ not contained in the diagonal of $\mathbb{P}^n \times \mathbb{P}^n$ we can likewise restrict the rational map j to Z to describe a subvariety $j(Z) \subset \mathcal{J}(X, Y)$; the union of the corresponding lines will again be a subvariety of $\mathbb{P}^n$. This notion will be taken up again in Examples 8.14 and 8.17.

Example 8.3. The Secant Plane Maps

This is just the special case $X = Y$ of the preceding construction: for any irreducible variety $X \subset \mathbb{P}^n$ (other than a point) we have a rational map

$$s: X \times X \dashrightarrow \mathbb{G}(1, n),$$

defined on the complement of the diagonal Δ in $X \times X$ by sending the pair (p, q) to the line $\overline{pq}$. This is called the *secant line map*; the image $\mathcal{S}(X)$ of this map is, naturally, called the *variety of secant lines* to X. This is a point of potential confusion in the terminology: when we say a line $L \in \mathbb{G}(1, n)$ is a *secant line* to X, it is meant that $L \in \mathcal{S}(X)$, not necessarily that L is spanned by its intersection with X. We will see in Lecture 15 how to characterize such lines.

Needless to say, there are analogues of this for any number $k + 1$ of points; for $X \subset \mathbb{P}^n$ irreducible and not contained in any $(k - 1)$-plane, we can define a rational map

$$s_k: X^{k+1} \dashrightarrow \mathbb{G}(k, n)$$

sending a general $(k + 1)$-tuple of points of X to the plane they span. (Note that the locus where this map fails to be regular need not be just the diagonal as it was in the first case. It is correspondingly harder in general to characterize the planes $\Lambda \in s_k(X)$ in its image other than those spanned by their intersection with X.) The notation is generalized in the obvious way: $s_k(X)$ is called the *secant plane map* and its image $\mathscr{S}_k(X)$ the *variety of secant k-planes* to X.

Exercise 8.4. Consider the case of the twisted cubic curve $C \subset \mathbb{P}^3$. Show that the image of the secant line map $s: C \times C \dashrightarrow \mathbb{G}(1, 3)$, viewed as a subvariety of $\mathbb{P}^5$ via the Plücker embedding $\mathbb{G}(1, 3) \hookrightarrow \mathbb{P}^5$, is in fact just the Veronese surface. (*)

Example 8.5. Secant Varieties

As in Example 8.1, we can use Proposition 6.13 in combination with Example 8.3 to create a new variety: we see that the union of the secant lines to a variety X is again a variety, called the *chordal variety* or *secant variety* of X, and denoted $S(X)$[2]. (Recall that a secant line to X is defined simply to be a point of the image of the secant line map $s: X \times X \dashrightarrow \mathbb{G}(1, n)$.) Similarly, the union of the secant k-planes to X is again a variety, denoted $S_k(X)$.

Exercise 8.6. Let $C \subset \mathbb{P}^3$ be a twisted cubic curve. Show that any point $p \in \mathbb{P}^3$ not on C lies on a unique line L in the image of the secant line map (so that in particular $S(C) = \mathbb{P}^3$). (*)

Exercise 8.7. Now let $C \subset \mathbb{P}^4$ be a rational normal curve. Find the equation(s) of its chordal variety. (*)

Exercise 8.8. Let $S \subset \mathbb{P}^5$ be the Veronese surface. Find the equations of its chordal variety and compare this result with that of Exercise 8.7.

Example 8.9. Trisecant Lines, etc

Again, let $X \subset \mathbb{P}^n$ be a variety, and let $\Delta \subset X \times X \times X$ be the big diagonal (that is, the locus of triples with two or more points equal). Then the locus $\tilde{V}_{1,3}(X)$ of triples of distinct points $(p, q, r) \in X \times X \times X$ such that p, q, and r are collinear is a subvariety of $X \times X \times X - \Delta$, and hence so is its closure $V_{1,3}(X)$ in $X \times X \times X$.

Exercise 8.10. Show that if $X \subset \mathbb{P}^n$ is a hypersurface of degree $d \geq 3$ with $n \geq 3$ the small diagonal in $X \times X \times X$ will be contained in the variety $V_{1,3}(X)$. (By contrast, the big diagonal will never be, as long as X is not itself a linear space. This is easy

[2] In general, we will try to follow the following convention: for varieties in $\mathbb{P}^n$, associated maps to Grassmannians will be denoted with lower-case letters, images of such maps with upper-case script letters, and the unions of the corresponding planes by upper-case Roman letters. We apologize in this case for the potential confusion (for example, we have already used $S(X)$ for the homogeneous coordinate ring of $X \subset \mathbb{P}^n$); it should be clear from the context what is meant in any case.

to see using the notion of projective tangent space introduced in Lecture 14; it is somewhat trickier, but still doable, with the techniques available to us now.)

More generally, we can for any k and l define a variety $V_{l,k}(X) \subset X^k$ to be the closure of the locus in X^k of k-tuples of distinct points of X contained in an l-plane.

We define the variety $\mathscr{V}_{1,3}(X) \subset \mathbb{G}(1, n)$ of *trisecant lines* to X to be the closure of the locus of lines $\overline{pqr}$ with $(p, q, r) \in \tilde{V}_{1,3}(X)$; likewise we define the variety $\mathscr{V}_{l,k}$ of k-secant l-planes to be the closure in $\mathbb{G}(l, n)$ of the locus of l-planes containing and spanned by k distinct points of X.

Exercise 8.11. Let $\Psi_{l,k} \subset X^k \times \mathbb{G}(l, n)$ be the incidence correspondence defined as the closure of the locus

$$\tilde{\Psi} = \{(p_1, \ldots, p_k; \Gamma): \Gamma = \overline{p_1, \ldots, p_k};\ p_i \text{ distinct}\}.$$

Show that the images of Ψ in X^k and $\mathbb{G}(l, n)$ are $V^{\circ}_{l,k}(X)$ and $\mathscr{V}_{l,k}(X)$, respectively.

Exercise 8.12. Find an example of a variety $X \subset \mathbb{P}^n$ and integers l and $k \geq l + 1$ such that $V_{k,l}(X)$ is *not* the closure of the locus of distinct k-tuples $p_1, \ldots, p_k \in X$ lying in an l-plane. (*)

Exercise 8.13. Consider the curves $C_{\alpha,\beta} \subset \mathbb{P}^3$ of Example 1.26. Describe the union of the trisecant lines to these curves. (*)

Example 8.14. Joins of Corresponding Points

Once more, let X and Y be subvarieties of $\mathbb{P}^n$, and suppose we are given a regular map $\varphi: X \to Y$ such that $\varphi(x) \neq x$ for all $x \in X$. We may then define a map

$$k_\varphi: X \to \mathbb{G}(1, n)$$

by sending a point $x \in X$ to the line $\overline{x, \varphi(x)}$ joining x to its image under φ. This is a regular map, so its image is a variety; it follows that the union

$$K(\varphi) = \bigcup_{x \in X} \overline{x, \varphi(x)}$$

is again a variety. As before, if the condition $\varphi(x) \neq x$ is violated for some (but not all) $x \in X$, we can still define a rational map k_φ, and define $K(\varphi)$ to be the union of the lines corresponding to its image; of course, we will no longer be able to describe $K(\varphi)$ naively as the union of the lines $\overline{x, \varphi(x)}$.

Exercise 8.15. Let X and Y be skew lines in $\mathbb{P}^3$, take $\varphi: X \to Y$ an isomorphism and describe the resulting $K(\varphi)$.

Exercise 8.16. Now let $X = Y \subset \mathbb{P}^3$ be a twisted cubic curve and φ an automorphism of projective space carrying X into itself. Again, what may $K(\varphi)$ look like?

Example 8.17. Rational Normal Scrolls

This is a case of the join of two varieties that represents a generalization of both rational normal curves and Exercise 8.15. Let k and l be positive integers with $k \leq l$ and $n = k + l + 1$, and let Λ and Λ' be complementary linear subspaces of dimensions k and l in $\mathbb{P}^n$ (that is, Λ and Λ' are disjoint and span $\mathbb{P}^n$). Choose rational normal curves $C \subset \Lambda$ and $C' \subset \Lambda'$, and an isomorphism $\varphi: C' \to C$; and let $S_{k,l}$ be the union of the lines $\overline{p, \varphi(p)}$ joining points of C to corresponding points of C'. $S_{k,l}$ is called a *rational normal scroll.* The lines $\overline{p, \varphi(p)}$ are called the *lines of the ruling* of $S_{k,l}$; as we will see in Proposition 8.20 they are the only lines lying on $S_{k,l}$ unless $k = 1$. Observe that $S_{k,l}$ is determined up to projective isomorphism by the integers k and l: we can move any pair of planes Λ; Λ' into any other, any rational normal curves C, C' into any others, and finally adjust φ by composing with an automorphism of Λ or Λ' inducing an automorphism of C or C'. Note also that we can think of the case where C is a point (and φ the constant map) as the degenerate case $k = 0$, so that a cone over a rational normal curve $C \subset \mathbb{P}^{n-1}$ may be thought of as the scroll $S_{0,n-1}$ in $\mathbb{P}^n$. In these terms, Exercise 8.15 simply says that the scroll $S_{1,1}$ is a quadric hypersurface in $\mathbb{P}^3$.

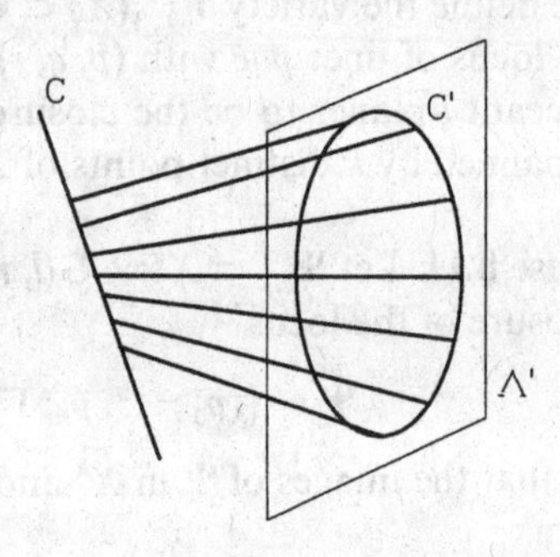

the scroll $S_{2,1} \subset \mathbb{P}^4$

Exercise 8.18. Let $H \subset \mathbb{P}^n$ be any hyperplane not containing a line of the ruling of the scroll $S = S_{k,l}$. Show that the hyperplane section $H \cap S$ is a rational normal curve in $\mathbb{P}^{n-1}$.

Exercise 8.19. Show that the image of a Veronese surface $S = \nu_2(\mathbb{P}^2) \subset \mathbb{P}^5$ under projection from a point $p \in S$ is the cubic scroll $S_{1,2} \subset \mathbb{P}^4$. In particular, show that $S_{1,2}$ is isomorphic to the plane $\mathbb{P}^2$ blown up at a point (see also Example 7.24).

The following exercises establish some basic facts about surface scrolls $S_{k,l}$; we summarize these facts in the following proposition.

Proposition 8.20. (a) *The scrolls $S_{k,l}$ and $S_{k',l'} \subset \mathbb{P}^n$ are projectively equivalent if and only if $k = k'$.*

(b) *In case $k < l$, the rational normal curve $C \subset S = S_{k,l}$ of degree k appearing in the construction of the rational normal scroll is the unique rational normal curve of degree $< l$ on S (other than the lines of the ruling of S); in particular, it is uniquely determined by S (it is called the* directrix *of S). This is not the case for the rational normal curve C' of larger degree l or for C itself in case $k = l$.*

(c) *The image of the scroll $S = S_{k,l}$ under projection from a point $p \in S$ is projec-*

tively equivalent to $S_{k-1,l}$ if p lies on the directrix of S; it is projectively equivalent to $S_{k,l-1}$ otherwise. In particular, all scrolls $S_{k,l}$ are rational.

Exercise 8.21. Show that any $k+1$ lines of the ruling of a scroll $S = S_{k,l}$ are independent (i.e., span a $\mathbb{P}^{2k+1}$), but that any $k+2$ lines are dependent (i.e., contained in a $\mathbb{P}^{2k+2}$). Deduce part (a) of Proposition 8.20.

Exercise 8.22. In case $k < l$, show that any $l = n - k - 1$ lines L_i of the ruling of the scroll $S = S_{k,l}$ span a hyperplane $H \subset \mathbb{P}^n$ and that the intersection of H with S consists of the lines L_i together with the curve C. Deduce part (b) of Proposition 8.20.

Exercise 8.23. A rational normal curve of degree l on the scroll $S = S_{k,l}$ that lies in a linear subspace $\mathbb{P}^l$ complementary to the span of the directrix, and that meets each line of the ruling once will be called a *complementary section* of S. Show that any complementary section can play the role of C' in the preceding construction. In case $k = l$, any rational normal curve $C \subset S$ of degree l meeting each line of the ruling once is called a complementary section; show that any two lie in complementary linear subspaces and can play the roles of C and C' in the basic construction.

Exercise 8.24. Show that any hyperplane $H \subset \mathbb{P}^n$ containing $n - l = k + 1$ or more lines of the ruling of S will intersect S in a union of lines of the ruling and the directrix C, while a hyperplane containing $n - l - 1 = k$ lines of the ruling will intersect S in either a union of lines of the ruling and the directrix or in the given k lines and a complementary section. Deduce that any point of the scroll S not lying on the directrix lies on a complementary section and deduce part (c) of Proposition 8.20.

Exercise 8.25. Let $S = S_{k,l} \subset \mathbb{P}^n$ be a scroll, with $k \geq 2$. Show that projection from a line of the ruling gives a biregular isomorphism of $S_{k,l}$ with $S_{k-1,l-1} \subset \mathbb{P}^{n-2}$. (It is conversely the case if $k, k' > 0$ that $S_{k,l}$ is abstractly isomorphic to $S_{k',l'}$ if and only if $l - k = l' - k'$, but we will not prove it here.)

Example 8.26. Higher-Dimensional Scrolls

A further generalization of this notion is to allow several rational normal curves; that is, for any collection $a_1, \ldots, a_k$ of natural numbers with $a_1 \leq \cdots \leq a_k$ and $\sum a_i = n - k + 1$ we can find complementary linear subspaces $\Lambda_i \cong \mathbb{P}^{a_i} \subset \mathbb{P}^n$ and rational normal curves $C_i \subset \Lambda_i$ in each. Choose isomorphisms $\varphi_i \colon C_1 \to C_i$ and let

$$S = \bigcup_{p \in C_1} \overline{p, \varphi_2(p), \ldots, \varphi_k(p)}.$$

S is called a *rational normal k-fold scroll* (or *rational normal scroll of dimension* k), and sometimes denoted $S_{a_1, \ldots, a_k}$. As before, S is determined up to projective equivalence by the integers a_i.

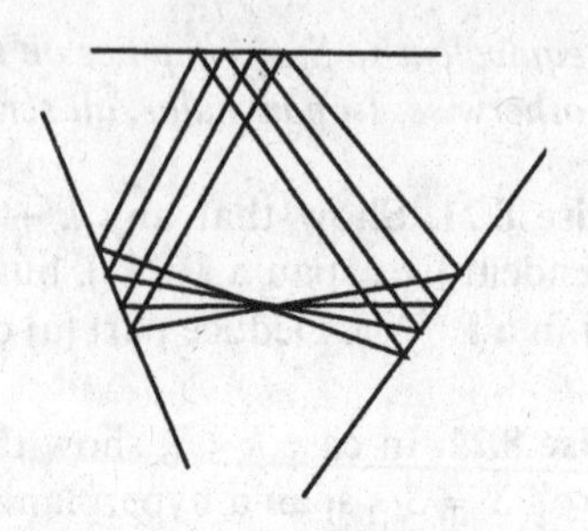

Exercise 8.27. Show that the Segre threefold, that is, the image $\Sigma_{2,1}$ of the Segre map

$$\sigma: \mathbb{P}^2 \times \mathbb{P}^1 \to \mathbb{P}^5,$$

is the rational normal threefold scroll $S_{1,1,1}$. More generally, the Segre variety $\Sigma_{k-1,1} \subset \mathbb{P}^{2k-1}$ is the rational normal k-fold scroll $S_{1,\ldots,1}$.

Exercise 8.28. Let $S = S_{1,1,1} \subset \mathbb{P}^5$ be the Segre threefold, $H \subset \mathbb{P}^5$ any hyperplane not containing a two-plane of the ruling of S. Show that the hyperplane section $H \cap S$ is the scroll $S_{1,2}$ in $\mathbb{P}^4$. (An interesting question to ask in general: what scrolls will appear as hyperplane sections of a given scroll $S_{a_1,\ldots,a_k}$?)

There are two special cases of this construction that are usually included in the class of rational normal scrolls. First, if some of the integers a_i are zero—say, $a_1 = \cdots = a_m = 0$—then the scroll we arrive at is the cone, with vertex the span of the points $C_1, \ldots, C_m$, over a rational normal scroll $S_{a_{m+1},\ldots,a_k}$. Also, if $k = 1$, we get simply a rational normal curve, which we will consider a one-dimensional rational normal scroll. With these provisos, we may state the following general theorem.

Theorem 8.29. *Let $S \subset \mathbb{P}^n$ be a rational normal k-fold scroll.*

(i) *If $p \in S$ any point, then the projection $\bar{S} = \pi_p(S)$ from p is again a rational normal scroll and*

(ii) *if $H \subset \mathbb{P}^n$ is any hyperplane not containing a $(k-1)$-plane of S, then the hyperplane section $Y = S \cap H$ is again a scroll.*

The first part of this statement is relatively straightforward; it is simply a more elaborate version of Exercise 8.21. (In particular, if $S = S_{a_1,\ldots,a_k}$, it is possible to say which scrolls occur as images of projections of S.) The second part will need a further characterization of scrolls, either as determinantal varieties (as in the following lecture) or by degree (as in Lecture 19).

Example 8.30. More Incidence Correspondences

There are a number of generalizations of the construction in Example 6.12. For instance, we can look in the product of two Grassmannians of planes in the same space $\mathbb{P}^n$ at the locus of pairs of planes that meet: for any k and l we set

$$\Omega = \{(\Lambda, \Lambda') : \Lambda \cap \Lambda' \neq \emptyset\} \subset \mathbb{G}(k, n) \times \mathbb{G}(l, n).$$

That this is a variety is clear: we can write

$$\Omega = \{([v_0 \wedge \cdots \wedge v_k], [w_0 \wedge \cdots \wedge w_l]):$$
$$v_0 \wedge \cdots \wedge v_k \wedge w_0 \wedge \cdots \wedge w_l = 0\}.$$

Similarly, we can for any k and $l \geq k$ consider the locus of nested pairs:

$$\mathbb{F}(k, l; n) = \{(\Lambda, \Lambda'): \Lambda \subset \Lambda'\} \subset \mathbb{G}(k, n) \times \mathbb{G}(l, n).$$

Note that both these constructions specialize to the construction of the incidence correspondence in Example 6.12 in case $k = 0$. A common generalization of them both in turn is the variety

$$\Psi_m = \{(\Lambda, \Lambda'): \dim(\Lambda \cap \Lambda') \geq m\} \subset \mathbb{G}(k, n) \times \mathbb{G}(l, n),$$

which gives Ω when $m = 0$ and $\mathbb{F}(k, l; n)$ when $m = k < l$.

Of course, over Ω, $\mathbb{F}(k, l; n)$, and Ψ_m there are "universal families" that play with respect to these varieties the same role as the original incidence correspondence Σ played with respect to $\mathbb{G}(k, n)$, i.e., we may set

$$\Xi_m = \{(\Lambda, \Lambda'; p): p \in \Lambda \cap \Lambda'\} \subset \Psi_m \times \mathbb{P}^n.$$

Exercise 8.31. Show that Ω, $\mathbb{F}(k, l; n)$ and Ψ_m are indeed varieties.

Exercise 8.32. Let $\Phi \subset \mathbb{G}(k, n)$ be any subvariety of the Grassmannian of k-planes in $\mathbb{P}^n$ and for any $l > k$ let $\Psi \subset \mathbb{G}(l, n)$ be the locus of l-planes containing some plane $\Lambda \in \Phi$. Show that Ψ is a variety.

Exercise 8.33. Let $Q \subset \mathbb{P}^3$ be a quadric surface. Describe the locus in $\mathbb{P}^{3*}$ of planes in $\mathbb{P}^3$ containing a line of Q. More generally, if $\Sigma_{k,1} = \sigma(\mathbb{P}^1 \times \mathbb{P}^k) \subset \mathbb{P}^{2k+1}$ is the Segre variety, describe the locus in $(\mathbb{P}^{2k+1})^*$ of hyperplanes in $\mathbb{P}^{2k+1}$ containing a k-plane of $\Sigma_{k,1}$.

Example 8.34. Flag Manifolds

The incidence correspondence $\mathbb{F}(k, l; n)$ introduced in Example 8.30 is a special case of what is called a *flag manifold*. Briefly, for any increasing sequence of integers $a_1 < a_2 < \cdots < a_k < n$ we can form the variety of *flags*

$$\begin{aligned} \mathbb{F}(a_1, \ldots, a_k; n) &= \{(\Lambda_1, \ldots, \Lambda_k): \Lambda_1 \subset \cdots \subset \Lambda_k\} \\ &\subset \mathbb{G}(a_1, n) \times \cdots \times \mathbb{G}(a_k, n). \end{aligned}$$

That this is a subvariety of the product of Grassmannians is easy to see (in particular, it is the intersection of the inverse images of the flag manifolds $\mathbb{F}(a_i, a_j; n) \subset \mathbb{G}(a_i, n) \times \mathbb{G}(a_j, n)$ under the corresponding projection map).

Example 8.35. More Joins and Intersections

In the preceding discussion, we considered the subvariety of the Grassmannian obtained by taking the joins of corresponding points on two subvarieties X, $Y \subset \mathbb{P}^n$. In fact, we can take the joins of arbitrary subspaces as well. Suppose now that $X \subset \mathbb{G}(k, n)$ and $Y \subset \mathbb{G}(l, n)$ are subvarieties of the Grassmannians of k- and l-planes in a projective space $\mathbb{P}^n$ and $\varphi: X \to Y$ is a regular map such that $\varphi(\Lambda)$ is disjoint from Λ for all Λ. Then we obtain a regular map

$$k_\varphi: X \to \mathbb{G}(k + l + 1, n)$$

by sending a point $\Lambda \in X$ to the join of the planes Λ and $\varphi(\Lambda)$; as usual the union of these $(k + l + 1)$-planes is again a variety $K(\varphi)$. We may define $K(\varphi)$ similarly if Λ fails to be disjoint from $\varphi(\Lambda)$ for some, but not all, $\Lambda \in X$.

Using this construction together with an isomorphism $\mathbb{G}(k, n) \cong \mathbb{G}(n - k - 1, n)$ between Grassmannians, we can likewise construct a variety as the union of the intersections of corresponding planes in two families. Specifically, let X, Y, and φ be as earlier, but suppose now that $k + l \geq n$ and that for each $\Lambda \in X$ the planes Λ and $\varphi(\Lambda)$ intersect transversely, i.e., span $\mathbb{P}^n$. Then we can define a map

$$l_\varphi: X \to \mathbb{G}(k + l - n, n)$$

by sending x to the plane $x \cap \varphi(x)$. The union

$$L(\varphi) = \bigcup_{\Lambda \in X} \Lambda \cap \varphi(\Lambda)$$

is thus again a variety. As before, analogous constructions may be made for several subvarieties of Grassmannians $\mathbb{G}(k_i, n)$ or in the case where Λ and $\varphi(\Lambda)$ fail to meet transversely for all Λ.

Example 8.36. Quadrics of Rank 4

The simplest example of the construction described in Example 8.35 occurs in $\mathbb{P}^3$; it is an analog of the construction given in Example 1.22 for conics (compare it also with Exercise 8.5). Start with two skew lines L and $M \subset \mathbb{P}^3$. The family of planes containing each one is parameterized by $\mathbb{P}^1$; choose such a parameterization for each (for example, introduce an auxiliary line N disjoint from each of L and M and to each point $R \in N$ let correspond the planes $\overline{LR}$ and $\overline{MR}$). For each $\lambda \in \mathbb{P}^1$, then, the two planes corresponding to λ will meet in a line; the union of these lines will be a quadric surface.

Note that this quadric will contain L and M and, if we use the parameterization suggested earlier, the auxiliary line N as well; in this way we may see again that through three mutually skew lines in $\mathbb{P}^3$ there passes a unique quadric surface. Note also that we did not need to assume that L and M were skew (but if they are not, we do need to know that the plane $\overline{LM}$ does not correspond to itself).

We can make this construction in projective spaces of higher dimensions as well, simply choosing two planes $L, M \cong \mathbb{P}^{n-2} \subset \mathbb{P}^n$, putting the families of hyperplanes through L and M in one-to-one correspondence, and taking the union of the intersections of corresponding planes. What we will get in general is a cone, with vertex $L \cap M$, over a quadric in $\mathbb{P}^3$ (or over a conic in $\mathbb{P}^2$, if L and M fail to span all of $\mathbb{P}^n$). These are called, in keeping with the terminology of Example 3.3, *quadrics of rank* 3 or 4; they may also be characterized as

(i) quadrics obtained in the preceding construction;
(ii) cones over irreducible quadrics in $\mathbb{P}^3$;
(iii) quadrics given, in suitable coordinates, as $Z_0Z_3 = Z_1Z_2$ or $Z_0Z_2 = Z_1^2$;

(iv) quadrics that are also rational normal scrolls (they are isomorphic to the scroll $X_{1,1,0,\dots,0}$ or $X_{2,0,\dots,0}$; or simply as

(v) irreducible quadrics that contain a plane $\mathbb{P}^{n-2}$.

Exercise 8.37. Verify that these conditions are indeed equivalent.

Exercise 8.38. Recall the construction of a general rational normal curve through $d + 3$ points given in Example 1.22. For each i and j let $Q_{i,j}$ be the quadric obtained by applying the construction to the families of hyperplanes containing Λ_i and Λ_j. Show that the $Q_{i,j}$ span the space of all quadrics containing the rational normal curve and in particular that their intersection is the rational normal curve.

Example 8.39. Rational Normal Scrolls II

A variant of the preceding construction gives us the general rational normal scroll. It is not hard to guess what to do. To construct a rational normal curve—a one-dimensional scroll—we took n families of hyperplanes in $\mathbb{P}^n$ and took the intersections of corresponding members, while to construct those quadric hypersurfaces that are scrolls we used two. In general, if we use an intermediate number, we will obtain a rational normal scroll.

To be specific, suppose that $\Lambda_1, \dots, \Lambda_{n-k+1}$ are planes of dimension $n - 2$ in $\mathbb{P}^n$ and as before we put the families of hyperplanes containing them in one-to-one correspondence, that is, we write the family of hyperplanes containing Λ_i as $\{\Gamma_i(\lambda)\}_{\lambda \in \mathbb{P}^1}$. Suppose moreover that for each value of $\lambda \in \mathbb{P}^1$, the corresponding hyperplanes are independent, i.e., intersect in a $(k - 1)$-plane. Then the union

$$X = \bigcup_{\lambda \in \mathbb{P}^1} \Gamma_1(\lambda) \cap \cdots \cap \Gamma_{n-k+1}(\lambda)$$

of their intersections is a rational normal scroll.

We will not justify this statement here; it follows from the description of rational normal scrolls as determinantal varieties given in Example 9.10.

LECTURE 9

Determinantal Varieties

In this lecture we will introduce a large and important class of varieties, those whose equations take the form of the minors of a matrix. We will see that many of the varieties we have looked at so far—Veronese varieties, Segre varieties, rational normal scrolls, for example—are determinantal.

Generic Determinantal Varieties

We will start with what is called the *generic determinantal variety*. Let M be the projective space $\mathbb{P}^{mn-1}$ associated to the vector space of $m \times n$ matrices. For each k, we let $M_k \subset M$ be the subset of matrices of rank k or less; since this is just the common zero locus of the $(k+1) \times (k+1)$ minor determinants, which are homogeneous polynomials of degree $k+1$ on the projective space M, M_k is a projective variety.

Note. More or less by definition, the $(k+1) \times (k+1)$ minors cut out the variety M_k set theoretically. The stronger statement—that they generate the homogeneous ideal $I(M_k)$ of M_k—is also true, but is nontrivial to prove.

Example 9.1. Segre Varieties

The simplest example of a generic determinantal variety, the case $k = 1$, is a variety we have already encountered. The basic observation here is that an $m \times n$ matrix $(Z_{i,j})$ will be of rank 1 if and only if it is expressible as a product $Z = {}^tU \cdot W$, where $U = (U_1, \ldots, U_m)$ and $W = (W_1, \ldots, W_n)$ are vectors. We see from this that the subvariety $M_1 \subset M = \mathbb{P}^{mn-1}$ is just the Segre variety, that is, the image of the Segre map

$$\sigma: \mathbb{P}^{m-1} \times \mathbb{P}^{n-1} \to \mathbb{P}^{mn-1}.$$

(In intrinsic terms, a homomorphism $A: K^n \to K^m$ of rank 1 is determined up to scalars by its kernel $\text{Ker}(A) \in (\mathbb{P}^{n-1})^*$ and its image $\text{Im}(A) \in \mathbb{P}^{m-1}$.) Thus, we can represent the quadric surface $\Sigma_{1,1} = \sigma(\mathbb{P}^1 \times \mathbb{P}^1) \subset \mathbb{P}^3$ as

$$\Sigma_{1,1} = \left\{[Z]: \begin{vmatrix} Z_0 & Z_1 \\ Z_2 & Z_3 \end{vmatrix} = 0\right\}$$

and similarly the Segre threefold $\Sigma_{2,1} = \sigma(\mathbb{P}^2 \times \mathbb{P}^1) \subset \mathbb{P}^5$ may be realized as

$$\Sigma_{2,1} = \left\{[Z]: \text{rank}\begin{pmatrix} Z_0 & Z_1 & Z_2 \\ Z_3 & Z_4 & Z_5 \end{pmatrix} \leq 1\right\}.$$

Example 9.2. Secant Varieties of Segre Varieties

We can in fact describe the other generic determinantal varieties $M_k \subset M$ in terms of the Segre variety M_1. We may observe, for example, that a linear map $A: K^n \to K^m$ has rank 2 or less if and only if it is expressible as a sum of two maps of rank 1. This is immediate in case $m = 2$; in general we can factor a rank 2 map

$$A: K^n \to \text{Im}(A) \hookrightarrow K^m$$

and apply the case $m = 2$ to the first of these maps. We conclude from this that the generic determinantal variety M_2 is the chordal variety of the Segre variety $M_1 \subset M$; more generally, the variety M_k may be realized as the union of the secant $(k-1)$-planes to M_1.

Linear Determinantal Varieties in General

Generalizing, suppose now that we have an $m \times n$ matrix of homogeneous linear forms $\Omega = (L_{i,j})$ on a projective space $\mathbb{P}^l$, not all vanishing simultaneously. The variety

$$\Sigma_k(\Omega) = \{[Z_0, \ldots, Z_l]: \text{rank}(\Omega(Z)) \leq k\} \subset \mathbb{P}^l$$

is also called a *linear determinantal variety*. Of course, this is just the pullback of M_k under the linear embedding $\mathbb{P}^l \to M$ given by the linear forms $(L_{i,j})$. (Extending this definition, for any subvariety $X \subset \mathbb{P}^l$ the locus of points $[Z] \in X$ such that $\text{rank}(L_{i,j}(Z)) \leq k$ is likewise called a determinantal subvariety of X.)

An $m \times n$ matrix Ω of linear forms on a projective space $\mathbb{P}^l = \mathbb{P}V$ may be thought of as a linear map

$$\omega: V \to \text{Hom}(U, W)$$

for $U \cong K^n$, $W \cong K^m$ vector spaces over K; equivalently, it may be thought of as an element of the triple tensor product

$$\omega \in V^* \otimes U^* \otimes W.$$

In view of the number of variations on this representation—for example, we can equally well think of ω as a map from U to $\mathrm{Hom}(V, W)$, or from $\mathrm{Hom}(U^*, V)$ to W, etc.—we will abandon any attempt to give separate symbols to these incarnations of the same object and call them all ω. (This is in contrast to the case of a bilinear object $\varphi \in \mathrm{Hom}(U, V) = U^* \otimes V$, whose one other representation as an element of $\mathrm{Hom}(V^*, U^*)$ may be called the *transpose* of φ and denoted ${}^t\varphi$.)

However we view the trilinear object ω given by an $m \times n$ matrix Ω of linear forms on $\mathbb{P}^k = \mathbb{P}V$, it is clear that three groups act on it: the automorphism groups $\mathrm{GL}(V)$, $\mathrm{GL}(U)$, and $\mathrm{GL}(W)$ of V, U, and W. We say that two such matrices Ω and Ω' are *conjugate* if one can be carried into the other under the combined action of these three groups—in more concrete terms, if Ω can be obtained from Ω' by a sequence of row and column operations, followed by a linear change of variables in the Z_i.

Here we should draw a fundamental and important distinction between the cases of bilinear and tri- or multilinear objects: while a bilinear object $\varphi \in V \otimes W$ is completely determined (up to the action of $\mathrm{GL}(V) \times \mathrm{GL}(W)$) by its one (discrete) invariant, i.e., the *rank*, the situation with trilinear objects $\omega \in U \otimes V \otimes W$ is much more subtle. The action of $\mathrm{GL}(U) \times \mathrm{GL}(V) \times \mathrm{GL}(W)$ on $U \otimes V \otimes W$ will not in general have a finite number of orbits (or even a single dense orbit), but rather (as we will see in Lecture 11) a continuously varying family of them, whose invariants are far from being completely understood. One exception to this is the special case where the dimension of one of the factors is 2; it is this special case that we will deal with in the examples that follow. We will also see in the final section of Lecture 22 an analysis of a related problem, the classification up to conjugation of elements of a product $U \otimes \mathrm{Sym}^2 V$ in case $\dim(U) = 2$.

The action of $\mathrm{GL}(U)$ and $\mathrm{GL}(W)$ obviously do not alter the determinantal varieties $\Sigma_k(\Omega)$ associated to Ω; the action of $\mathrm{GL}(V)$ simply carries $\Sigma_k(\Omega)$ into a projectively equivalent variety. The determinantal varieties $\Sigma_k(\Omega)$ may thus be thought of as giving invariants under conjugation of the trilinear object Ω. Indeed, in some cases, such as the rational normal curve in Example 9.3, a determinantal variety $\Sigma_k(\Omega)$ will actually determine the matrix Ω up to conjugation, so that understanding the geometry of determinantal varieties becomes equivalent to understanding trilinear algebra.

Example 9.3. Rational Normal Curves

In Example 1.16 we observed that the rational normal curve $C \subset \mathbb{P}^n$ given as the image of the Veronese map $v_n\colon \mathbb{P}^1 \to \mathbb{P}^n$ sending $[X, Y]$ to $[X^n, X^{n-1}Y, \ldots, Y^n]$ could be realized as the rank 1 determinantal variety $\Sigma_1(\Omega)$ associated to the matrix of linear forms

$$\Omega = \begin{pmatrix} Z_0 & Z_1 & \ldots & Z_{n-1} \\ Z_1 & Z_2 & \ldots & Z_n \end{pmatrix}.$$

We would now like to expand on those observations, and in particular to show that conversely any $2 \times n$ matrix of linear forms satisfying a nondegeneracy hypothesis is conjugate to Ω.

To begin with, recall the intrinsic description of the Veronese map $v: \mathbb{P}^1 \to \mathbb{P}^n$ given on page 25: if $\mathbb{P}^1 = \mathbb{P}V$ is the projective space associated to a two-dimensional vector space V, we can realize $\mathbb{P}^n$ as the projectivization of the nth symmetric power $W = \mathrm{Sym}^n V$, and the Veronese map as the map sending a point $[v] \in \mathbb{P}V$ to the hyperplane $H_v \subset W^*$ of polynomials vanishing at $[v]$ (equivalently, to the linear functional on the space W^* of polynomials given by evaluation at v).

With this said, consider for any pair of positive integers k, l with $k + l = n$ the ordinary multiplication map on polynomials on V

$$\mathrm{Sym}^k V^* \otimes \mathrm{Sym}^l V^* \to \mathrm{Sym}^n V^* = W^*.$$

Dualizing, we get a map

$$\mu: W \to \mathrm{Sym}^k V \otimes \mathrm{Sym}^l V,$$

or, equivalently, a $(k + 1) \times (l + 1)$ matrix Ω_k of linear forms on $\mathbb{P}W$. We claim now that the rank 1 locus of the matrix Ω_k is the rational normal curve $C \subset \mathbb{P}W$. This amounts to saying that for any linear functional $\varphi: \mathrm{Sym}^n V^* \to K$ on W^*, the bilinear form $B_\varphi: \mathrm{Sym}^k V^* \times \mathrm{Sym}^l V^* \to K$ defined as the composition

$$\mathrm{Sym}^k V^* \times \mathrm{Sym}^l V^* \to \mathrm{Sym}^n V^* \to K$$

has rank 1 if and only if φ is given by evaluation at a point $p \in \mathbb{P}V$. Certainly if φ is evaluation at a point p, B_φ will have rank 1, with kernel in either factor $\mathrm{Sym}^k V^*$ or $\mathrm{Sym}^l V^*$ the hyperplane of polynomials vanishing at p. Conversely, if B_φ has rank 1, its kernel in $\mathrm{Sym}^k V^*$ will be a subspace U such that the image of $U \times \mathrm{Sym}^l V^*$ in $\mathrm{Sym}^n V^*$ is contained in a hyperplane. By Lemma 9.8 below it follows that U must be the subspace of polynomials vanishing at p for some point $p \in \mathbb{P}^1$, that the image of $U \times \mathrm{Sym}^l V^*$ in $\mathrm{Sym}^n V^*$ is therefore the subspace of polynomials vanishing at p, and hence that φ is evaluation at p.

Note that the representations of C as the rank 1 loci of the matrices Ω_k are the various determinantal descriptions of the rational normal curve given in Example 1.16; in particular, we see that C is the determinantal variety associated to the $2 \times n$ matrix Ω_1 dual to the multiplication map

$$V^* \otimes \mathrm{Sym}^{n-1} V^* \to \mathrm{Sym}^n V^*,$$

which is the matrix Ω above.

We also asserted in Exercise 1.25 that conversely any $2 \times n$ matrix of linear forms

$$\Omega = \begin{pmatrix} L_1 & L_2 & \cdots & L_n \\ M_1 & M_2 & \cdots & M_n \end{pmatrix}$$

satisfying the condition that for all $[\lambda, \mu] \in \mathbb{P}^1$ the linear forms $\lambda L_1 + \mu M_1, \ldots, \lambda L_n + \mu M_n$ are independent has determinantal variety a rational normal curve. The argument was simply that for each $[\lambda, \mu] \in \mathbb{P}^1$ the equations

$$\lambda L_1 + \mu M_1 = \cdots = \lambda L_n + \mu M_n = 0$$

determined a point $p_{[\lambda,\mu]} \in \mathbb{P}^n$; and the map $\mathbb{P}^1 \to \mathbb{P}^n$ sending $[\lambda, \mu]$ to $p_{[\lambda,\mu]}$ was seen to be given by a basis of polynomials of degree n on $\mathbb{P}^1$. Note that we can express the independence condition more simply by saying just that no matrix of linear forms conjugate to Ω has a zero entry. This condition may be applied more generally to matrices of arbitrary size and is expressed by saying that the matrix is 1-*generic*. By way of terminology, we define a *generalized row* of a matrix Ω of linear forms to be a row of a matrix conjugate to Ω (that is, any linear combination of the rows of Ω) so that, for example, the points of the rational normal curve C given earlier as the determinantal variety of the 1-generic matrix Ω are the zeros of the generalized rows of Ω.

We define a *generalized column* or a *generalized $a \times b$ submatrix* of a matrix similarly. In particular, the condition of 1-genericity can be strengthened; we say that a matrix Ω of linear forms is *l-generic* if every generalized $l \times l$ submatrix of Ω has independent entries. The condition of genericity of a matrix defined in this way is closely related to the geometry of the associated determinantal variety; see, for example, [E1].

Finally, suppose that Ω is any 1-generic $2 \times n$ matrix of linear forms on $\mathbb{P}^n$ and $C \cong \mathbb{P}^1$ its determinantal variety. As we have observed, the generalized rows of Ω correspond to the points of C; for any $[\lambda, \mu] \in \mathbb{P}^1$ the common zeros of the linear forms $\lambda L_j + \mu M_j$ on $\mathbb{P}^n$ give a point $p_{[\lambda,\mu]} \in C$. We claim now that the common zero locus of any generalized column of C is an $(n-1)$-tuple of points on C; or, more precisely, that the entries of a generalized column of Ω are polynomials of degree n on $C \cong \mathbb{P}^1$ having $n-1$ common roots.

To see this, observe that an entry L_j of Ω will be a linear form on $\mathbb{P}^n$—that is, a polynomial of degree n on $C \cong \mathbb{P}^1$—and that the zeros of this polynomial will include the point $p_{[1,0]}$. On the other hand, we see that the remaining $n-1$ zeros $p_{[\lambda,\mu]}$ of L_j with $\mu \neq 0$ must be zeros of the other entry M_j in the column of L_j as well, i.e., the two entries of any column of Ω correspond to polynomials on C having $n-1$ roots in common. (An additional argument is necessary in case $p_{[1,0]}$ is a multiple root of L_j.)

In this way we see that the generalized columns of Ω correspond bijectively to polynomials of degree $n-1$ on C (up to scalars). In particular, if we are given coordinates $[X, Y]$ on $C \cong \mathbb{P}^1$ we can take the rows of the matrix Ω (after multiplying on the left by a 2×2 matrix of scalars) to correspond to the points $X = 0$ and $Y = 0$, and the columns (after multiplying on the right by an $n \times n$ matrix of scalars) to correspond to the polynomials $X^{n-1}, X^{n-2}Y, \ldots, Y^{n-1}$. If we then choose our homogeneous coordinates $Z_0, \ldots, Z_n$ on $\mathbb{P}^n$ to correspond to the polynomials $X^n, X^{n-1}Y, \ldots, Y^n$, we see that the matrix Ω will have the standard form $L_i = Z_{i-1}$, $M_i = Z_i$ above. We summarize this conclusion as the following proposition.

Proposition 9.4. *Any 1-generic $2 \times n$ matrix of linear forms on $\mathbb{P}^n$ is conjugate to the matrix*

$$\Omega(Z) = \begin{pmatrix} Z_0 & Z_1 & \dots & Z_{n-1} \\ Z_1 & Z_2 & \dots & Z_n \end{pmatrix}.$$

Exercise 9.5. Show that if a $2 \times n$ matrix of linear forms on $\mathbb{P}^n$ is not 1-generic, but satisfies the weaker condition that not every generalized row is dependent then its determinantal variety is a union of a rational normal curve C contained in a proper linear subspace of $\mathbb{P}^n$ and some lines meeting C. Use this description to show in particular that, despite the fact that by Proposition 9.4 an open subset of all $2 \times n$ matrices of linear forms on $\mathbb{P}^n$ are conjugate, it is not that case that there are only finitely many conjugacy classes of such matrices.

Example 9.6. Secant Varieties to Rational Normal Curves

We can generalize Example 1.16 to give a determinantal representation of the secant varieties to a rational normal curve. This will be very useful in Example 9.10.

Proposition 9.7. *For any $l \le \alpha$, $d - \alpha$ the rank l determinantal variety associated to the matrix*

$$M_\alpha = \begin{bmatrix} Z_0 & Z_1 & Z_2 & \dots & Z_{d-\alpha} \\ Z_1 & Z_2 & Z_3 & \dots & Z_{d-\alpha+1} \\ \vdots & & & & \\ Z_\alpha & Z_{\alpha+1} & . & \dots & Z_d \end{bmatrix}$$

is the l-secant variety $S_{l-1}(C)$ of the rational normal curve $C \subset \mathbb{P}^d$.

PROOF. We will prove this in the case $l = 2$, leaving the general case as Exercise 9.9. A key ingredient is the following lemma.

Lemma 9.8. *Let S_d be the vector space of polynomials of degree d on $\mathbb{P}^1$, and $V \subsetneqq S_d$ a proper linear subspace without common zeros. Let $W = S_1 \cdot V$ be the space of polynomials of degree $d + 1$ generated by all products of $F \in V$ with linear forms. Then*

$$\dim(W) \ge \dim(V) + 2.$$

PROOF. A basic observation is that if $V \subset S_d$ is k-dimensional and $p \in \mathbb{P}^1$ any point and we let $\text{Ord}_p(V)$ be the set of orders of vanishing of (nonzero) elements of V at p, then the cardinality of $\text{Ord}_p(V)$ is exactly k. In particular, if the dimension of $W = S_1 \cdot V$ is less than $k + 2$, (and V has no common zeros) then since

$$\text{Ord}_p(W) \supset \text{Ord}_p(V) \cup (\text{Ord}_p(V) + 1),$$

it follows that we must have

$$\text{Ord}_p(V) = \{0, 1, \dots, k - 1\}$$

and

$$\mathrm{Ord}_p(W) = \{0, 1, 2, \ldots, k-1, k\}$$

The first of these statements says that we can find a basis $\{F_1, F_2, \ldots, F_k\}$ for V with $\mathrm{ord}_p(F_\alpha) = k - \alpha$ for all α, where we take p to be the zero of X. But now the three polynomials XF_1, YF_1, and $XF_2 \in W$ all vanish to order at least $k-1$ at p, so that by the second statement there must be at least one linear relation

$$a_1 \cdot XF_1 + b_1 \cdot YF_1 + a_2 \cdot XF_2 \equiv 0$$

among them. Since XF_1 and YF_1 are independent, the coefficient a_2 must be nonzero; we deduce that F_1 and F_2 must have a common divisor of degree $d-1$. Similarly, since XF_1, YF_1, XF_2, YF_2, and XF_3 all vanish to order at least $k-2$ at p there must be at least two linear relations among them, and since there can be at most one linear relation among XF_1, YF_1, XF_2, and YF_2, there must be a relation

$$a_1 \cdot XF_1 + b_1 \cdot YF_1 + a_2 \cdot XF_2 + b_2 \cdot YF_2 + a_3 \cdot XF_3 \equiv 0$$

with $a_3 \neq 0$. As before, we deduce that F_1, F_2, and F_3 must have a common divisor of degree at least $d-2$. Proceeding in this way we deduce that $F_1, \ldots, F_\alpha$ will have at least $d - \alpha + 1$ zeros in common; if V has no common zeros we conclude that V is the vector space of all polynomials of degree d. □

Now suppose we have a point $[Z_0, \ldots, Z_d] \in \mathbb{P}^d$ such that the matrix

$$M_\alpha = \begin{pmatrix} Z_0 & Z_1 & Z_2 & \cdots & Z_{d-\alpha} \\ Z_1 & Z_2 & Z_3 & \cdots & Z_{d-\alpha+1} \\ \vdots & & & & \\ Z_\alpha & Z_{\alpha+1} & \cdot & \cdots & Z_d \end{pmatrix}$$

has rank at most 2. If we realize $\mathbb{P}^d = \mathbb{P}\,\mathrm{Sym}^d V$ as the space of linear functionals on the space S_d of polynomials of degree d on $\mathbb{P}^1 = \mathbb{P}V$, $(Z_0, \ldots, Z_d)$ is a linear functional

$$\varphi\colon S_d \to K$$

on S_d such that the composition

$$S_\alpha \times S_{d-\alpha} \xrightarrow{\ m\ } S_d \xrightarrow{\ \varphi\ } K$$

(where m is multiplication) has rank 2. This means that there exists subspaces $V_1 \subset S_\alpha$ and $V_2 \subset S_{d-\alpha}$, each of codimension 2, such that the products $V_1 \cdot S_{d-\alpha}$ and $S_\alpha \cdot V_2$ both lie in the kernel $V \subset S_d$ of φ. If either $V_1 \cdot S_{d-\alpha}$ or $S_\alpha \cdot V_2$ is equal to V, then by our lemma it follows that V has a common zero, i.e., $[\varphi]$ is a point on the rational normal curve. On the other hand, if both $W_1 = V_1 \cdot S_{d-\alpha}$ and $W_2 = S_\alpha \cdot V_2$ have codimension 2 in S_d, then they must each have a common divisor P_i of degree 2; indeed they must each be the space of polynomials of degree d divisible by P_i.

Now, since the W_i do not span S_d, either they are equal or their intersection is a subspace of codimension 3 in S_d and they together span V. In the latter case, the polynomials P_i must have a common root, which is then a common zero of V, so that once more $[\varphi]$ will be a point on the rational normal curve.

In the former case, φ will be a linear form on S_d vanishing on the space of polynomials divisible by $P = P_1 = P_2$, which is to say, if P has distinct roots q and r, a linear combination

$$\varphi(F) = a \cdot F(q) + b \cdot F(r)$$

of the linear forms given by evaluation at q and r. $[\varphi]$ is thus a point on the secant variety of the rational normal curve. Lastly, in case P has a double root q, it is not hard to check that $[\varphi]$ lies on the closure of the linear combinations given earlier as r ranges over $\mathbb{P}^1$. □

Note as a corollary that for any point $[Z_0, \dots, Z_d] \in \mathbb{P}^d$, the rank l of the matrix

$$M_\alpha = \begin{pmatrix} Z_0 & Z_1 & Z_2 & \dots & Z_{d-\alpha} \\ Z_1 & Z_2 & Z_3 & \dots & Z_{d-\alpha+1} \\ \vdots & & & & \\ Z_\alpha & Z_{\alpha+1} & \cdot & \dots & Z_d \end{pmatrix}$$

does not depend on α, as long as $l \leq \alpha \leq d - l$: in general, l is simply one greater than the dimension of the smallest secant plane to the rational normal curve containing $[Z_0, \dots, Z_d]$.

Exercise 9.9. Give a similar argument to establish Proposition 9.7 in general.

Example 9.10. Rational Normal Scrolls III

Consider next a 1-generic $2 \times k$ matrix

$$\Omega(Z) = \begin{pmatrix} L_1(Z) & \dots & L_k(Z) \\ M_1(Z) & \dots & M_k(Z) \end{pmatrix}$$

of linear forms on $\mathbb{P}^n$, and the corresponding determinantal variety

$$\Psi = \{[Z]: \operatorname{rank} \Omega(Z) = 1\}.$$

To describe Ψ, note that for any homogeneous 2-vector $[\lambda, \mu] \in \mathbb{P}^1$, the common zero locus of a generalized row of Ω

$$\Lambda_{[\lambda,\mu]} = \{[Z]: \lambda L_1(Z) + \mu M_1(Z) = \cdots = \lambda L_k(Z) + \mu M_k(Z) = 0\}$$

is a linear space $\mathbb{P}^{n-k}$ in $\mathbb{P}^n$, and that the variety Ψ is just the union, over all $[\lambda, \mu] \in \mathbb{P}^1$, of these linear spaces. Given that in case $k = n$ the determinantal variety is a rational normal curve and in case $k = 2$ it is a quadric of rank 3 or 4, it is not perhaps surprising that the variety Ψ is in general a rational normal scroll of dimension $n - k + 1$, a fact that we will now proceed to establish.

We will focus on the case $k = n - 1$ in the following; the general case is only notationally more complicated. We start by showing that conversely every rational normal surface scroll is a determinantal variety.

Exercise 9.11. Show that the determinantal variety

$$\Psi = \left\{[Z]\colon \operatorname{rank}\begin{pmatrix} Z_0 & \cdots & Z_{l-1} & Z_{l+1} & \cdots & Z_{n-1} \\ Z_1 & \cdots & Z_l & Z_{l+2} & \cdots & Z_n \end{pmatrix} \le 1\right\}$$

is the rational normal scroll $X_{l,n-l-1} \subset \mathbb{P}^n$.

With this said, our main statement is the following proposition.

Proposition 9.12. *Any 1-generic $2 \times n-1$ matrix Ω of linear forms on $\mathbb{P}^n$ is conjugate for some l to the matrix*

$$\Omega_0 = \begin{pmatrix} Z_0 & \cdots & Z_{l-1} & Z_{l+1} & \cdots & Z_{n-1} \\ Z_1 & \cdots & Z_l & Z_{l+2} & \cdots & Z_n \end{pmatrix}$$

PROOF. We start by observing that if $H \cong \mathbb{P}^{n-1} \subset \mathbb{P}^n$ is a general hyperplane, then the restriction of Ω to H will again be 1-generic; this is equivalent to saying that H will not contain any of the lines $\Lambda_{[\lambda,\mu]}$ coming from the generalized rows of Ω. It follows, if H is given by the linear form W, that modulo W the matrix Ω can be put in the normal form of Proposition 9.4; i.e., Ω is conjugate to a matrix of the form

$$\begin{pmatrix} Z_1 + a_{1,1}W & Z_2 + a_{1,2}W & \cdots & Z_{n-1} + a_{1,n-1}W \\ Z_2 + a_{2,1}W & Z_3 + a_{2,2}W & \cdots & Z_n + a_{2,n-1}W \end{pmatrix}.$$

After simply relabeling the variables Z_i, we can take this to be

$$\begin{pmatrix} Z_1 + a_1 W & Z_2 + a_2 W & \cdots & Z_{n-1} + a_{n-1}W \\ Z_2 & Z_3 & \cdots & Z_n \end{pmatrix}.$$

We now want to multiply this matrix on the right by an invertible $(n-1)\times(n-1)$ matrix $B = (b_{i,j})$ of scalars to put it in the desired form, i.e., so that the $(k+1)$st entry in the first row equals the kth entry in the second row for all but one value of k between 1 and $n-2$. Each such requirement imposes linear conditions on the $b_{i,j}$; for example, the first says that

$$(*_1) \qquad \begin{gathered} b_{1,1} = b_{2,2}, b_{2,1} = b_{3,2}, \ldots, b_{n-2,1} = b_{n-1,2} \\ b_{1,2} = b_{n-1,1} = 0 \quad \text{and} \\ a_1 b_{1,2} + a_2 b_{2,2} + \cdots + a_{n-1} b_{n-1,2} = 0 \end{gathered}$$

and the kth that

$$(*_k) \qquad \begin{gathered} b_{1,k} = b_{2,k+1}, b_{2,k} = b_{3,k+1}, \ldots, b_{n-2,k} = b_{n-1,k+1} \\ b_{1,k+1} = b_{n-1,k} = 0 \quad \text{and} \\ a_1 b_{1,k+1} + a_2 b_{2,k+1} + \cdots + a_{n-1} b_{n-1,k+1} = 0. \end{gathered}$$

The condition that the relations $(*_k)$ hold for all but one value l of k says that the matrix B is constant along diagonals (except between the lth and $(l+1)$st columns), zero on the top and bottom row (except in the first and $(l+1)$st entry of the top row and lth and last entry of the bottom row), and its columns (except for the first and the $(l+1)$st) are orthogonal to $(a_1, \ldots, a_{n-1})$. The picture of B is thus

$$\begin{bmatrix} b_1 & 0 & 0 & 0 & 0 & c_1 & 0 & 0 \\ b_2 & b_1 & 0 & 0 & 0 & c_2 & c_1 & 0 \\ b_3 & b_2 & b_1 & 0 & 0 & c_3 & c_2 & c_1 \\ b_4 & b_3 & b_2 & b_1 & 0 & c_4 & c_3 & c_2 \\ 0 & b_4 & b_3 & b_2 & b_1 & c_5 & c_4 & c_3 \\ 0 & 0 & b_4 & b_3 & b_2 & c_6 & c_5 & c_4 \\ 0 & 0 & 0 & b_4 & b_3 & 0 & c_6 & c_5 \\ 0 & 0 & 0 & 0 & b_4 & 0 & 0 & c_6 \end{bmatrix}.$$

In particular, B will be determined by a pair of vectors $b = (b_1, \ldots, b_{n-l})$ and $c = (c_1, \ldots, c_{l+1})$ appearing in the first and $(l+1)$st columns (in this example, $n = 9$ and $l = 5$). Now, for any m, let A_m be the matrix

$$A_m = \begin{bmatrix} a_2 & a_3 & \cdots & a_{m+1} \\ a_3 & a_4 & \cdots & a_{m+2} \\ \vdots & & & \\ a_{n-m} & \cdot & \cdots & a_{n-1} \end{bmatrix}.$$

The relations $(*_k)$ then say that the vectors b and c must lie in the kernels of the matrices A_{n-l} and A_{l+1}, respectively. Thus, we can take l to be the smallest integer such that A_{l+1} has a kernel (by simple size considerations l will always be less than or equal to $(n-1)/2$) and choose $c \in \mathrm{Ker}(A_{l+1})$ any nonzero vector; we can then take b a general element of $\mathrm{Ker}(A_{n-l})$, which for reasons of size will have kernel of dimension at least $n - 2l - 1 \geq 2$ (in the "balanced" case $l + 1 = n - l$ we simply have to choose b and c a basis for $\mathrm{Ker}(A_{l+1})$).

The question now is why the matrix B is nonsingular, given these choices. To see this, we go back to Proposition 9.7, which describes in more detail the ranks of the matrices A_m in terms of the location of the point $A = [a_2, \ldots, a_{n-1}]$ in relation to the secant varieties to the rational normal curve

$$C = \{[1, t, t^2, \ldots, t^{n-3}]: t \in \mathbb{P}^1\} \subset \mathbb{P}^{n-3}.$$

Specifically, Proposition 9.7 identifies the integer l as the smallest integer such that A lies on an l-secant plane $\overline{p_1, \ldots, p_l}$ to C; indeed the proof shows that the vector b is just the vector of coefficients of the polynomial P of degree l with roots at the points p_i. Now, for m between l and $n - 2 - l$, we know that A can lie on no m-secant plane to C except one containing the plane $\overline{p_1, \ldots, p_l}$ (consider the projection of C from the plane $\overline{p_1, \ldots, p_{l-1}}$: the image will again be a rational normal curve $\overline{C}$ in $\mathbb{P}^{n-l-2}$, with A mapping to p_l, which we know cannot lie on any secant plane $\overline{q_1, \ldots, q_m}$ to $\overline{C}$ for $m \leq n - l - 2$ unless $q_i = p_l$ for some i). By the same token, however, for $m = n - l - 1$, A will lie on many m-secant planes $\overline{q_1, \ldots, q_m}$ with $\{q_1, \ldots, q_m\}$ disjoint from $\{p_1, \ldots, p_l\}$; so that the general element of the kernel of the matrix A_{n-l} will be the vector of coefficients of a polynomial Q with roots $q_1, \ldots, q_{n-l-1}$ disjoint from $\{p_1, \ldots, p_l\}$. By Lemma 3.6, then, we see that the matrix B is nonsingular. □

Exercise 9.13. State and prove the analogs of Exercise 9.11 and Proposition 9.12 for $2 \times k$ matrices of linear forms on $\mathbb{P}^n$ and rational normal scrolls in general.

Exercise 9.14. Use Exercise 9.13 to verify the description of rational normal scrolls given in Example 8.39.

Observe also that, inasmuch as all determinantal varieties $\Psi \subset \mathbb{P}^n$ given by matrices of linear forms are intersections of the generic determinantal variety $M_k \subset M$ with linear subspaces of M, Exercise 9.11 and Proposition 9.12 subsume Exercises 8.18 and 8.28. Indeed, the general statement we can make is that every rational normal scroll is the intersection of the Segre variety $\Sigma_{k,1} = \sigma(\mathbb{P}^k \times \mathbb{P}^1) \subset \mathbb{P}^{2k+1}$ with a linear subspace $\mathbb{P}^n \subset \mathbb{P}^{2k+1}$.

There is an alternative proof of Proposition 9.12 (or rather the generalization of it to $2 \times k$ matrices for arbitrary k), communicated to me by Frank Schreyer. As before, we consider a 1-generic $2 \times k$ matrix

$$\Omega(Z) = \begin{pmatrix} L_1(Z) & \cdots & L_k(Z) \\ M_1(Z) & \cdots & M_k(Z) \end{pmatrix}$$

of linear forms on $\mathbb{P}^n$. Assume that the entries of the matrix Ω together span the vector space W of linear forms on $\mathbb{P}^n$. We claim that for some sequence of integers $a_1, \ldots, a_l$ $(l = n - k)$ this will be conjugate to the matrix

$$\Omega_a = \left(\begin{array}{ccc|ccc|ccc|ccc} Z_0 & \cdots & Z_{a_1-1} & Z_{a_1+1} & \cdots & Z_{a_2-1} & Z_{a_2+1} & \cdots & Z_{a_l-1} & Z_{a_l+1} & \cdots & Z_{n-1} \\ Z_1 & \cdots & Z_{a_1} & Z_{a_1+2} & \cdots & Z_{a_2} & Z_{a_2+2} & \cdots & Z_{a_l} & Z_{a_l+2} & \cdots & Z_n \end{array}\right)$$

i.e., to a matrix consisting of $l + 1$ blocks of size $2 \times a_1, \ldots, 2 \times (n - a_l - 1)$, with each block a *catalecticant*, that is, a matrix A with $a_{i,j+1} = a_{i+1,j}$ for all i, j. (Note that as an immediate corollary in case the entries of Ω span a subspace of W of dimension $m + 1 < n + 1$ we can put Ω in the same form, with n replaced by m and l by $m - k$.)

PROOF. Let $U_1 \subset W$ be the span of the entries in the first row of Ω and U_2 similarly the span of the entries in the second row; by the condition that Ω is 1-generic the dimension of each U_i must be k. It follows that the intersection $U_1 \cap U_2$ will have dimension exactly $2k - n - 1$, and multiplying Ω on the right by a suitable $k \times k$ matrix we can assume that this intersection $U_1 \cap U_2$ is exactly the span of the entries $M_1, \ldots, M_{2k-n-1}$. Let Ω' be the submatrix consisting of the first $2k - n - 1$ columns of Ω. Ω' is again 1-generic, and so by induction on k we can make a change of variables in the Z_i and multiply Ω on the left and right by scalar matrices to put Ω' in the form

$$\left(\begin{array}{ccc|ccc|ccc|ccc} Z_0 & \cdots & Z_{b_1-1} & Z_{b_1+1} & \cdots & Z_{b_2-1} & Z_{b_2+1} & \cdots & Z_{b_{l'}-1} & Z_{b_{l'}+1} & \cdots & Z_{n'-1} \\ Z_1 & \cdots & Z_{b_1} & Z_{b_1+2} & \cdots & Z_{b_2} & Z_{b_2+2} & \cdots & Z_{b_{l'}} & Z_{b_{l'}+2} & \cdots & Z_{n'} \end{array}\right)$$

where $n' + 1$ is the dimension of the span of the entries of Ω', and $l' = n' - (2k - n - 1)$.

We now look at the remaining $n + 1 - k$ columns of Ω. In particular, consider the variables Z_{b_i}, $i = 1, \ldots, l'$ and $Z_{n'}$: by construction, these are in the span U_1 of the upper-row entries of Ω, but their span is visibly independent from the span of the first row of Ω'. It follows that after multiplying Ω on the right by a suitable scalar matrix, Ω' will be unchanged and these variables will appear as the upper entries in the next $l' + 1$ columns of Ω.

What are the entries in the lower row of these columns? All we know is that they are independent modulo the span of the entries of Ω', so that we can take them to be among the coordinates Z_α with $\alpha > n'$. We can then insert these columns at the right-hand ends of the corresponding $l' + 1$ blocks of Ω'; after renumbering the variables the submatrix Ω'' consisting of the first $(2k - n - 1) + (l' + 1) = n' + 1$ columns of Ω will have the form

$$\left(\begin{array}{ccc|ccc|ccc|ccc} Z_0 & \cdots & Z_{a_1-1} & Z_{a_1+1} & \cdots & Z_{a_2-1} & Z_{a_2+1} & \cdots & Z_{a_{l'}-1} & Z_{a_{l'}+1} & \cdots & Z_{n'+l'} \\ Z_1 & \cdots & Z_{a_1} & Z_{a_1+2} & \cdots & Z_{a_2} & Z_{a_2+2} & \cdots & Z_{a_{l'}} & Z_{a_{l'}+2} & \cdots & Z_{n'+l'+1} \end{array}\right)$$

where $a_i = b_i + i$.

Finally, we have $k - (n' + 1)$ columns of Ω left. But the entries of the first $n' + 1$ span a space of dimension $n' + l' + 2$ in the $(n + 1)$-dimensional space W, and by hypothesis the entries of Ω altogether span W. Since

$$\begin{aligned} n + 1 - (n' + l' + 2) &= n + 1 - (n' + (n' - (2k - n - 1)) + 2) \\ &= 2(k - (n' + 1)) \end{aligned}$$

it follows that the entries of these last columns are independent modulo the entries of Ω''. We can thus take them to be the coordinates $Z_{n'+l'+2}, \ldots, Z_{n'}$ and thinking of these columns as

$$l - l' = k - (n' + 1)$$

blocks of width 1 we see we have put Ω in the desired form. □

To give a geometric interpretation to this argument, what we are doing here is: i. assuming the rank 1 locus associated to the matrix Ω is not a cone; ii. projecting this determinantal variety, which is swept out by a one-parameter family of $(n - k)$-planes, from one of those planes; and iii. if the resulting variety turns out to be a cone, projecting from the vertex of that cone.

Example 9.15. Rational Normal Scrolls IV

The preceding description of rational normal scrolls gives rise to yet another classical construction of scrolls (in fact, it is equivalent to it). This is analogous to the description of a twisted cubic as the residual intersection of two quadrics containing a line $L \subset \mathbb{P}^3$. (Given a collection of varieties $X_i \subset \mathbb{P}^n$ containing a variety $Z \subset \mathbb{P}^n$ and such that Z is an irreducible component of their intersection, we call the union of the remaining irreducible components of the intersection the *residual intersection* of the X_i.)

To start, suppose we have a rational normal k-fold scroll $X \subset \mathbb{P}^n$, given to us as the determinantal variety associated to a matrix of linear forms

$$\Gamma = \begin{pmatrix} L_1(Z) & \dots & L_{n-k+1}(Z) \\ M_1(Z) & \dots & M_{n-k+1}(Z) \end{pmatrix}$$

Let $\Lambda \subset \mathbb{P}^n$ be the $(n-2)$-plane described by the vanishing of the first column of this matrix. Then for any $[Z] \in \mathbb{P}^n - \Lambda$, the matrix Γ will have rank 1 if and only if the remaining columns of Γ are dependent on the first, that is, if and only if the $(1, j)$th minor of Γ vanishes for $j = 2, \dots, n-k+1$. Thus, if we let Q_j be the quadric hypersurface defined by the $(1, j)$th minor, we see that

$$Q_2 \cap Q_3 \cap \dots \cap Q_{n-k+1} = \Lambda \cup X.$$

In English, then, the scroll X is the residual intersection of $n-k$ quadric hypersurfaces containing an $(n-2)$-plane $\Lambda \subset \mathbb{P}^n$.

We have a converse to this statement as well. Suppose now that $\Lambda \subset \mathbb{P}^n$ is a codimension 2 subspace, and $Q_2, \dots, Q_{n-k+1}$ quadric hypersurfaces containing Λ. Since the ideal of Λ is generated by two linear forms—call them $L_1(Z)$ and $M_1(Z)$—we can write each Q_j as a linear combination of L_1 and M_1 with homogeneous linear coefficients:

$$Q_j(Z) = M_j(Z) \cdot L_1(Z) - L_j(Z) \cdot M_1(Z).$$

It follows then that, away from Λ, the common zero locus of the Q_j is the rank 1 locus of the matrix

$$\begin{pmatrix} L_1(Z) & \dots & L_{n-k+1}(Z) \\ M_1(Z) & \dots & M_{n-k+1}(Z) \end{pmatrix},$$

which, as long as it satisfies the independence condition, will be a rational normal scroll.

Exercise 9.16. Show that if $Q_2, \dots, Q_{n-k+1}$ are general quadrics containing Λ, then the matrix Γ does satisfy the independence condition; conclude that the residual intersection of general quadrics containing an $(n-2)$-plane $\Lambda \subset \mathbb{P}^n$ is a rational normal scroll.

We can use a similar idea to give a cute solution to Exercise 1.11, that is, to show that conversely any two quadrics Q_1, Q_2 containing a twisted cubic curve $C \subset \mathbb{P}^3$ intersect in the union of C and a line. To see this, represent the twisted cubic as the zero locus of the 2×3 matrix

$$\begin{pmatrix} Z_0 & Z_1 & Z_2 \\ Z_1 & Z_2 & Z_3 \end{pmatrix}.$$

Now, each quadric Q_μ containing the twisted cubic can be written as a linear combination of the 2×2 minors of this matrix, that is, as the determinant of a 3×3 matrix

$$\begin{bmatrix} Z_0 & Z_1 & Z_2 \\ Z_1 & Z_2 & Z_3 \\ \mu_1 & \mu_2 & \mu_3 \end{bmatrix}$$

(the coefficients μ_i here do not correspond exactly to those of Exercise 1.11). The locus outside of C where two such both vanish is thus the rank ≤ 2 locus of the 4×3 matrix

$$\begin{bmatrix} Z_0 & Z_1 & Z_2 \\ Z_1 & Z_2 & Z_3 \\ \mu_1 & \mu_2 & \mu_3 \\ \nu_1 & \nu_2 & \nu_3 \end{bmatrix},$$

which, if the vectors μ and ν are independent, is the same as the locus of the two 3×3 minors

$$\begin{vmatrix} Z_0 & Z_1 & Z_2 \\ \mu_1 & \mu_2 & \mu_3 \\ \nu_1 & \nu_2 & \nu_3 \end{vmatrix} = \begin{vmatrix} Z_1 & Z_2 & Z_3 \\ \mu_1 & \mu_2 & \mu_3 \\ \nu_1 & \nu_2 & \nu_3 \end{vmatrix} = 0,$$

which is a line.

More General Determinantal Varieties

To generalize one step further, suppose now that $\{F_{i,j}\}$ is an $m \times n$ matrix of homogeneous polynomials on a variety $X \subset \mathbb{P}^k$, with $\deg(F_{i,j}) = d_{i,j}$, and assume that for some pair of sequences of integers $\{e_1, \ldots, e_m\}$ and $\{f_1, \ldots, f_n\}$ we have $d_{i,j} = e_i - f_j$ (as will be the case, for example, if the degrees of the $F_{i,j}$ are constant along either rows or columns). The minor determinants of the matrix $(F_{i,j})$ are then homogeneous polynomials, and the subvariety $Y \subset X$ they cut out is again called a determinantal subvariety of X. (For those familiar with the notion of vector bundles, this is still a special case of a more general definition; we can define a determinantal subvariety of X to be the locus where a map $\varphi: E \to F$ between two vector bundles E, F on X has rank at most k.)

Exercise 9.17. Let $S \cong \mathbb{P}^1 \times \mathbb{P}^1 \subset \mathbb{P}^3$ be the Segre surface $Z_0Z_3 - Z_1Z_2 = 0$. Let $C \subset S$ be the curve given as the locus of a bihomogeneous polynomial $F(X, Y)$ of bidegree $(m, m - 1)$ in the variables X, Y on $\mathbb{P}^1 \times \mathbb{P}^1$. Show that $C \subset \mathbb{P}^3$ may be described as a determinantal variety of the form

$$C = \left\{[Z]: \operatorname{rank}\begin{pmatrix} Z_0 & Z_1 & G(Z) \\ Z_2 & Z_3 & H(Z) \end{pmatrix} \leq 1\right\}$$

where G and H are homogeneous polynomials of degree $m - 1$; and conversely.

It may be that having generalized the notion of determinantal variety to this extent, the suspicion will enter the reader's mind that every variety $X \subset \mathbb{P}^n$ is determinantal. This is true in a trivial sense; if X is the zero locus of homogeneous polynomials $F_1, \dots, F_k$, then it is the determinantal variety associated to the $1 \times k$ matrix $(F_1, \dots, F_k)$. It is not, however, the case that every variety is nontrivially determinantal, as the following example shows.

Exercise 9.18. Show that for d sufficiently large not every surface of degree d in $\mathbb{P}^3$ is expressible as the determinant of a $d \times d$ matrix of linear forms.

Symmetric and Skew-Symmetric Determinantal Varieties

To mention one other class of determinantal varieties, there are analogous notions of symmetric and skew-symmetric determinantal varieties. We mention here a couple of such varieties we have already seen in other contexts.

Exercise 9.19. Show that the quadratic Veronese variety $v_2(\mathbb{P}^n) \subset \mathbb{P}^{(n+1)(n+2)/2-1}$ is the locus where a certain symmetric $(n+1) \times (n+1)$ matrix of linear forms on $\mathbb{P}^{(n+1)(n+2)/2-1}$ has rank 1.

Exercise 9.20. Show that the Grassmannian $\mathbb{G}(1, n) \subset \mathbb{P}^{n(n+1)/2-1}$ is the locus where a certain skew-symmetric $(n+1) \times (n+1)$ matrix of linear forms on $\mathbb{P}^{n(n+1)/2-1}$ has rank 2.

Just as in the case of the Segre varieties, the determinantal descriptions of the Veronese varieties and the Grassmannians allow us to describe their secant varieties as determinantal.

Note finally that Exercise 9.19, in combination with Exercise 2.9, allows us to conclude that any projective variety is isomorphic to a determinantal variety given by a matrix with linear entries.

Example 9.21. Fano Varieties of Determinantal Varieties

We come now to the interesting problem of describing linear spaces lying on determinantal varieties, in other words, vector spaces of matrices of low rank. To be precise, we let M_k be the variety of $m \times n$ matrices of rank at most k in the projective space M of all $m \times n$ matrices and we ask for a description of the variety $F_l(M_k)$ of l-planes on M_k.

Start with the simplest case, the variety M_1 of matrices of rank 1. As we saw, this is the Segre variety $\mathbb{P}^{m-1} \times \mathbb{P}^{n-1} \subset \mathbb{P}^{mn-1}$; under the Segre map the fibers of the projections of $\mathbb{P}^{m-1} \times \mathbb{P}^{n-1}$ onto its two factors are carried into linear subspaces $\mathbb{P}^{n-1}$ and $\mathbb{P}^{m-1} \subset \mathbb{P}^{mn-1}$, respectively. Call these families of linear spaces the *rulings* of the Segre variety M_1; we then have the following theorem.

Theorem 9.22. *Any linear subspace $\Lambda \subset M_1 \subset M \cong \mathbb{P}^{mn-1}$ is contained in a ruling of M_1.*

PROOF. Translated into the language of linear algebra, this amounts to saying that if $V \subset \text{Hom}(K^m, K^n)$ is any vector space of matrices such that all $A \in V$ have rank at most 1, then either all nonzero $A \in V$ have the same kernel or all $A \neq 0 \in V$ have the same image. To see this, note that if any two nonzero elements $A, B \in V$ had both distinct images and distinct kernels, we could find vectors $v, w \in K^m$ with $v \in \text{Ker}(A)$ and $w \in \text{Ker}(B)$ but not vice versa; then $B(v)$ and $A(w)$ would be independent and any linear combination $\alpha \cdot A + \beta \cdot B$ with nonzero coefficients α, β would have rank 2. Thus for any two elements $A, B \in V$, either the kernel of $L_{\alpha,\beta} = \alpha \cdot A + \beta \cdot B$ is constant and $\text{Im}(L_{\alpha,\beta})$ varies with $[\alpha, \beta]$ or vice versa. In the former case, any third element $C \in V$ must have its kernel in common with all the $L_{\alpha,\beta}$, since it can have its image in common with at most one; in the latter case, the image of C must be that of $L_{\alpha,\beta}$. □

If $m > n$, the theorem implies in particular that the Fano variety $F_{m-1}(M_1)$ is isomorphic to $\mathbb{P}^{n-1}$ and if $m = n$ to two disjoint copies of $\mathbb{P}^{n-1}$.

Exercise 9.23. Generalize Exercise 6.8 to show that if $m > n$ the Plücker embedding of the Grassmannian $\mathbb{G}(m-1, mn-1)$ into projective space carries the subvariety $F_{m-1}(M_1) \cong \mathbb{P}^{n-1}$ into a Veronese variety.

The linear spaces lying on M_k with $k \geq 2$ are less uniformly behaved. In M_2, for example, there are linear spaces defined analogously to the rulings of M_1: we can choose a subspace $V \cong K^{m-2} \subset K^m$ and look at the space of matrices A with $\text{Ker}(A) \supset V$, we can choose $W \cong K^2 \subset K^n$ and look at the space of A with $\text{Im}(A) \subset W$, or we can choose a hyperplane $V \subset K^m$ and a line $W \subset K^n$ and look at the space of matrices A such that $A(V) \subset W$. These all yield maximal linear subspaces of M_2. Unlike the case of M_1, however, these are not all the maximal linear subspaces: consider, for example, in the case $m = n = 3$, the space of skew-symmetric matrices. The situation for $k = 2$, 3, and 4 is pretty well understood, but the general picture is not.

Exercise 9.24. Consider the case $m = n = 3$ and $k = 2$, (i) Show that any linear space $\Lambda \cong \mathbb{P}^l \subset M_2$ of dimension $l \geq 3$ either consists of matrices having a common kernel or of matrices having a common image. (ii) Show that any linear subspace $\Lambda \cong \mathbb{P}^2 \subset M_2$ is either one of these types or, after a suitable change of basis, the space of skew-symmetric matrices.

LECTURE 10
Algebraic Groups

As with most geometric categories, we have the notion of a group object in classical algebraic geometry. An *algebraic group* is defined to be simply a variety X with regular maps

$$m: X \times X \to X$$

and

$$i: X \to X$$

satisfying the usual rules for multiplication and inverse in a group (e.g., there exists a point $e \in X$ with $m(e, p) = m(p, e) = p$ and $m(p, i(p)) = e \ \forall p \in X$). By the same token, a *morphism* of algebraic groups is a map $\varphi: G \to H$ that is both a regular map and a group homomorphism.

The fundamental examples are the following classical groups.

Example 10.1. The General Linear Group GL_nK

The set of invertible $n \times n$ matrices is just the complement, in the vector space $\mathbb{A}^{n^2}$ of all $n \times n$ matrices, of the hypersurface given by the determinant; it is thus a distinguished open subset of $\mathbb{A}^{n^2}$ and therefore an affine variety in its own right. The multiplication map $GL_nK \times GL_nK \to GL_nK$ is clearly regular; that the inverse map is so follows from Cramer's rule: for A an invertible $n \times n$ matrix,

$$(A^{-1})_{i,j} = (-1)^{i+j} \det(M_{j,i})/\det(A)$$

where $M_{j,i}$ is the submatrix of A obtained by deleting the jth row and ith column.

The subgroup $SL_nK \subset GL_nK$ is a subvariety, closed under the group operations, and so an algebraic subgroup; the quotient group $PGL_nK \subset \mathbb{P}^{n^2-1}$ (this is the quotient of GL_nK by its center, the subgroup of scalar matrices) is seen to be an algebraic group the same way GL_nK is.

Example 10.2. The Orthogonal Group SO_nK (in characteristic $\neq 2$)

This is a priori defined as a subgroup of SL_nK. Specifically, assume that the characteristic of K is not 2 and let Q be a nondegenerate symmetric bilinear form

$$Q: V \times V \to K$$

on an n-dimensional vector space V, and consider the subgroup of the group $SL(V)$ given by

$$SO(V, Q) = \{A \in SL(V): Q(Av, Aw) = Q(v, w) \ \forall v, w \in V\}.$$

This is clearly a subvariety of $SL(V)$ closed under composition and inverse, and hence an algebraic group. Actually, the condition that A preserve Q implies that the determinant of A is ± 1, so that if we allow $A \in GL(V)$ rather than $SL(V)$ we get a $\mathbb{Z}/2$ extension of $SO(V, Q)$, denoted simply $O(V, Q)$.

Note that we could equivalently have defined $SO(V, Q)$ as the subgroup of $SL(V)$ preserving the quadratic polynomial $q(v) = Q(v, v)$ on V. To see this, observe that for any $v, w \in V$

$$Q(v, w) = \frac{q(v + w) - q(v) - q(w)}{2}$$

so that q determines Q. Thus, if $A \in SL(V)$ and $q(Av) = q(v)$ for all $v \in V$, then $Q(Av, Aw) = Q(v, w)$ for all $v, w \in V$, i.e., $A \in SO(V, Q)$.

If we take $V = K^n$ and the form Q to be given simply by dot product

$$Q(v, w) = {}^t v \cdot w,$$

then the subgroup $SO(V, Q)$ is denoted SO_nK and may be described as

$$SO_nK = \{A \in SL_nK: {}^tA \cdot A = I\}.$$

As before, if we allow $A \in GL_nK$, we obtain a $\mathbb{Z}/2$ extension of SO_nK with two connected components, called O_nK. Given any nondegenerate symmetric bilinear form Q on a vector space V we may find a basis for V—that is, an identification of V with K^n—in terms of which Q is given in this way. Thus, all groups $SO(V, Q)$ with Q a nondegenerate symmetric bilinear form on the n-dimensional vector space V over K are in fact isomorphic to SO_nK.

SO_nK has a nontrivial center $\{\pm I\}$ if n is even; the quotient by this center is naturally a subgroup of PGL_nK denoted PSO_nK (if n is odd, $PSO_nK = SO_nK$). Similarly, we can for any n form the quotient $PO_nK = O_nK/\{\pm I\}$; if n is odd this is irreducible, while for n even this still has two disjoint irreducible components. Lastly (though we won't establish it here), it is also the case that for any n there is an irreducible algebraic group, called $Spin_nK$, that admits a map of degree 2 to the group SO_nK.

Example 10.3. The Symplectic Group $\mathrm{Sp}_{2n}K$

This is defined in the same way as $\mathrm{SO}_n K$, except that the nondegenerate bilinear form Q is taken to be skew-symmetric rather than symmetric, and the dimension of the vector space must correspondingly be even.

Group Actions

By an *action* of an algebraic group G on an algebraic variety X, we mean a regular map

$$\varphi: G \times X \to X$$

satisfying the usual rules of composition, that is, $\varphi(g, \varphi(h, x)) = \varphi(gh, x)$. (Given such an action, we will often suppress the φ and write $g(p)$ for $\varphi(g, p)$.) By a *projective action* on a variety $X \subset \mathbb{P}^n$ we mean an action of G on $\mathbb{P}^n$ such that $\varphi(g, X) = X$ for all $g \in G$. An even stronger condition is to require that the action of G on X lift to an action of G on the homogeneous coordinate ring $S(X)$ of X in $\mathbb{P}^n$; in this case we say that the action of G on X is *linear*.

Example 10.4. $\mathrm{PGL}_{n+1}K$ acts on $\mathbb{P}^n$

Clearly, the basic group action of the subject is the action of $\mathrm{PGL}_{n+1}K$ on $\mathbb{P}^n$. We should remark here that this action is not linear. On the other hand, if $X = v_d(\mathbb{P}^n) \subset \mathbb{P}^N$ is the image of $\mathbb{P}^n$ under the Veronese map of degree d, then the action of $\mathrm{PGL}_{n+1}K$ on X is projective (we will see this explicitly in many of the following examples) and is linear iff $n + 1$ divides d.

One other aspect of the action of $\mathrm{PGL}_{n+1}K$ on the projective space $\mathbb{P}^n$ (which we will prove in Lecture 18) is that $\mathrm{PGL}_{n+1}K$ is the entire group of automorphisms of $\mathbb{P}^n$.

Here is a classical and easy fact about the action of $\mathrm{PGL}_{n+1}K$ on $\mathbb{P}^n$.

Exercise 10.5. Show that if $\{p_1, \ldots, p_{n+2}\} \subset \mathbb{P}^n$ is any set of $n + 2$ points in general position (i.e., with no $n + 1$ lying on a hyperplane) and $\{q_1, \ldots, q_{n+2}\}$ any other such set, there is a unique element of $\mathrm{PGL}_{n+1}K$ carrying p_i to q_i for all i.

Exercise 10.6. Let $X \subset \mathbb{P}^n$ be any variety. Show that the subgroup

$$\mathrm{Aut}(X, \mathbb{P}^n) = \{A \in \mathrm{PGL}_{n+1}K: A(X) = X\}$$

is an algebraic subgroup of $\mathrm{PGL}_{n+1}K$. This is called the *group of projective motions* of $X \subset \mathbb{P}^n$.

Note. $\mathrm{Aut}(X, \mathbb{P}^n)$ is a subgroup of the group $\mathrm{Aut}(X)$ of all automorphisms of the variety X. It's not hard to come up with examples where it's a proper subgroup, i.e., where there exist automorphisms of X not induced by motions of the projective space, as in Exercise 10.7. There also exist varieties X such that $\mathrm{Aut}(X)$ is not an algebraic group at all, though examples of these are somewhat harder to come by given our limited techniques. (For those with the necessary background, the simplest I can think of is the blow-up X of the plane $\mathbb{P}^2$ at the nine points of intersection of two general cubic curves; X is then a family of elliptic curves with eight sections, generating a group $\mathbb{Z}^8$ of automorphisms of X.)

Exercise 10.7. Give examples of projective varieties $X \subset \mathbb{P}^n$ such that $\mathrm{Aut}(X, \mathbb{P}^n) \subsetneqq \mathrm{Aut}(X)$. (*)

Group actions play an important role in algebraic geometry, for a number of reasons. At an elementary level (as we will see in later examples), many of the varieties we have encountered so far—Veronese and Segre varieties, Grassmannians, generic determinantal varieties, rational normal scrolls, for example—admit nontrivial projective actions by subgroups of classical groups; an understanding of these actions is essential to understanding the geometry of the varieties in question. In fact, it would probably be fair to say that a characteristic of the sort of varieties we consider here—those that can be studied for the most part without the aid of modern tools such as sheaves or scheme theory—is the presence of a group action.

On a somewhat deeper level, we have already referred to the existence of parameter spaces, that is, varieties whose points correspond to members of a family of varieties in a projective space $\mathbb{P}^n$, such as the variety of hypersurfaces of a given degree d in $\mathbb{P}^n$. Since $\mathrm{PGL}_{n+1}K$ acts on $\mathbb{P}^n$, it will act as well on these parameter spaces; this action is clearly central to the study of these families of varieties up to projective equivalence. This leads to the subject of *geometric invariant theory*, about which we can only hint at this stage (see Lecture 21 for a brief discussion of this topic).

Example 10.8. PGL_2K Acts on $\mathbb{P}^2$

Assume $\mathrm{char}(K) \neq 2$. If we realize GL_2K as the group of automorphisms of a two-dimensional vector space V, then it also acts on the symmetric square Sym^2V; since the center of GL_2K still acts by scalar multiplication on Sym^2V, the quotient group PGL_2K acts on the projective plane $\mathbb{P}(\mathrm{Sym}^2V) \cong \mathbb{P}^2$. Alternatively (and equivalently) we may consider the action of PGL_2K on the dual space $\mathbb{P}(\mathrm{Sym}^2V^*)$, which we think of as the space of quadratic polynomials on $\mathbb{P}^1$ up to scalars; the action of PGL_2K on $\mathbb{P}^1$ naturally induces an action on this space.

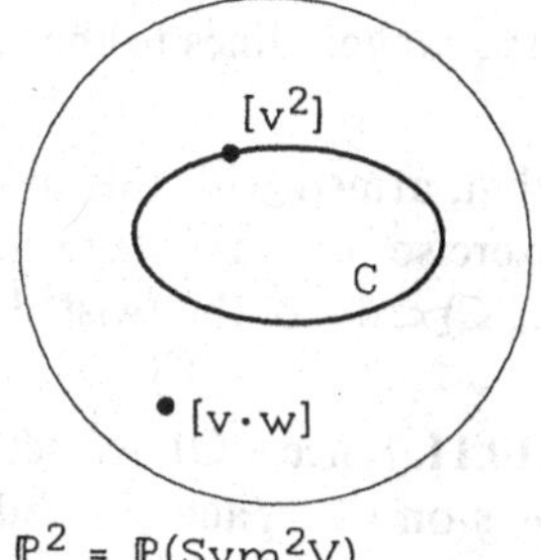

Observe that this action preserves the subset in $\mathbb{P}(\mathrm{Sym}^2 V) \cong \mathbb{P}^2$ of squares, that is, vectors of the form $v \cdot v$. On the other hand, this subset is just the conic curve $C = v_2(\mathbb{P}^1)$ given as the image of the quadratic Veronese map. If we accept the statement that $\mathrm{Aut}(\mathbb{P}^1) = \mathrm{PGL}_2 K$, we conclude that

$$\mathrm{PGL}_2 K = \mathrm{Aut}(C, \mathbb{P}^2),$$

i.e., the group of projective motions of a plane conic is exactly the action of $\mathrm{PGL}(V)$ on $\mathbb{P}(\mathrm{Sym}^2 V)$.

Note also that any element of $\mathrm{Sym}^2 V$ that is not a square is the product of two independent linear factors, and any such product can be carried into any other by an element of $\mathrm{PGL}_2 K$. There are thus exactly two orbits of the action of $\mathrm{PGL}_2 K$ on $\mathbb{P}^2$, the conic curve C and the complement $\mathbb{P}^2 - C$.

Example 10.9. $\mathrm{PGL}_2 K$ acts on $\mathbb{P}^3$

Assume now that $\mathrm{char}(K) \neq 2, 3$. In the same vein as the preceding example, if V is a two-dimensional vector space then $\mathrm{PGL}(V)$ acts on the projective space $\mathbb{P}(\mathrm{Sym}^3 V) \cong \mathbb{P}^3$ (or equivalently, on the dual space of polynomials of degree 3 on $\mathbb{P}^1$ modulo scalars). Analogously, it preserves the subset $C \subset \mathbb{P}^3$ of cubes, which we have seen is the twisted cubic curve; the group of projective motions of the twisted cubic curve is thus the action of $\mathrm{PGL}_2 K$ on $\mathbb{P}^3$.

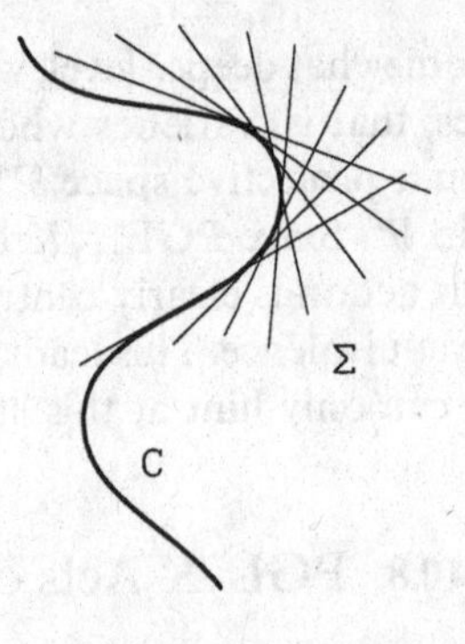

As in the previous case, there are finitely many orbits of this action; the set C of cubes, the locus Σ of products of squares with linear factors (i.e., points $[v^2 \cdot w]$, with v, w independent), and the set Φ of products of pairwise independent linear factors.

Exercise 10.10. Show that the union $\Sigma \cup C$ is a quartic surface (that is, a hypersurface of degree 4) in $\mathbb{P}^3$. (In fact, it is the *tangential surface* of C, that is, the union of the tangent lines to the twisted cubic curve C).

Note that, armed with this description of the orbits of $\mathrm{Aut}(C, \mathbb{P}^3)$, the second half of Exercise 3.8 (that there are, up to projective equivalence, only two projections $\pi_p(C) \subset \mathbb{P}^2$ of the twisted cubic from points $p \notin C$) follows immediately.

Exercise 10.11. Since $\mathrm{PGL}_2 K$ acts on $\mathbb{P}^3$ preserving the twisted cubic curve C, it thereby acts on the space of quadric surfaces in $\mathbb{P}^3$ containing C, which we have seen is parametrized by $\mathbb{P}^2$. Show that this action of $\mathrm{PGL}_2 K$ on $\mathbb{P}^2$ is the same as that described in Example 10.8, and identify the two orbits. (*)

Example 10.12. PGL$_2K$ Acts on $\mathbb{P}^n$ (and on $\mathbb{P}^4$ in particular)

Assume that char(K) is either 0 or greater than n. The action of $\mathrm{PGL}_2K = \mathrm{PGL}(V)$ on the projective space $\mathbb{P}(\mathrm{Sym}^n V) \cong \mathbb{P}^n$ may be described similarly as the group of projective motions of the rational normal curve $C \subset \mathbb{P}(\mathrm{Sym}^n V)$ consisting of nth powers $[v \cdot \ldots \cdot v]$. What becomes much more interesting in the general case is the space of orbits of this action; we will describe this in the (relatively) simple case $n = 4$.

To begin with, there are four orbits of PGL_2K on $\mathbb{P}^4$ not consisting of products of four independent factors: we have the rational normal quartic curve

$$C = \{[v^4], v \in V\};$$

in addition, there are the loci

$$\Sigma = \{[v^3 \cdot w], v, w \text{ independent} \in V\}$$
$$\Phi = \{[v^2 \cdot w^2], v, w \text{ independent} \in V\}$$

and

$$\Psi = \{[v^2 \cdot u \cdot w], v, u, w \text{ pairwise independent} \in V\}.$$

Observe that C is the only closed orbit; Σ and Φ contain C in their closure and are contained in turn in the closure of Ψ. In terms of constructions we have not yet introduced, Σ is the tangential surface of C, Ψ the union of the osculating 2-planes to C and Φ the union of the pairwise intersections of distinct osculating 2-planes.

Things get more interesting when we look at the orbits corresponding to products $[v \cdot u \cdot w \cdot z]$ of pairwise independent factors. The point is, these are not all conjugate: such an element of $\mathbb{P}(\mathrm{Sym}^4 V)$ corresponds to a fourtuple of distinct points on $\mathbb{P}^1$ (actually $(\mathbb{P}^1)^*$, if we're keeping track of these things), and not every collection of four points $z_1, \ldots, z_4 \in \mathbb{P}^1$ can be carried into any other four. What we can do is to carry three of the points z_2, z_3, and z_4 to 0, ∞, and 1 in that order; the remaining point will then be sent to the point λ, where λ is the cross-ratio

$$\lambda(z_1, \ldots, z_4) = \frac{(z_1 - z_2)\cdot(z_3 - z_4)}{(z_1 - z_3)\cdot(z_2 - z_4)}.$$

Now, permuting the four points z_i has the effect of changing the cross-ratio from λ to either $1 - \lambda$, $1/\lambda$, $1/(1 - \lambda)$, $(\lambda - 1)/\lambda$, or $\lambda/(\lambda - 1)$. Thus, two fourtuples can be carried into each other if and only if the subsets $\{\lambda, 1 - \lambda, 1/\lambda, 1/(1 - \lambda), (\lambda - 1)/\lambda, \lambda/(\lambda - 1)\} \subset K - \{0, 1\}$ coincide. To characterize when this is the case, we introduce the celebrated j-function

$$j(\lambda) = 256 \cdot \frac{(\lambda^2 - \lambda + 1)^3}{\lambda^2 \cdot (\lambda - 1)^2}$$

(the factor of 256 is there for number-theoretic reasons). We have the following exercise.

Exercise 10.13. Show that two subsets $\{0, \infty, 1, \lambda\}$ and $\{0, \infty, 1, \lambda'\} \subset \mathbb{P}^1$ are congruent if and only if the values $j(\lambda) = j(\lambda')$.

From this exercise it follows that we have one orbit Ω_j for each value of j in K, completing the list of orbits.

We will mention here a few facts about the orbits Ω_j; they can all be worked out directly, though we will not do that here. To begin with, for most values of j, the closure of the orbit Ω_j is a hypersurface of degree 6. Indeed, this is true for all values except $j = 0$ and 1728. As it turns out, the closure $\overline{\Omega}_{1728}$ is a variety we have run into before: it is the chordal variety of the rational normal quartic, which is a cubic hypersurface in $\mathbb{P}^4$. The variety $\overline{\Omega}_0$ is also an interesting one; it is a quadric hypersurface containing the rational normal quartic. This yields an interesting fact: there is a canonically determined quadric containing the rational normal quartic $C \subset \mathbb{P}^4$, i.e., one that is carried into itself by any motion of $\mathbb{P}^4$ carrying C into itself. This can also be seen by classical representation theory, which says that as representations of SL_2K,

$$\mathrm{Sym}^2(\mathrm{Sym}^4 V) = \mathrm{Sym}^8 V \oplus \mathrm{Sym}^4 V \oplus K.$$

In this decomposition, the last two factors correspond to the space $I(C)_2$ of quadrics in the ideal $I(C)$ of C, while the last factor corresponds to the quadric $\overline{\Omega}_0$.

Exercise 10.14. Which orbits of the action of PGL_2K on $\mathbb{P}^4$ lie in the closure of a given orbit Ω_j? (*)

In general, an explicit description of the space of orbits of the action of PGL_2K on $\mathbb{P}^n$ was worked out classically for n up to 6 (this was known as the theory of "binary quantics"); a number of qualitative results are known in general.

Example 10.15. PGL_3K Acts on $\mathbb{P}^5$

Assume $\mathrm{char}(K) \neq 2$. We will consider next the action of the group $\mathrm{GL}(V) \cong \mathrm{GL}_3K$ of automorphisms of a three-dimensional vector space V on the space $\mathrm{Sym}^2 V$. This induces an action of PGL_3K on the associated projective space $\mathbb{P}(\mathrm{Sym}^2 V) \cong \mathbb{P}^5$. Equivalently, we may look at the action on the dual space $\mathbb{P}(\mathrm{Sym}^2 V^*) \cong \mathbb{P}^5$, which we may view as the space of conics in $\mathbb{P}^2$; an automorphism of $\mathbb{P}^2$ naturally induces an automorphism of the space $\mathbb{P}^5$ of conics in $\mathbb{P}^2$.

The question of what orbits there are in this action was thus already answered in Example 3.3: there is a closed orbit Φ corresponding to double lines (i.e., squares $v \cdot v \in \mathrm{Sym}^2 V$), another orbit Ψ corresponding to unions of distinct lines (i.e., products $v \cdot w$ of independent linear factors), and finally an open orbit corresponding to smooth conics (i.e., those projectively equivalent to the Veronese image $v_2(\mathbb{P}^1)$). To describe these orbits, observe that the first—the locus of squares $[v^2]$—is just the Veronese surface $v_2(\mathbb{P}^2)$ in $\mathbb{P}^5$. Next, it's not hard to see that a quadric polynomial on $\mathbb{P}^2$ factors if and only if it can be written as the sum of two squares, $x^2 + y^2 = (x + iy)(x - iy)$, while conversely vw can be written as $((v + w)/2)^2 + (i(v - w)/2)^2$. The closure $\overline{\Psi} = \Psi \cup \Phi$ is thus just the chordal variety of the Veronese surface.

We can deduce from this that there are, up to projective equivalence, only two varieties in $\mathbb{P}^4$ obtained as regular projections $\pi_p(S)$ of the Veronese surface $S = v_2(\mathbb{P}^2) \subset \mathbb{P}^5$ (that is, projections from a point $p \in \mathbb{P}^5$ not lying on S).

Example 10.16. $\mathrm{PGL}_3 K$ Acts on $\mathbb{P}^9$

Assume now that $\mathrm{char}(K) \neq 2, 3$. As was the case when we passed from the action of $\mathrm{PGL}_2 K$ on $\mathbb{P}^2$ and $\mathbb{P}^3$ to its action on $\mathbb{P}^n$ for $n \geq 4$, the situation changes dramatically when we look at the action of $\mathrm{PGL}_3 K$ on the space $\mathbb{P}(\mathrm{Sym}^3 V) \cong \mathbb{P}^9$ (or, equivalently, on the space of cubic polynomials on $\mathbb{P}^2$ up to scalars). We will simply state here some facts about this action, without going through the verifications.

To begin with, there are a number of orbits of $\mathrm{PGL}_3 K$ corresponding to reducible cubics (or ones with multiple components): triple lines (i.e., tensors of the form $[v^3]$) form a single orbit, as do sums of double lines and lines ($[u \cdot v^2]$ with u, $v \in V$ independent). There are two orbits corresponding to cubics that are unions of three distinct lines; one where the lines are not concurrent (i.e., $[u \cdot v \cdot w]$ with u, v, and w independent) and one where they are ($[u \cdot v \cdot w]$ with u, v, and w pairwise independent and spanning a two-plane in V). There are similarly two orbits consisting of cubics that are unions of conics and lines, one where the conic meets the line in two points, and one where it is tangent at one.

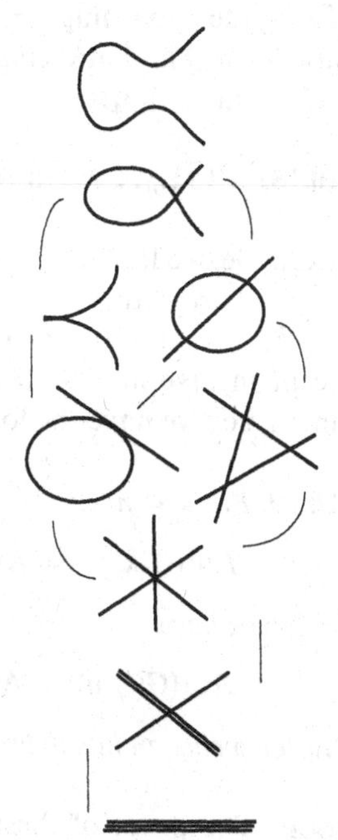

Among the irreducible cubics, there are first of all the two cubics described in Exercise 3.8; these form two more orbits. Finally, there are the smooth cubics. Any smooth cubic can be, after a projective motion, brought into a cubic C_λ of the form

$$Y^2 Z = X \cdot (X - Z) \cdot (X - \lambda Z),$$

and two such curves C_λ, $C_{\lambda'}$ are projectively equivalent if and only if the two subsets $\{0, 1, \infty, \lambda\}$ and $\{0, 1, \infty, \lambda'\}$ can be carried into one another by a projective motion of $\mathbb{P}^1$—that is, if and only if the j-invariants

$$j(\lambda) = 256 \cdot \frac{(\lambda^2 - \lambda + 1)^3}{\lambda^2 \cdot (\lambda - 1)^2}$$

coincide. The situation is thus very similar to that of the action of $\mathrm{PGL}_2 K$ on $\mathbb{P}^4$.

The diagram above indicates some of the obvious closure relationships among the orbits, especially those corresponding to singular cubics. There are some relatively subtle ones, however, that are not indicated; for example, for every value of $j \in K$, the orbit consisting of smooth cubics with j-invariant j contains in its closure the locus of cuspidal cubics (i.e., the orbit of cubics projectively equivalent to $Y^2Z - X^3$), but not the locus of triangles.

Example 10.17. $\mathrm{PO}_n K$ Acts on $\mathbb{P}^{n-1}$

Suppose $\mathrm{char}(K) \neq 2$. We will consider next the action of the orthogonal group $\mathrm{O}_n K$ on K^n and the corresponding action of $\mathrm{PO}_n K$ on $\mathbb{P}^{n-1}$. The only point we have to make here is that the action of $\mathrm{PO}_n K$ preserves the quadric hypersurface

$$X = \{[v]\}: Q(v, v) = 0\};$$

conversely $\mathrm{PO}_n K$ is the entire group of projective motions of X, i.e., $\mathrm{PO}_n K = \mathrm{Aut}(X, \mathbb{P}^{n-1})$.

Note in particular that both $\mathrm{PGL}_2 K$ and $\mathrm{SO}_3 K$ may in this way be identified with the group of projective motions of a conic curve $X \subset \mathbb{P}^2$, showing that these two groups are isomorphic. Similarly, the isomorphism of a quadric surface $X \subset \mathbb{P}^3$ with $\mathbb{P}^1 \times \mathbb{P}^1$ induces an isomorphism of $\mathrm{PSO}_4 K$ with $\mathrm{PGL}_2 K \times \mathrm{PGL}_2 K$. Note that the actual group of projective motions of a quadric surface X is in fact a $\mathbb{Z}/2$ extension of $\mathrm{PGL}_2 K \times \mathrm{PGL}_2 K$; $\mathrm{PGL}_2 K \times \mathrm{PGL}_2 K$ is the subgroup fixing the two rulings of the quadric (the automorphism of K^4 preserving the quadratic form Q and interchanging the rulings of X has determinant -1, and so does not appear in $\mathrm{SO}_4 K$).

Example 10.18. $\mathrm{PGL}_n K$ Acts on $\mathbb{P}(\wedge^k K^n)$

As a final example, we look at the action of $\mathrm{GL}_n K$ on $\wedge^k K^n$ and the induced action on the associated projective space. Clearly, this action preserves the Grassmannian $G(k, n) \subset \mathbb{P}(\wedge^k K^n)$, that is, the locus of totally decomposable vectors $[v_1 \wedge \cdots \wedge v_k]$. In fact, except in case $n = 2k$, this is the entire group of automorphisms of the Grassmannian, i.e., we have the following theorem.

Theorem 10.19. *For $k < n/2$,*

$$\mathrm{Aut}(G(k, n)) = \mathrm{Aut}(G(k, n), \mathbb{P}(\wedge^k K^n)) = \mathrm{PGL}_n K.$$

In case $n = 2k$ we have

$$\mathrm{Aut}(G(k, n)) = \mathrm{Aut}(G(k, n), \mathbb{P}(\wedge^k K^n)) \supset \mathrm{PGL}_n K$$

where the latter inclusion has index 2.

(Half-Proof). The proof of these facts consists of two parts, the first of which is to verify the first equality in each case, that is, to show that every automorphism of $G(k, n)$ is projective. This is not deep—it comes down to the assertion that every codimension 1 subvariety of $G(k, n)$ is the intersection of $G(k, n)$ with a hypersurface in $\mathbb{P}(\wedge^k K^n)$—but is beyond our means at present. Instead, we will here assume it and prove the remaining equality.

The key to seeing the second equality is the observation made in Lecture 6 that any linear space $\Lambda \cong \mathbb{P}^l$ lying on the Grassmannian $G(k, n) \subset \mathbb{P}(\wedge^k K^n)$ is a sub-Grassmannian of the form $G(1, l + 1)$ or $G(l, l + 1)$—that is, either the locus of k-planes containing a fixed $(k - 1)$-plane and lying in a fixed $(k + l)$-plane, or the locus of k-planes contained in a fixed $(k + 1)$-plane and containing a fixed $(k - l)$-plane. In particular, if $k < n/2$, then every linear subspace $\Lambda \cong \mathbb{P}^{n-k} \subset G(k, n)$ is of the former type; explicitly, it is the locus of k-planes containing a fixed $(k - 1)$-plane. In other words, as stated in Exercise 6.9, the Fano variety

$$F_{n-k}(G(k, n)) \cong G(k - 1, n).$$

A projective motion of $G(k, n)$ thus induces an automorphism of $G(k - 1, n)$; given the statement that such an automorphism is always projective, it in turn will induce one on $G(k - 2, n)$, and so on. We arrive in this way at an automorphism of $G(1, n) = \mathbb{P}^{n-1}$; this is given by an element of $\mathrm{PGL}_n K$ that in turn induces the given projective motion of $G(k, n) \subset \mathbb{P}(\wedge^k K^n)$.

In case $n = 2k$ the Fano variety of $(k - 1)$-planes on $G(k, n)$ will consist of two disjoint copies of $G(k - 1, n)$ (to be pedantic, one copy of $G(k - 1, n)$ and one copy of $G(k + 1, n)$); the argument here shows that the subgroup fixing these components individually is $\mathrm{PGL}_n K$. It remains thus just to show that there is indeed an automorphism of $G(k, n)$ exchanging them; we leave this as an exercise. □

Given this last statement, we can give another identification along the lines of the note on page 122: the connected component of the identity in automorphism group of the Grassmannian $G(2, 4)$ is $\mathrm{PGL}_4 K$. However, since the Grassmannian $G(2, 4)$ is just a quadric hypersurface in $\mathbb{P}^5$ we deduce that

$$\mathrm{PGL}_4 K \cong \mathrm{PSO}_6 K.$$

Quotients

One natural and important question to ask in the context of an action of a group G on a variety X is about the existence of quotients. What a quotient should be, naively, is a variety Y whose points correspond one to one to the orbits of G on X, and such that the corresponding map $X \to Y$ is regular; or equivalently, a surjective map $\pi: X \to Y$ of varieties such that $\pi(p) = \pi(q)$ iff there exists $g \in G$ with $g(p) = q$. In fact, we want to require something a little stronger, namely, a quotient should be a variety Y and map $\pi: X \to Y$ such that any regular map $\rho: X \to Z$ to another variety Z factors through π if and only if $\rho(p) = \rho(g(p))$ for all $p \in X$ and $g \in G$. (This prevents us from doing

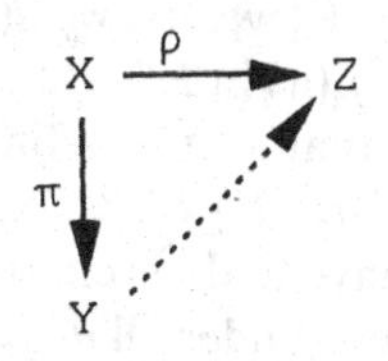

something stupid like composing the map π with a map $\varphi\colon Y \to Y'$ that is one to one but not an isomorphism, and in particular clearly makes the quotient unique if it exists.) When there is a variety Y and map $\pi\colon X \to Y$ satisfying this condition, we say that the quotient of X by the action of G exists.

That such a quotient need not exist in general is clear, even in the simplest cases. For example, $K^* = \mathrm{GL}_1 K$ acts on $\mathbb{A}^1$ by multiplication, with only two orbits, $\{0\}$ and $\mathbb{A}^1 - \{0\}$; but there does not exist a surjective morphism from $\mathbb{A}^1$ onto a variety with two points.

In general, the first problem we face in constructing a quotient is exactly that of the example: one orbit of G may lie in the closure of another, so that the natural topology on the set X/G will be pathological. We see more serious instances of this in Examples 10.12 and 10.16. In the latter case, we would naturally like to have a variety parametrizing projective equivalence classes of plane cubic curves, and the most natural candidate for such a variety is the quotient of $\mathbb{P}^9$ by $\mathrm{PGL}_3 K$; but the quotient does not exist. As was stated there, for example, the orbit corresponding to cuspidal cubics is contained in the closure of every orbit of smooth cubics.

What we can do is restrict our attention to an open subset U of the variety X invariant under G and try to take the quotient of U by G. Thus, for instance, in Example 10.16 we could take U the union of the locus of smooth cubics together with the orbit corresponding to irreducible cubics with a node; as it turns out, the quotient of U by $\mathrm{PGL}_3 K$ would then exist and be isomorphic to $\mathbb{P}^1$. The study of when such quotients exist and when they are compact is a deep one, called *geometric invariant theory*.

One circumstance in which nice quotients always exist is the case of finite group actions, which we will now discuss.

Quotients of Affine Varieties by Finite Groups

We start with the case of a finite group G acting on an affine variety X, say the zero locus of polynomials $\{f_\alpha(x_1, \ldots, x_n)\}$ in K^n; we claim that the quotient of X by G, in the sense above, exists.

To see this, let $I(X)$ be the ideal of X, and $A(X) = K[x_1, \ldots, x_n]/I(X)$ its coordinate ring. By the basic requirement for a quotient applied to $Z = \mathbb{A}^1$, the regular functions on Y must be exactly the functions on X invariant under the action of G, i.e., the coordinate ring of Y must be the subring $A(Y) = A(X)^G \subset A(X)$. The first thing to check, then, is that the ring $A(Y)$ is finitely generated over K; so we can write $A(Y) = K[w_1, \ldots, w_m]/(g_1(w), \ldots, g_l(w))$ and take Y the zero locus of the polynomials $g_\alpha(w)$ in K^m.

To prove that $A(X)^G$ is finitely generated over K, note first that we can write $A(X)$ in the form $K[x_1, \ldots, x_m]/I$, where G acts on the generators x_i by permutation; all we have to do is enlarge the given set $\{z_i\}$ of generators by throwing in the images of the z_i under all $g \in G$. Next, we observe that we have a surjection

$$K[x_1, \ldots, x_m]^G \twoheadrightarrow (K[x_1, \ldots, x_m]/I(X))^G:$$

an element $\bar{\alpha} \in (K[x_1, \ldots, x_m]/I)^G$ corresponds to an element $\alpha \in K[x_1, \ldots, x_m]$ congruent to each of its images $g^*(\alpha)$ mod I, and given such an element the average

$$\frac{1}{|G|} \sum_{g \in G} g^*(\alpha) \in K[x_1, \ldots, x_m]$$

will be invariant under G and will map to $\bar{\alpha}$. Thus it is enough to show that $K[x_1, \ldots, x_m]^G$ is finitely generated. But this is sandwiched between two polynomial rings:

$$K[x_1, \ldots, x_m]^{\mathfrak{S}_m} \subset K[x_1, \ldots, x_m]^G \subset K[x_1, \ldots, x_m],$$

where $\mathfrak{S}_m$ is the symmetric group on m letters. Now, the ring of invariants $K[x_1, \ldots, x_m]^{\mathfrak{S}_m}$ is just the polynomial ring $K[y_1, \ldots, y_m]$, where the y_i are the elementary symmetric functions of the x_j; and since $K[x_1, \ldots, x_m]$ is a finitely generated module over $K[y_1, \ldots, y_m]$ it follows that $K[x_1, \ldots, x_m]^G$ is as well.

Next, we have to see that the points of Y correspond to orbits of G on X. This is not hard. For one thing, suppose that two points p and $q \in X$ are not in the same orbit under G. To see that $\pi(p) \neq \pi(q)$, observe that we can find a function $f \in A(X)$ on X vanishing at p but not at $g(q)$ for any $g \in G$; the product $\prod g^*(f)$ of the images of f under G is then a function $f \in A(Y)$ vanishing at $\pi(p)$ but not at $\pi(q)$. To see that π is surjective, suppose that $\mathfrak{m} = (h_1, \ldots, h_k)$ is an ideal in $A(Y)$; we have to show that $\mathfrak{m} \cdot A(X) \neq A(X)$ unless $\mathfrak{m} = A(Y)$ is itself the unit ideal in $A(Y)$. But now if $\mathfrak{m} \cdot A(X)$ is the unit ideal, we can write

$$1 = a_1 \cdot h_1 + \cdots + a_k \cdot h_k$$

for some $a_1, \ldots, a_k \in A(X)$. Summing the images of this equation under all $g \in G$, then, we have

$$|G| = (\Sigma g^*(a_1)) \cdot h_1 + \cdots + (\Sigma g^*(a_k)) \cdot h_k;$$

since the coefficients of the right-hand side of this equation are invariant under G—that is, lie in $A(Y)$—we deduce that indeed $\mathfrak{m} = A(Y)$.

We leave as an exercise the verification that the variety Y, together with the map $X \to Y$, satisfies the universality property described earlier. We remark also that while the argument here seems to require that the field K have characteristic 0 (or at least characteristic prime to the order of G), this is not necessary for the result. In fact, if we did this correctly, using the notion of integral extensions of rings, the "averaging" process would not be invoked; see [S1] for a proper discussion.

Exercise 10.20. Show that the action of the symmetric group on four letters on four-tuples of distinct points in $\mathbb{P}^1$ induces an action of $\mathfrak{S}_4$ on $\mathbb{A}^1 - \{0, 1\}$, with quotient $\mathbb{A}^1$ and map $\mathbb{A}^1 - \{0, 1\} \to \mathbb{A}^1$ given by the j-function (cf. Exercise 10.13).

Example 10.21. Quotients of Affine Space

Probably the simplest examples come from linear actions on vector spaces. For instance, consider the involution $(z, w) \mapsto (-z, -w)$ on $\mathbb{A}^2$. The subring of $K[z, w]$

invariant under this involution is generated by $a = z^2$, $b = zw$, and $c = w^2$, which satisfy the one relation $ac = b^2$; thus the quotient of $\mathbb{A}^2$ by this involution is a quadric cone $Q \subset \mathbb{A}^3$.

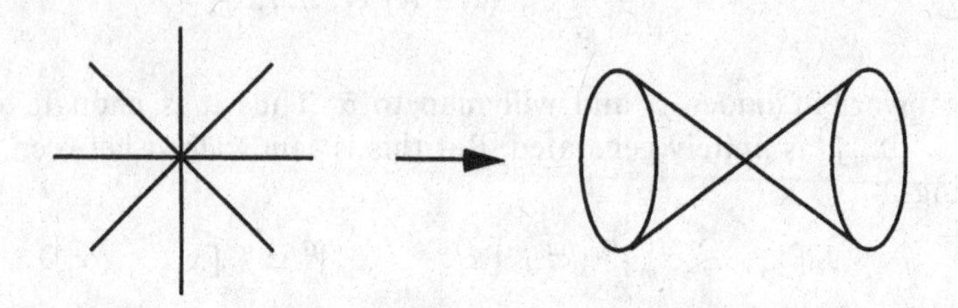

For another example, consider the action of $\mathbb{Z}/3$ on $\mathbb{A}^2$ given by sending (z, w) to $(\zeta z, \zeta w)$, where ζ is a cube root of unity. The ring of invariants is now generated by the monomials z^3, z^2w, zw^2, and w^3, from which we see that the quotient Y is a cone in $\mathbb{A}^4$ over a twisted cubic curve in $\mathbb{P}^3$. By contrast, if the action of $\mathbb{Z}/3$ on $\mathbb{A}^2$ is given by sending (z, w) to $(\zeta z, \zeta^2 w)$, the ring of invariants is generated by $a = z^3$, $b = zw$, and $c = w^3$, which satisfy the equation $ac = b^3$, giving a very different quotient.

Exercise 10.22. Show that the last two quotients are indeed not isomorphic as varieties. (*)

Example 10.23. Symmetric Products

Another example of finite group quotients is the symmetric product $X^{(n)}$ of a variety X with itself. This is defined to be the quotient of the ordinary n-fold product X^n of X with itself by the action of the symmetric group on n letters, that is, the variety whose points are unordered n-tuples of points of X. For example, consider the simplest case, that of $X = \mathbb{A}^1$. Here the ordinary n-fold product $X^n = \mathbb{A}^n$ has coordinate ring $K[z_1, \ldots, z_n]$, and the nth symmetric product has coordinate ring the subring of symmetric polynomials in $z_1, \ldots, z_n$. But this subring is freely generated by the elementary symmetric polynomials $\sigma_1, \ldots, \sigma_n$; so the symmetric product $(\mathbb{A}^1)^{(n)}$ is itself isomorphic to $\mathbb{A}^n$. Put another way, the space of monic polynomials of degree n in one variable is the n-fold symmetric product of $\mathbb{A}^1$ with itself.

Exercise 10.24. Describe the symmetric product $(\mathbb{A}^2)^{(2)}$ of $\mathbb{A}^2$ with itself, and more generally, the symmetric square of $\mathbb{A}^n$.

Quotients of Projective Varieties by Finite Groups

Just as the quotient of an affine variety by a finite group always exists and is an affine variety, the quotient of a projective variety $X \subset \mathbb{P}^n$ by a finite group always exists and is a projective variety, though this requires slightly more work to establish.

To begin with, observe that any action of a finite group on a projective variety $X \subset \mathbb{P}^n$ can be made projective; i.e., after reembedding X in a projective space $\mathbb{P}^N$ we may assume that G acts on $\mathbb{P}^N$ carrying X into itself. This is easy: to do this, simply embed the original projective variety X in a product of projective spaces $\mathbb{P}^n \times \mathbb{P}^n \times \cdots \times \mathbb{P}^n$ indexed by the elements $g \in G$ of the group, by the map whose gth factor is the automorphism g of X followed by the inclusion of X in $\mathbb{P}^n$. The action of G on X is then just the restriction to X of the action of G on $(\mathbb{P}^n)^{|G|}$ given by permuting the factors; after embedding $(\mathbb{P}^n)^{|G|}$ in projective space of dimension $(n+1)^{|G|} - 1$ by the Segre map, this will be a projective action. Finally, composing this embedding with a Veronese map, we can take the action to be linear.

Now let $S(X)$ be the homogeneous coordinate ring of X in $\mathbb{P}^n$ and consider the subring $B = S(X)^G \subset S(X)$ invariant under the action of G. B is again a graded ring, though it may not be generated by its first graded piece B_1; for that matter, B_1 may be zero. This is no problem: we let

$$B^{(k)} = \bigoplus_{n=0}^{\infty} B_{nk}$$

and give $B^{(k)}$ a grading by declaring B_{nk} its nth graded piece; since B is finitely generated, for some k $S(Y) = B^{(k)}$ will be generated by its first graded piece. Thus we can write

$$S(Y) = K[Z_0, \ldots, Z_m]/(F_1(Z), \ldots, F_l(Z))$$

where the F_α are homogeneous polynomials of the Z_i. We claim, finally, the subvariety $Y \subset \mathbb{P}^m$ defined by the equations $F_\alpha(Z) = 0$ is the quotient of X by G, in the preceding sense.

Exercise 10.25. Verify this last statement. (Hint: use the fact that the definition of quotient is local, in the sense that if $U \subset X$ is an affine open subset preserved by G then the image of U in the quotient of X by G is the quotient of U by G.)

The same argument, incidentally, shows the existence of finite group quotients of quasi-projective varieties in general.

Exercise 10.26. Show that the nth symmetric product of $\mathbb{P}^1$ is $\mathbb{P}^n$.

Example 10.27. Weighted Projective Spaces

Probably the most basic examples of quotients of projective varieties by finite groups, weighted projective spaces are quotients of projective space by the action of abelian groups acting diagonally. Specifically, let $a_0, \ldots, a_n$ be any positive integers, and consider the action of the group $(\mathbb{Z}/a_0) \times \cdots \times (\mathbb{Z}/a_n)$ on $\mathbb{P}^n$ generated by the automorphisms

$$[Z_0, \ldots, Z_n] \mapsto [Z_0, \ldots, \zeta \cdot Z_i, \ldots, Z_n]$$

where ζ is a primitive (a_i)th root of unity. The quotient of $\mathbb{P}^n$ by this action is called a *weighted projective space* and is denoted $\mathbb{P}(a_0, \ldots, a_n)$.

The simplest examples of weighted projective spaces (in view of Exercise 10.29) are the weighted planes $\mathbb{P}(1, 1, 2)$ and $\mathbb{P}(1, 2, 2)$. The first of these is just the quotient of $\mathbb{P}^2$ by the action of $\mathbb{Z}/2$ with generator

$$[Z_0, Z_1, Z_2] \mapsto [Z_0, Z_1, -Z_2].$$

The subring B of invariant polynomials is thus generated by the monomials Z_0, Z_1, and Z_2^2. We accordingly pass to the ring $B^{(2)}$, which is generated by the monomials

$$W_0 = Z_0^2, \qquad W_1 = Z_0 Z_1, \qquad W_2 = Z_1^2, \quad \text{and} \quad W_3 = Z_2^2.$$

We conclude that the plane $\mathbb{P}(1, 1, 2)$ is isomorphic to the surface in $\mathbb{P}^3$ with equation $W_0 W_2 = W_1^2$—that is, a cone over a conic curve (it is interesting to compare this with Example 10.21).

A less interesting example, as it turns out, is the weighted plane $\mathbb{P}(1, 2, 2)$. Here the ring of invariants is generated by Z_0, Z_1^2, and Z_2^2; passing to $B^{(2)}$ we see that it is generated by Z_0^2, Z_1^2, and Z_2^2 with no relations among them, and hence that the plane $\mathbb{P}(1, 2, 2) \cong \mathbb{P}^2$. In fact, both of these examples are special cases of the following.

Exercise 10.28. Show that any weighted projective space of the form $\mathbb{P}(1, \ldots, 1, k, \ldots, k)$ is isomorphic to a cone over a Veronese variety $v_k(\mathbb{P}^l)$ (and in particular $\mathbb{P}(1, k, \ldots, k) \cong \mathbb{P}^n$).

Exercise 10.29. Show that any weighted projective line $\mathbb{P}(a, b)$ is in fact isomorphic to $\mathbb{P}^1$.

As an example not covered by this, consider the weighted plane $\mathbb{P}(1, 2, 3)$. This is the quotient of $\mathbb{P}^2$ by the action of $\mathbb{Z}/6$ with generator

$$[Z_0, Z_1, Z_2] \mapsto [Z_0, -Z_1, \omega Z_2]$$

where ω is a cube root of unity. This is already a linear action; the subring $B \subset K[Z_0, Z_1, Z_2]$ invariant under $\mathbb{Z}/6$ is generated by the powers Z_0, Z_1^2 and Z_2^3. We accordingly pass to the ring $B^{(6)}$, which is generated by the monomials

$$W_0 = Z_0^6, \qquad W_1 = Z_0^4 Z_1^2, \qquad W_2 = Z_0^2 Z_1^4, \qquad W_3 = Z_1^6,$$

$$W_4 = Z_0^3 Z_2^3, \qquad W_5 = Z_2^6, \qquad \text{and} \qquad W_6 = Z_0 Z_1^2 Z_2^3.$$

We may thus conclude that the space $\mathbb{P}(1, 2, 3)$ is isomorphic to the projective variety $X \subset \mathbb{P}^6$ given by the equations

$$W_0 W_2 = W_1^2, \qquad W_0 W_3 = W_1 W_2, \qquad W_1 W_3 = W_2^2,$$

$$W_0 W_5 = W_4^2, \qquad W_2 W_5 = W_6^2, \qquad W_1 W_5 = W_4 W_6,$$

$$W_2 W_6 = W_3 W_4, \qquad W_1 W_6 = W_2 W_4, \qquad W_0 W_6 = W_1 W_4.$$

Exercise 10.30. Show that any weighted projective space is rational. Can you find an explicit birational isomorphism of $\mathbb{P}(1, 2, 3)$ with $\mathbb{P}^2$? (In fact, $\mathbb{P}(1, 2, 3) \subset \mathbb{P}^6$ is a special case of a *del Pezzo surface*; in particular, it is expressible as the image of a rational map from $\mathbb{P}^2$ to $\mathbb{P}^6$ given by cubic polynomials on $\mathbb{P}^2$.)

Exercise 10.31. Show that, just as ordinary projective space may be realized as the quotient of $K^{n+1} - \{0\}$ by the action of K^* acting by scalar multiplication, weighted projective space $\mathbb{P}(a_0, \ldots, a_n)$ may be viewed as the quotient of $K^{n+1} - \{0\}$ by K^*, where $\lambda \in K^*$ acts by

$$(Z_0, \ldots, Z_n) \mapsto (\lambda^{a_0} Z_0, \ldots, \lambda^{a_n} Z_n).$$

PART II

ATTRIBUTES OF VARIETIES

Having now accumulated a reasonable vocabulary of varieties and maps, we are prepared to begin in earnest developing the tools for studying them, which at this stage means finding more of the invariants and properties that distinguish one variety from another. Of course, as these are introduced, we will be able to define and describe additional varieties as well.

The characteristics of a variety that we shall introduce are its dimension; its smoothness or singularity, and in either case its Zariski tangent spaces (and in the latter case its tangent cones as well); the resolution of its ideal, with the attendant Hilbert function and Hilbert polynomial; and its degree. In each case, after giving the definitions, we will mention the main theorems describing the behavior of these invariants under some of the constructions introduced in the last section and then apply these to compute the invariants of the various varieties we know.

LECTURE 11

Definitions of Dimension and Elementary Examples

We will start by giving a number of different definitions of dimension and we will try to indicate how they relate to one another. All of our definitions initially apply to an irreducible variety X; the dimension of an arbitrary variety will be defined to be the maximum of the dimensions of its irreducible components.

The first definition is illegitimate on several counts, but is perhaps the most intuitive. Suppose that X is an irreducible variety over the complex numbers. If X is actually a complex submanifold of $\mathbb{P}^n$ then it is connected and we define the dimension of X to be the dimension of X as a complex manifold. More generally, it is the case (though we will not prove this until Exercise 14.3) that the smooth points of X are dense—in fact, a Zariski open subset $U \subset X$ is a connected complex manifold. We define the dimension of X to be the dimension of this manifold.

This is illicit for four reasons: it invokes the notion of smoothness, which we have not formally defined; it quotes an unproved theorem; it involves the complex numbers specifically; and it presumes a preexisting notion of dimension in a different context. Illegitimate as it may be, however, it does represent our a priori concept of what dimension should be; our goal in what follows will be to find an algebraic definition that conveys this notion.

Let's start with something we can agree on: the dimension of $\mathbb{P}^n$ is n. A second point we can go with is that if X and Y are irreducible varieties and $\varphi: X \to Y$ a dominant map all of whose fibers are finite, then the dimension of X should equal the dimension of Y. Combining these, we can give the following preliminary definition.

Definition 11.1. The *dimension* of an irreducible variety X is k if X admits a finite-to-one dominant map to $\mathbb{P}^k$.

Clearly this definition needs some work; it is not immediately clear either that such a map can always be found or that if it can the dimension k of the target projective space $\mathbb{P}^k$ is the same for all such maps. The latter problem will be rectified later, when we give other equivalent definitions; the former we can deal with now.

To do this, begin by taking $X \subset \mathbb{P}^n$ projective. Given an irreducible variety $X \subset \mathbb{P}^n$, the natural place to look for a finite map to projective space is among projections. Now, when we project X from a point $p \in \mathbb{P}^n$ not on X, the fibers are necessarily finite—since $p \notin X$, no line through p can meet X in more than a finite number of points. If the image $\overline{X} = \pi_p(X)$ is not all of $\mathbb{P}^{n-1}$, we can repeat the process, continuing to project X from points until we arrive at a finite surjective map of X to $\mathbb{P}^k$. To put it another way, projection of X from any linear subspace $\Lambda \cong \mathbb{P}^l \subset \mathbb{P}^n$ disjoint from X will be surjective if and only if no subspace strictly containing Λ is disjoint from X, so we have simply to find a maximal linear space disjoint from X. In particular, we may characterize the dimension of a projective variety $X \subset \mathbb{P}^n$ as the smallest integer k such that there exists an $(n-k-1)$-plane $\Lambda \subset \mathbb{P}^n$ disjoint from X.

To extend this definition to nonprojective varieties requires one twist. To begin with, recall that for any l the variety of l-planes meeting X is a closed subvariety of the Grassmannian $\mathbb{G}(l, n)$, so we could equally well have characterized the dimension of a projective $X \subset \mathbb{P}^n$ as the smallest integer k such that the general $(n-k-1)$-plane $\Lambda \subset \mathbb{P}^n$ is disjoint from X. Now, by Exercise 6.15, for any quasi-projective variety $X \subset \mathbb{P}^n$ the closure in $\mathbb{G}(l, n)$ of the variety of l-planes meeting X is the variety of l-planes meeting the closure $\overline{X}$. We may therefore make the following definition.

Definition 11.2. The *dimension* of an irreducible quasiprojective variety $X \subset \mathbb{P}^n$ is the smallest integer k such that a general $(n-k-1)$-plane $\Lambda \subset \mathbb{P}^n$ is disjoint from X, or if X is projective, such that there exists a subspace of dimension $n-k-1$ disjoint from X.

Next, note that a general $(n-k-1)$-plane contained in a general $(n-k)$-plane is a general $(n-k-1)$-plane. It follows that in the preceding situation, the general $(n-k)$-plane in $\mathbb{P}^n$ will meet X in a finite set of points. Moreover, by the same reasoning the general $(n-k+1)$-plane will meet X in a variety consisting of more than finitely many points, since otherwise the general $(n-k)$-plane would be disjoint from X. Thus we make the following definition.

Definition 11.3. The *dimension* of an irreducible quasi-projective variety $X \subset \mathbb{P}^n$ is that integer k such that the general $(n-k)$-plane in $\mathbb{P}^n$ intersects X in a finite set of points.

Note that from either of these definitions a seemingly obvious fact follows: that if X is irreducible and $Y \subset X$ is a proper closed subvariety, then $\dim(Y) < \dim(X)$ (or, equivalently, a closed subvariety of an irreducible variety X having the same dimension as X is equal to X).

Here is an observation that is just a restatement of what we have said but that is useful enough to merit the status of a proposition.

Proposition 11.4. *If $X \subset \mathbb{P}^n$ is a k-dimensional variety and $\Lambda \cong \mathbb{P}^l$ any linear space of dimension $l \geq n - k$, then Λ must intersect X.*

Note that this is a direct extension of Corollary 3.15. In Exercise 11.38 we will see how to generalize this to the statement: if $X, Y \subset \mathbb{P}^n$ are any subvarieties of dimensions k and l with $k + l \geq n$, then $X \cap Y \neq \emptyset$.

The arguments preceding the last two definitions certainly show that any variety admits a finite-to-one map to a projective space, but they do not imply that the dimension of that projective space is the same for all such maps. To see this, we need another characterization of dimension that depends only on an invariant of X (as opposed to the particular embedding in projective space)—the function field $K(X)$.

The key observation is Proposition 7.16, which says that for any generically finite dominant rational map $\varphi: X \to Y$, the induced inclusion $K(Y) \hookrightarrow K(X)$ of function fields expresses $K(X)$ as a finite algebraic extension of $K(Y)$. In our present circumstances, this says that if a variety X has dimension k in the sense of any of the preceding definitions—i.e., if X admits a finite map $\pi: X \to \mathbb{P}^k$—then we have an inclusion

$$\pi^* : K(\mathbb{P}^k) = K(x_1, \ldots, x_k) \hookrightarrow K(X)$$

expressing $K(X)$ as a finite extension of $K(x_1, \ldots, x_k)$. This gives us the following definition.

Definition 11.5. The *dimension* of an irreducible variety X is the transcendence degree of its function field $K(X)$ over K.

This definition of course has the virtue of making dimension visibly an invariant of the variety X, and not of the particular embedding of X in projective space. It also says that dimension is essentially a local property, since the function field $K(X)$ of an irreducible variety X is the quotient field of its local ring $\mathcal{O}_{X,p}$ at any point $p \in X$. (In particular, we can define the dimension of an irreducible quasi-projective variety $X \subset \mathbb{P}^n$ in general in this way; observe that the dimension of X so defined coincides with the dimension of its closure.) We could also apply any of the other standard definitions of the dimension of a local ring to the ring $\mathcal{O}_{X,p}$ (length of a maximal chain of prime ideals, minimal number of generators of an $\mathfrak{m}$-primary ideal, etc.) to arrive at a definition of the dimension of a variety; see [E] or [AM] for a thorough discussion of these various characterizations of the dimension of local rings.

Note in particular that the length of a maximal chain of prime ideals in $\mathcal{O}_{X,p}$ is the same as the length of a maximal chain of irreducible subvarieties

$$\{p\} \subset X_1 \subset X_2 \subset \cdots \subset X_{k-1} \subset X_k = X$$

through p in X. This characterization of dimension can be directly seen to be equivalent to Definition 11.1, which provides another verification that the dimension k of that definition is unique.

To extend the definition of dimension to possibly reducible varieties, we define the dimension of an arbitrary variety to be the maximum of the dimensions of its irreducible components; we say that X has *pure dimension* k (or simply is *of pure dimension*) if every irreducible component of X has the same dimension k. We may also define the *local dimension* of a variety X at a point $p \in X$, denoted $\dim_p(X)$, to be the minimum of the dimensions of all open subsets of X containing p (equivalently, the maximum of the dimensions of the irreducible components of X containing p, or the dimension of the quotient ring of the local ring $\mathcal{O}_{X,p}$). Thus, for example, X is of pure dimension if and only if $\dim_p(X) = \dim(X)\ \forall p \in X$.

One basic fact worth mentioning is the following.

Exercise 11.6. Using any of the characterizations of dimension, show that if $X \subset \mathbb{P}^n$ is any quasi-projective variety and $Z \subset \mathbb{P}^n$ any hypersurface not containing an irreducible component of X, then

$$\dim(Z \cap X) = \dim(X) - 1.$$

(Similarly, if f is any regular function on X not vanishing on any irreducible component of X, then the zero locus $Y \subset X$ of f has dimension $\dim(X) - 1$.) This will be used to prove the more general Theorem 17.24 on the subadditivity of codimensions of intersections.

We will see one more characterization of dimension in Lecture 13; it is in some senses the cleanest, but also the least intuitive of all our definitions. For now, though, it's time to discuss properties of dimension and to give some examples.

Example 11.7. Hypersurfaces

In Lecture 1, we defined a hypersurface X in $\mathbb{P}^n$ to be a variety expressible as the zero locus of a single homogeneous polynomial $F(Z)$. By what we have said, X will be of pure dimension $n - 1$. It is worth pointing out the converse here: that a variety $X \subset \mathbb{P}^n$ of pure codimension 1 is a hypersurface. This follows from the fact that a minimal prime ideal in the ring $K[Z_0, \ldots, Z_n]$ is principal. It also follows that the homogeneous ideal of a hypersurface is generated by a single element.

Example 11.8. Complete Intersections

Definition. We say that a k-dimensional variety $X \subset \mathbb{P}^n$ is a *complete intersection* if its homogeneous ideal $I(X) \subset K[Z_0, \ldots, Z_n]$ is generated by $n - k$ elements.

Note that by Exercise 11.6, this is the smallest number of generators the ideal of a k-dimensional subvariety $\mathbb{P}^n$ can have.

Complete intersections are often studied simply because they are, in a sense, ubiquitous; as a consequence of Theorem 17.16, if $F_\alpha(Z)$ is a general homogeneous polynomial of degree d_α for $\alpha = 1, \ldots, n - k$, the common zero locus of the F_α will be a complete intersection. (We will implicitly use this fact in computing the dimension of the family of complete intersections later.) It is thus easy to write down many complete intersection varieties. At the same time, we should counter this with the observation that complete intersections form a very small subclass of all varieties; though we cannot prove any of it here, there are numerous restrictions on the topology of complete intersections (e.g., if they have dimension two or more they must be simply connected) and on their numerical invariants. Even among varieties with given topology and numerical invariants, the complete intersections may form a small subfamily; for example, let $\Gamma \subset \mathbb{P}^2$ consist of three noncollinear points. Since the homogeneous ideal $I(\Gamma)$ contains no elements of degree 1 but the dimension of $I(\Gamma)_2$ is three, $I(\Gamma)$ cannot have fewer than three generators; thus Γ is not a complete intersection. More generally, we have the following.

Exercise 11.9. Let $\Gamma \subset \mathbb{P}^n$ be a general collection of d points. It is then the case that Γ is a complete intersection if and only if $n = 1$, $d = 1$ or 2, or $d = 4$ and $n = 2$. Prove this for $n = 2$ and $d \leq 7$.

The example of three points in $\mathbb{P}^2$ illustrates the distinction between the notion of complete intersection and *set-theoretic complete intersection.* A subvariety $X \subset \mathbb{P}^n$ of dimension k is said to be a set-theoretic complete intersection if it is the common zero locus of a collection of $n - k$ homogeneous polynomials; for example, if $\Gamma = \{p, q, r\} \subset \mathbb{P}^2$ is three noncollinear points, then Γ is the intersection of any irreducible conic containing it with the cubic curve $\overline{pq} \cup \overline{pr} \cup \overline{qr}$. Indeed, by contrast with Exercise 11.9, we have the following.

Exercise 11.10. Let $\Gamma \subset \mathbb{P}^2$ be a general collection of d points. Show that Γ is a set-theoretic complete intersection.

Another famous example of this distinction is the twisted cubic, which is not a complete intersection (for the same reason as in the case of three noncollinear points in $\mathbb{P}^2$), but which is a set-theoretic complete intersection (this is harder to see than for three points in the plane). It is not known whether every space curve $C \subset \mathbb{P}^3$ is a set-theoretic complete intersection; it is not even known whether the curves $C_{\alpha,\beta}$ introduced in Example 1.26 are.

Exercise 11.11. Find a homogeneous quadratic polynomial $Q(Z_0, \ldots, Z_3)$ and a homogeneous cubic polynomial $P(Z_0, \ldots, Z_3)$ whose common zero locus is a rational normal curve $C \subset \mathbb{P}^3$. (Hint: think determinantal!) (*)

It might appear, based on the discussion in Lecture 5, that there is a notion intermediate between complete intersection and set-theoretic complete intersection: what if $X \subset \mathbb{P}^n$ is a k-dimensional variety and $F_1, \ldots, F_{n-k}$ are homogeneous polynomials that generate the ideal of X locally? In fact, in this case it turns out that X is a complete intersection, with $I(X) = (F_1, \ldots, F_{n-k})$. To see this requires a little commutative algebra, though: it amounts to the assertion that because $F_1, \ldots, F_{n-k}$ is a regular sequence in the ring $K[Z_0, \ldots, Z_n]$, the ideal $(F_1, \ldots, F_{n-k})$ cannot have any embedded primes; but if $(F_1, \ldots, F_{n-k})$ were properly contained in its saturation $I(X)$, it would have the irrelevant ideal $(Z_0, \ldots, Z_n)$ as an associated prime.

One further important notion is a *local complete intersection.* We say that a quasi-projective variety $X \subset \mathbb{P}^n$ having local dimension k at a point $p \in X$ is a local complete intersection at p if there is some affine neighborhood U of p in $\mathbb{P}^n$ such that the ideal of $U \cap X$ in $A(U)$ is generated by $n - k$ elements; equivalently, if $X \cup Y$ is a complete intersection for some variety $Y \subset \mathbb{P}^n$ of dimension k not containing p. (Interestingly, this is a property purely of the variety X and point p; it does not depend on the particular embedding.) We say that a variety X is a local complete intersection if it is so at every point $p \in X$. The simplest example of a variety that is not a local complete intersection is the union of the three coordinate axes in $\mathbb{A}^3$.

Immediate Examples

Next, there a number of varieties whose dimension is clear from one or more of the preceding definitions. For example,

- The Grassmannian $G(k, n)$ contains, as a Zariski open subset, affine space $\mathbb{A}^{k(n-k)}$, and thus has dimension $k(n - k)$.
- The product $X \times Y$ of varieties of dimensions d and e has dimension $d + e$.
- The cone $\overline{pX}$ with vertex p over an irreducible variety $X \subset \mathbb{P}^n$ of dimension d has dimension $d + 1$, unless X is a linear space and $p \in X$.
- The projection $\overline{X} = \pi_p(X)$ of a variety $X \subset \mathbb{P}^n$ of dimension d from a point $p \in \mathbb{P}^n$ has dimension d again, unless X is a cone with vertex p.
- The algebraic groups $\mathrm{PGL}_n K$, $\mathrm{PO}_n K$, and $\mathrm{PSp}_{2n} K$ have dimensions $n^2 - 1$, $n(n - 1)/2$, and $2n^2 + n$, respectively.

To find the dimensions of other varieties, we need to know a little more about dimension, for example, how it behaves in maps. The basic fact in this regard is the following.

Theorem 11.12. *Let X be a quasi-projective variety and $\pi: X \to \mathbb{P}^n$ a regular map; let Y be closure of the image. For any $p \in X$, let $X_p = \pi^{-1}(\pi(p)) \subset X$ be the fiber of π through p, and let $\mu(p) = \dim_p(X_p)$ be the local dimension of X_p at p. Then $\mu(p)$ is an upper-semicontinuous function of p, in the Zariski topology on X—that is, for any m the locus of points $p \in X$ such that $\dim_p(X_p) \geq m$ is closed in X. Moreover, if $X_0 \subset X$ is any irreducible component, $Y_0 \subset Y$ the closure of its image and μ the minimum value of $\mu(p)$ on X_0, then*

$$\dim(X_0) = \dim(Y_0) + \mu.$$

The standard example of the upper semicontinuity of $\mu(p)$ is the blow-up map $\tilde{\mathbb{P}}^2 \to \mathbb{P}^2$, which has point fibers over each point of $\mathbb{P}^2$ except the point p being blown up, and fiber $\mathbb{P}^1$ over that point. A simple example of the sort of behavior that is to be expected in the context of $\mathscr{C}^\infty$ maps but that the theorem explicitly precludes in the category of varieties is the projection of the two-sphere S^2 onto a line, where the fibers range from empty, to a point, to a circle, and back again.

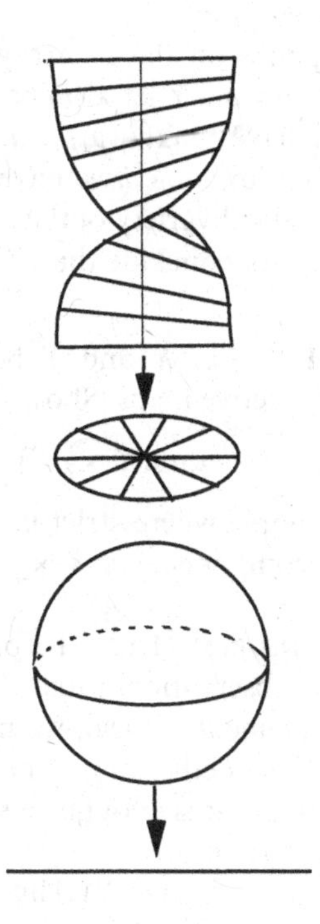

Note also that when X is projective we can express this in terms of the fiber dimension as a function on the image.

Corollary 11.13. *Let X be a projective variety and $\pi: X \to \mathbb{P}^n$ any regular map; let $Y = \pi(X)$ be its image. For any $q \in Y$, let $\lambda(q) = \dim(\pi^{-1}(q))$. Then $\lambda(q)$ is an upper-semicontinuous function of q in the Zariski topology on Y. Moreover, if $X_0 \subset X$ is any irreducible component, $Y_0 = \pi(X_0)$ its image, and λ the minimum value of $\lambda(p)$ on Y_0, we have*

$$\dim(X_0) = \dim(Y_0) + \lambda.$$

The condition that X (and, correspondingly, its image Y) be projective can be weakened; X can be taken to be quasi-projective, as long as we add the assumptions that X is irreducible and that Y is a variety as well. The latter is not a serious restriction, since we can always find an open subset $U \subset Y \subset \overline{Y}$ and replace X by $\pi^{-1}(U)$. The former is just to avoid such idiocies as taking X the disjoint union of $\mathbb{A}^1 \times (\mathbb{A}^1 - \{0\})$ and a point p, with $\pi: X \to \mathbb{A}^1$ the projection on the second factor on $\mathbb{A}^1 \times (\mathbb{A}^1 - \{0\})$ and $\pi(p) = 0$. We can also extend the corollary to the case of rational maps $f: X \to Y$ by applying it to the graph of the map.

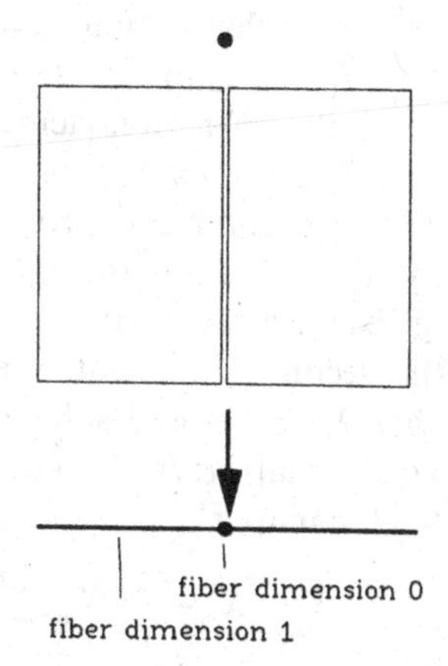

Finally, before we prove Theorem 11.12, note that as a consequence we have the following characterization of irreducibility.

Theorem 11.14. *Let $\pi: X \to Y$ be a regular map of projective varieties, with Y irreducible. Suppose that all fibers $\pi^{-1}(p)$ of π are irreducible of the same dimension d. Then X is irreducible.*

PROOF. Suppose on the contrary that X is the union of irreducible closed subvarieties X_i; for $p \in Y$ let $\lambda_i(p)$ be the dimension of the fiber of $\pi_i = \pi|_{X_i}$. For each p, we must have $\max(\lambda_i(p)) = d$, and since each λ_i is upper semicontinuous we conclude that for some i we have $\lambda_i \equiv d$. But then for each $p \in Y$ the fiber $\pi_i^{-1}(p)$, being a closed subvariety of the fiber $\pi^{-1}(p)$ and having the same dimension, must equal $\pi^{-1}(p)$; we conclude that $X_i = X$. □

Exercise 11.15. Let X and Y be irreducible projective varieties, $f: X \to Z$ and $g: Y \to Z$ surjective maps. Show that the dimension of the fiber product satisfies

$$\dim(X \times_Z Y) \geq \dim(X) + \dim(Y) - \dim(Z).$$

Find an example where strict inequality holds. Also, show by example that some irreducible components of $X \times_Z Y$ may have strictly smaller dimension.

PROOF OF THEOREM 11.12. The proof consists of several reductions plus one relatively trivial observation. First, it is enough to prove it for X irreducible. Secondly, since the statement is local, we may take X to be affine and the map $\pi: X \to \mathbb{A}^n$. Finally, by the Noetherianness of the space X, to establish the upper semicontinuity of the function μ it is enough to show that it achieves its minimum μ_0 on an open set (i.e., there is an open $U \subset X$ with $\mu \equiv \mu_0$ on U; we do not have to show that $\{p: \mu(p) = \mu_0\}$ is open in X). Thus, it will be enough to establish the two statements:

(i) $\mu(p) \geq \dim(X) - \dim(Y)$ everywhere and
(ii) there is an open subset $U \subset X$ such that $\mu(p) = \dim(X) - \dim(Y)\ \forall p \in U$.

The proof of the first statement is just a direct application of Exercise 11.6. Using this, we see that if $\dim(Y) = e$ then we can find for any $p \in X$ a collection of polynomials $f_1, \ldots, f_e$ on $\mathbb{A}^n$ whose common zero locus on Y is a finite collection of points including $\pi(p)$. Replacing Y by an affine open subset V including $\pi(p)$ but none of the other zeros of $f_1, \ldots, f_e$, and X by the inverse image of V, we may assume that the common zero locus of the f_α is exactly $\{\pi(p)\}$. But then the subvariety $X_p \subset X$ is just the common zero locus of the functions $\pi^*f_1, \ldots, \pi^*f_e$, and using Exercise 11.6 again we deduce that $\mu(p) = \dim_p(X_p) \geq \dim(X) - e$.

For the second statement, we apply a trick we have used before: we may assume that X is a closed subvariety of $\mathbb{A}^{n+m}$, and the map π the restriction of the projection map $\pi: \mathbb{A}^{n+m} \to \mathbb{A}^n$ given by $(z_1, \ldots, z_{n+m}) \mapsto (z_1, \ldots, z_n)$. We may then factor the map π

$$X = X_m \xrightarrow{\pi_m} X_{m-1} \xrightarrow{\pi_{m-1}} \cdots \xrightarrow{\pi_2} X_1 \xrightarrow{\pi_1} X_0 = Y \subset \mathbb{A}^n$$

where $X_k \subset \mathbb{A}^{n+k}$ is the closure of the image of X under the projection $\mathbb{A}^{n+m} \to \mathbb{A}^{n+k}$ given by $(z_1, \ldots, z_{n+m}) \mapsto (z_1, \ldots, z_{n+k})$, and π_k the restriction of the projection map $\mathbb{A}^{n+k} \to \mathbb{A}^{n+k-1}$.

Now, it is easy to see that Theorem 11.12 holds for the individual maps π_k. Clearly the local fiber dimension $\mu_k(p)$ of the map π_k can be only 0 or 1; if $X_k \subset \mathbb{A}^{n+k}$ is given by a collection of polynomials f_α, which we may write as

$$f_\alpha(z_1, \ldots, z_{n+k}) = \sum a_{\alpha,i}(z_1, \ldots, z_{n+k-1}) \cdot (z_{n+k})^i$$

then the locus of points p with $\mu(p) = 1$ will be just the intersection of X_{n+k} with the common zero locus of the polynomials $a_{\alpha,i}$. Thus (bearing in mind that X_k will be irreducible since X is), either all the coefficients $a_{\alpha,i}$ are zero on X_{k-1}, μ_k is identically 1 on X_k, and $\dim(X_k) = \dim(X_{k-1}) + 1$; or the common zero locus of the coefficients $a_{\alpha,i}$ is a proper subvariety Z_{k-1} of X_{k-1}, μ_k is 0 outside the inverse image of this subvariety, the map π_k is generically finite on X_k, and $\dim(X_k) = \dim(X_k)$.

Note that the number of values of k such that $\mu_k \equiv 1$ is exactly the difference in dimensions $\dim(X) - \dim(Y)$. For each of the remaining values of k, let $U_k \subset X$ be the inverse image of the open subset $X_{k-1} - Z_{k-1}$; let U be the intersection of these open subsets of X. Let $p \in U$ be any point. Applying the succession of maps π_k to X_p, we conclude that

$$\dim(X_p) \leq \dim(X) - \dim(Y)$$

(we cannot deduce directly that we have equality, since the image of X_p may not be dense in the fiber of X_k over $\pi(p)$). But we have already established the opposite inequality at every point p; thus

$$\dim_p(X_p) = \dim(X) - \dim(Y). \qquad \square$$

Exercise 11.16. (a) Show by example that in general the function μ is not the sum of the pullbacks of the functions μ_k (so that we cannot deduce upper semicontinuity of μ directly from the corresponding statement for the maps π_k). (b) Justify the third reduction step: that in order to show that μ is upper semicontinuous in general it is enough to show that for any map π there is an open subset $U \subset X$ where it attains its minimum.

To conclude this discussion, note that as a corollary to Theorem 11.12 we can prove Proposition 5.15: if $\pi: X \to Y$ is any morphism and $Z \subset X$ a locally closed subset, then for general points $p \in Y$

$$\overline{(Z_p)} = X_p \cap \bar{Z},$$

i.e., the closure of the fiber Z_p of $\pi|_Z$ is the fiber $X_p \cap \bar{Z}$ of the closure $\bar{Z}$ over p.

First, we can restrict the map π to the closure of Z, i.e., we can assume $X = \bar{Z}$, or in other words, that Z is an open subset of X (we can also assume X is irreducible). Next, we can assume the image of π is dense, since otherwise both sides of the presumed inequality are empty. The result now follows immediately from applying Theorem 11.12 to the map π on X and on $W = X - Z$: if the dimensions of Y, X, and W are k, n, and $m < n$ respectively, then the local dimension of X_p is at least $n - k$ everywhere, while for general p the dimension of W_p must be $m - k < n - k$. It follows that W_p cannot contain an irreducible component of X_p; thus W_p lies in the closure of Z_p. $\square$

We will now use Theorem 11.12 (together with some common sense) to determine the dimension of some of the varieties introduced in Part I.

Example 11.17. The Universal k-Plane

In Lecture 6, we defined the incidence correspondence $\Sigma \subset \mathbb{G}(k, n) \times \mathbb{P}^n$ by setting

$$\Sigma = \{(\Lambda, x): x \in \Lambda\}.$$

As we said there, Σ is the "universal family" of k-planes; that is, it is simply the subvariety of the product whose fiber over a given point $\Lambda \in \mathbb{G}(k, n)$ is the k-plane $\Lambda \subset \mathbb{P}^n$ itself. We can thus apply Theorem 11.12 to the projection map $\pi_1: \Sigma \to \mathbb{G}(k, n)$ to deduce that Σ is irreducible of dimension

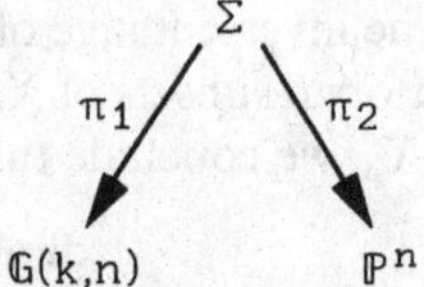

$$\begin{aligned}\dim(\Sigma) &= k + \dim(\mathbb{G}(k, n)) \\ &= nk - k^2 + n.\end{aligned}$$

Note that we could also compute its dimension (and establish its irreducibility) by looking at the projection map π_2 to $\mathbb{P}^n$. The fiber of π_2 over a point $p \in \mathbb{P}^n$ is just the subvariety of $\mathbb{G}(k, n)$ of k-planes containing p, which is isomorphic to the Grassmannian $\mathbb{G}(k-1, n-1)$. Thus, the fibers of π_2 are all irreducible of dimension $k(n-k)$, and Σ correspondingly irreducible of dimension $k(n-k) + n$.

Example 11.18. Varieties of Incident Planes

Let $X \subset \mathbb{P}^n$ be an irreducible variety, and for any $k \leq n - \dim(X)$ consider the subvariety $\mathscr{C}_k(X) \subset \mathbb{G}(k, n)$ of k-planes meeting X, introduced in Example 6.14. We may compute its dimension by realizing $\mathscr{C}_k(X)$, as we did there, as

$$\mathscr{C}_k(X) = \pi_1(\pi_2^{-1}(X)) \subset \mathbb{G}(k, n)$$

where π_1, π_2 are the projection maps on the incidence correspondence Σ introduced in Example 11.17. As in that example, the fibers of π_2 over X are all isomorphic to $\mathbb{G}(k-1, n-1)$; that is, they are irreducible of dimension $k(n-k)$, and $\Psi = (\pi_2)^{-1}(X)$ correspondingly irreducible of dimension $k(n-k) + \dim(X)$. Finally, since the map $\Psi \to \mathscr{C}_k(X) \subset \mathbb{G}(k, n)$ is generically finite (in fact it is generically one-to-one; see Exercise 11.23), we conclude that $\mathscr{C}_k(X)$ is likewise irreducible of dimension $k(n-k) + \dim(X)$. To express the conclusion in English: the codimension of $\mathscr{C}_k(X)$ in $\mathbb{G}(k, n)$ is the difference between the codimension of X and k.

The simplest example of this construction already exhibits interesting behavior. This is the variety $\Phi \subset \mathbb{G}(1, 3)$ of lines $l \subset \mathbb{P}^3$ meeting a given line $X = l_0$. Here, if we let $\Psi \subset \Sigma \subset \mathbb{G}(1, 3) \times \mathbb{P}^3$ be the incidence correspondence

$$\Psi \xrightarrow{\pi_1} \mathbb{G}(1,3) \supset \Phi, \qquad \Psi \xrightarrow{\pi_2} l_0 \subset \mathbb{P}^3$$

$$\Psi = \{(l, p): p \in l\} \subset \Phi \times l_0$$

we see that Ψ maps to l_0 with fibers all isomorphic to $\mathbb{P}^2$; the map to Φ, on the other hand, is one to one except over the point $l_0 \in \Phi$, where the fiber is isomorphic to $\mathbb{P}^1$.

Exercise 11.19. For another view of the variety Φ of lines meeting a given line l_0 let $l_0^* \subset \mathbb{P}^{3*}$ be the locus of planes $H \subset \mathbb{P}^3$ containing l_0 and consider the incidence correspondence

$$\Omega = \{(l, H): l \subset H\} \subset \Phi \times l_0^*.$$

Show that, just as in the case of the incidence correspondence Ψ, the fibers of Ω over l_0^* are isomorphic to $\mathbb{P}^2$, while the map $\pi_1: \Omega \to \Phi$ is one to one except over l_0, where the fiber is $\mathbb{P}^1$. Show that the two incidence correspondences Ψ and Ω are isomorphic (in fact, they are both isomorphic to the rational normal scroll $X_{2,2,1}$), but not via an isomorphism commuting with the maps to Φ. (*)

We will encounter the variety Φ a number of times and in various roles in this book. Indeed, we have in a sense already met in it Example 3.1; Φ is a quadric hypersurface in $\mathbb{P}^4$ of rank 4. (See also Example 6.3.)

Exercise 11.20. Let $X \subset \mathbb{P}^n$ be an irreducible nondegenerate k-dimensional variety. For $l < n - k$, find the dimension of the closure of the locus of l-planes meeting X in at least two points. Show by example that no analogous formula exists if we replace "two" by "three," even if we require $l \geq 2$.

Exercise 11.21. Let $X \subset \mathbb{P}^n$ be an irreducible nondegenerate k-dimensional variety. Show that the plane spanned by $n - k + 1$ general points of X meets X in only finitely many points, and use this to compute the dimension of the closure of the locus of l-planes spanned by their intersection with X when $l \leq n - k$.

Example 11.22. Secant Varieties

In Lecture 8 we introduced the chordal, or secant variety of a variety $X \subset \mathbb{P}^n$; we consider here its dimension. To begin with, recall that we define the secant line map

$$s: X \times X - \Delta \to \mathbb{G}(1, n),$$

on the complement of the diagonal Δ in $X \times X$ by sending the pair (p, q) to the line $\overline{pq}$; the chordal variety $S(X)$ is defined to be the union of the lines corresponding to points of the image $\mathscr{S}(X) \subset \mathbb{G}(1, n)$ of s. Now, unless X is a linear subspace of $\mathbb{P}^n$, the map s is generically finite; the fiber over a general point $l = \overline{pq}$ in the image will be positive-dimensional if and only if $l \subset X$ and the only variety that contains the line joining any two of its points is a linear subspace of $\mathbb{P}^n$. Thus, the dimension of the variety $\mathscr{S}(X)$ of secant lines is always equal to the dimension of $X \times X$, that is, twice the dimension of X.

Exercise 11.23. Counting dimensions, show that the variety $\mathscr{S}(X) \subset \mathbb{G}(1, n)$ of secant lines to a variety $X \subset \mathbb{P}^n$ of dimension $k < n - 1$ is a proper subvariety of the variety $\mathscr{C}_1(X)$ of incident lines. Use this to deduce (in characteristic 0) that the general projection $\pi_\Lambda: X \to \mathbb{P}^{k+1}$ of X from an $(n-k-2)$-plane $\Lambda \subset \mathbb{P}^n$ is birational onto its image; and in particular that every variety is birational to a hypersurface (cf Lecture 7).

Now, consider the incidence correspondence

$$\Sigma = \{(l, p): p \in l\} \subset \mathscr{S}(X) \times \mathbb{P}^n \subset \mathbb{G}(1, n) \times \mathbb{P}^n$$

whose image $\pi_2(\Sigma) \subset \mathbb{P}^n$ is the secant variety $S(X)$. The projection map π_1 on the first factor is surjective, with all fibers irreducible of dimension 1; thus Σ is irreducible of dimension $2 \cdot \dim(X) + 1$. The following proposition then follows.

Proposition 11.24. *The variety $\mathscr{S}(X) \subset \mathbb{G}(1, n)$ of secant lines to an irreducible k-dimensional variety $X \subset \mathbb{P}^n$ is irreducible of dimension $2k$. The secant variety $S(X) \subset \mathbb{P}^n$ is irreducible of dimension at most $2k + 1$, with equality holding if and only if the general point $p \in \overline{qr}$ lying on a secant line to X lies on only a finite number of secant lines to X.*

(Note that in this statement we need only consider points p lying on honest secant lines, that is, secant lines actually spanned by points of X.)

Exercise 11.25. Show that if $X \subset \mathbb{P}^n$ is an irreducible curve then the chordal variety $S(X)$ is three-dimensional unless X is contained in a plane.

By way of contrast, let $X = v_2(\mathbb{P}^2) \subset \mathbb{P}^5$ be the Veronese surface. We claim that the chordal variety $S(X)$ is only four-dimensional. Indeed, we can see this in two (apparently) different ways. To begin with, suppose that $r \in \mathbb{P}^5$ is a general point lying on a secant line to X; we may write the secant line as $\overline{v(p), v(q)}$ for some pair of points $p, q \in \mathbb{P}^2$. Now, the line $L = \overline{pq} \subset \mathbb{P}^2$ is carried, under the Veronese map v, to a plane conic curve $C \subset X \subset \mathbb{P}^5$, and since $r \in \overline{v(p), v(q)}$, r will lie on the plane Λ spanned by C. But then every line through r in Λ will be a secant line to C, and hence to X. In particular, it follows that a general point lying on a secant line to X lies on a one-dimensional family of secant lines to X. We may deduce from this that the dimension of $S(X)$ is at most 4; since it is clear on elementary grounds that $S(X)$ cannot have dimension less than 4 (for example, the cones $\overline{p, X}$ over X with vertex a point $p \in X$ would all have to coincide), we conclude that $\dim(S(X)) = 4$.

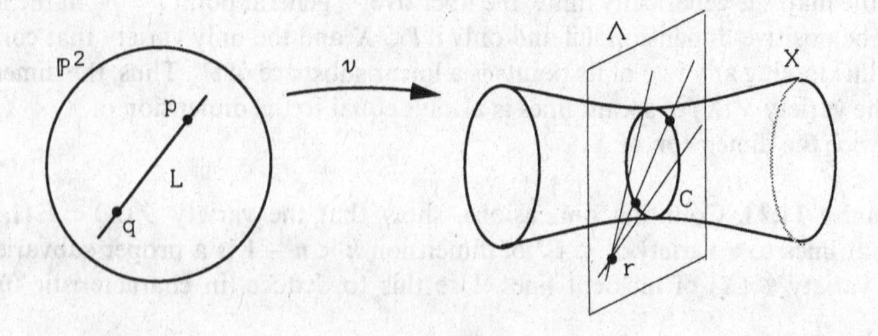

Another way to show that $\dim(S(X)) = 4$ is via the determinantal form of the equations of the Veronese surface. Recall that the equations of the Veronese surface may be expressed as the 2×2 minors of the matrix

$$M = \begin{bmatrix} Z_0 & Z_3 & Z_4 \\ Z_3 & Z_1 & Z_5 \\ Z_4 & Z_5 & Z_2 \end{bmatrix};$$

in other words, in the space $\mathbb{P}^5$ of symmetric 3×3 matrices, the locus of rank 1 matrices is a Veronese surface. But a linear combination of two rank 1 matrices can have rank at most 2, from which it follows that the chordal variety $S(X)$ of the Veronese surface will be contained in the cubic hypersurface in $\mathbb{P}^5$ with equation $\det(M) = 0$. (To rephrase this, we have seen in Examples 1.20 and 4.8 that the set of conics in $\mathbb{P}^2$ is parametrized by $\mathbb{P}^5$, with the subset of double lines the Veronese surface; the chordal variety to the Veronese surface is then simply the locus of conics of rank 1 or 2.)

In general, we call a variety $X \subset \mathbb{P}^n$ *nondegenerate* if it spans $\mathbb{P}^n$, i.e., is not contained in any hyperplane. If $X \subset \mathbb{P}^n$ is an irreducible nondegenerate k-dimensional variety whose chordal variety $S(X)$ has dimension strictly less than $\min(2k + 1, n)$, we say that X has *deficient secant variety*; in these circumstances we define the *deficiency* $\delta(X)$ of X to be the difference

$$\delta(X) = 2k + 1 - \dim(S(X)).$$

Here are some more examples of varieties with deficient secant varieties.

Exercise 11.26. Let $\Sigma = \sigma(\mathbb{P}^2 \times \mathbb{P}^2) \subset \mathbb{P}^8$ be the Segre variety. By arguments analogous to each of the preceding arguments, show (twice) that the chordal variety of Σ has dimension 7, so that Σ has deficiency $\delta(\Sigma) = 2$.

Exercise 11.27. Let $\mathbb{G} = \mathbb{G}(1, n) \subset \mathbb{P}(\wedge^2 K^{n+1}) = \mathbb{P}^{n(n+1)/2-1}$ be the Grassmannian of lines in $\mathbb{P}^n$. Again using arguments analogous to each of the preceding two, show that the chordal variety of $\mathbb{G}$ has dimension $4n - 7$, so that $\delta(\mathbb{G}) = 4$.

Exercise 11.28. Let $X \subset \mathbb{P}^n$ be an irreducible nondegenerate variety with deficient secant variety.

(i) Show that the cone $\overline{X} = \overline{p, X} \subset \mathbb{P}^{n+1}$ over X has deficiency $\delta(\overline{X}) = \delta(X) + 1$.

(ii) Show that the general hyperplane section $Y = X \cap H$ of X has deficiency $\delta(Y) = \delta(X) - 1$ (or simply has nondeficient secant variety in case $\delta(X) = 1$). (We will show in Proposition 18.10 that the general hyperplane section of an irreducible nondegenerate variety of dimension at least 2 is again irreducible and nondegenerate.)

The first part of the preceding exercise is in particular a source of examples of varieties with arbitrarily large deficiency. Here is another less trivial example.

Exercise 11.29. Let M be the projective space of $m \times n$ matrices, and let $M_k \subset M$ be the subvariety of matrices of rank at most k introduced in Lecture 9; take k so that $2k < \min(m, n)$. Show that the chordal variety $S(M_k)$ is equal to the subvariety $M_{2k} \subset M$ of matrices of rank at most $2k$. Indeed, show that for a map $A: K^m \to K^n$ of rank $2k$ and any pair of complementary k-dimensional subspaces $\Lambda, \Xi \subset \mathrm{Im}(A)$ the composition of A with the projections of Λ and Ξ give an expression of A as a sum of two matrices of rank k. Deduce that in the case $\mathbb{P}^n = M$, $X = M_k$ the fiber of the incidence correspondence Σ introduced in Example 11.22 over a general point $S(X)$ has dimension $2k^2$, and hence that

$$\dim(S(M_k)) \leq 2 \cdot \dim(M_k) + 1 - 2k^2.$$

A couple of remarks are in order here. First, it is known that the Veronese surface is the only example of a smooth surface with deficient secant variety. On the other hand, much less is known about the behavior of chordal varieties of varieties of higher dimension; for example, it is not even known whether the deficiency of smooth varieties is bounded (as we will see when we discuss smoothness in Lecture 14, the examples in Exercise 11.29 are in general singular).

Example 11.30. Secant Varieties in General

We begin with a basic calculation.

Exercise 11.31. Let $X \subset \mathbb{P}^n$ be an irreducible nondegenerate variety of dimension k. Use Exercise 11.21 to deduce that for any $l \leq n - k$ the general fiber of the secant plane map

$$s_l: X^{l+1} \to \mathbb{G}(l, n)$$

is finite.

From this we see that the variety of secant l-planes—defined to be the image $\mathcal{S}_l(X)$ of s_l—will always have dimension $(l + 1)k$. By the incidence correspondence argument then, we may also deduce that the secant variety $S_l(X)$, defined to be the union of the secant l-planes to X, will have dimension at most $lk + l + k$, with equality holding if and only if the general point $p \in S_l(X)$ lies on only finitely many secant l-planes.

Of course, we know even less about when this condition holds than we do in the case $l = 1$ discussed in the preceding example. For example, while it is true that the secant variety $S_l(X)$ of an irreducible nondegenerate curve $X \subset \mathbb{P}^n$ will have dimension $2l + 1$ whenever this number is less than or equal to n, it is less elementary to prove this (this will be a consequence of exercises 16.16 and 16.17). One very special case where we do have an elementary proof is the case of the rational normal curve.

Proposition 11.32. *Let $X \subset \mathbb{P}^n$ be a rational normal curve. The secant variety $S_l(X)$ has dimension* $\min(2l + 1, n)$.

Proof (in case $2l + 1 \le n$). Let $U \subset \mathcal{S}_l(X)$ be an open subset consisting of secant l-planes spanned by $l + 1$ distinct points of X (since the image of the complement of the diagonals $\Delta \subset X^{l+1}$ under s_l is dense and constructible, we can find such a U). It will suffice to show that for $l \le (n-1)/2$, a general point lying on a general secant l-plane spanned by points $p_1, \ldots, p_{l+1} \in X$ lies on no other secant l-plane Λ' to X with $\Lambda' \in U$. But now the statement that any $n + 1 \ge 2(l + 1)$ points on a rational normal curve are independent (Example 1.14) implies that the intersection of Λ with any other $\Lambda' \in U$ is contained in the subspace of Λ spanned by a proper subset of the points p_i. □

Exercise 11.33. Give the proof of Proposition 11.32 in case $2l = n$ either by bounding the dimension of the family of secant l-planes containing a general point of $S_l(X)$ or by using the preceding and arguing that $S_l(X)$ strictly contains $S_{l-1}(X)$ (and therefore has higher dimension). Either do the case $2l > n$ similarly, or simply deduce it from the preceding.

We mention this example largely because it has a nice interpretation in case $\text{char}(K) = 0$. As we saw in the remark following Exercise 2.10 and again in Example 9.3, if we view the space $\mathbb{P}^n$ as the projective space of homogeneous polynomials of degree n on $\mathbb{P}^1$, then the set of nth powers is a rational normal curve. A general point of the secant variety is thus a polynomial expressible as a sum of $l + 1$ nth powers, and so we have established the first half of the following corollary.

Corollary 11.34. *Assume K has characteristic 0. A general polynomial*

$$f(x) = a_n x^n + a_{n-1}x^{n-1} + \cdots + a_1 x + a_0$$

is expressible as a sum of d nth powers

$$f(x) = \Sigma(\alpha_i x - \beta_i)^n$$

if and only if $2d - 1 \ge n$. Moreover, if $2d - 1 = n$, it is uniquely so expressible.

Exercise 11.35. (i) Verify the second half of this statement. (ii) what is the smallest number m such that every polynomial of degree n is expressible as a sum of m nth powers? (*)

There is naturally an analog of this for polynomials in any given number of variables; you can, in similar fashion, work out the expected dimension of the secant varieties of the Veronese varieties in general, and so say when you expect to be able to express a general polynomial of degree n in l variables as a sum of m nth powers. It is not known (to me) when this number suffices, though there are cases when it doesn't. For example, a naive dimension count would lead you to expect that every quadratic polynomial in two variables (or homogeneous quadric in three variables) could be expressed as a sum of two squares; but the fact that the secant variety to the Veronese surface in $\mathbb{P}^5$ is deficient means it can't (you can see this directly by observing that a sum of two squares is reducible).

Example 11.36. Joins of Varieties

In Examples 6.17 and 8.1 we introduced the join $J(X, Y)$ of two varieties $X, Y \subset \mathbb{P}^n$ (of which the secant variety may be viewed as the special case $X = Y$). We may observe that by the same computation, the expected dimension of the join $J(X, Y)$ is $\dim(X) + \dim(Y) + 1$. Indeed, it is much harder for the join of two disjoint varieties to fail to have this dimension. We have the following.

Proposition 11.37. *Let X and $Y \subset \mathbb{P}^n$ be two disjoint varieties. The join $J(X, Y)$ will have dimension exactly $d = \dim(X) + \dim(Y) + 1$ whenever $d \leq n$.*

Proof. This is immediate if X and Y lie in disjoint linear subspaces of $\mathbb{P}^n$; in this case no two lines joining points of X and Y can meet, so every point of the join lies on a unique line joining X and Y.

To prove the proposition in general, we reembed X and Y as subvarieties $\tilde{X}$, $\tilde{Y}$ of disjoint linear subspaces $\Lambda_0, \Lambda_1 \cong \mathbb{P}^n \subset \mathbb{P}^{2n+1}$, where Λ_0 is the plane $Z_0 = \cdots = Z_n = 0$ and Λ_1 the plane $Z_{n+1} = \cdots = Z_{2n+1} = 0$. Let $\tilde{J} = J(\tilde{X}, \tilde{Y})$ be the join of these two, so that we have $\dim(\tilde{J}) = \dim(X) + \dim(Y) + 1$. Note that projection $\pi_L\colon \mathbb{P}^{2n+1} \to \mathbb{P}^n$ from the plane $L \cong \mathbb{P}^n \subset \mathbb{P}^{2n+1}$ defined by $Z_0 = Z_{n+1}, \ldots, Z_n = Z_{2n+1}$ carries $\tilde{X}$ and $\tilde{Y}$ to X and Y, respectively; moreover, the hypothesis that $X \cap Y = \emptyset$ implies that the join $\tilde{J}$ is disjoint from L, so the projection π_L is a regular map from $\tilde{J}$ onto $J(X, Y)$. But it is a general fact that the projection π_L of a projective variety from a linear space L disjoint from it is finite; if the fiber $(\pi_L)^{-1}(p)$ contained a curve C, C would be contained in the plane $\overline{p, L}$ spanned by p and L and so would necessarily meet the hyperplane $L \subset \overline{p, L}$. Thus $\pi_L\colon \tilde{J} \to J(X, Y)$ is finite and $\dim(J(X, Y)) = \dim(\tilde{J}) = \dim(X) + \dim(Y) + 1$. □

Exercise 11.38. Let X, Y, Λ_0, Λ_1, $\tilde{J}$, and L be as in the proof of Proposition 11.37, but do not assume X and Y are disjoint. Show that if $\dim(X) + \dim(Y) \geq n$, then $\tilde{J}$ must meet L; deduce that any two closed subvarieties $X, Y \subset \mathbb{P}^n$ the sum of whose dimensions is at least n must intersect.

The following is an amusing and useful corollary of Exercise 11.38.

Exercise 11.39. Show that for $m > n$ there does not exist a nonconstant regular map $\varphi\colon \mathbb{P}^m \to \mathbb{P}^n$.

Example 11.40. Flag Manifolds

We can think of the incidence correspondence of Example 11.17 as the special case $\mathbb{F}(0, k, n)$ of the flag manifold

$$\mathbb{F} = \mathbb{F}(k, l, n) = \{(\Gamma, \Lambda)\colon \Gamma \subset \Lambda\} \subset \mathbb{G}(k, n) \times \mathbb{G}(l, n)$$

introduced in Example 8.34. The dimension of the more general flag manifolds may be found in the same way; for example, the projection π_1 maps $\mathbb{F}$ onto $\mathbb{G}(k, n)$ with fibers $\mathbb{G}(l-k-1, n-k-1)$ of dimension $(l-k)(n-l)$, so that $\mathbb{F}$ is irreducible of dimension

$$\dim \mathbb{F}(k, l, n)$$
$$= \dim \mathbb{G}(k, n) + (l-k)(n-l)$$
$$= (k+1)(n-k) + (l-k)(n-l);$$

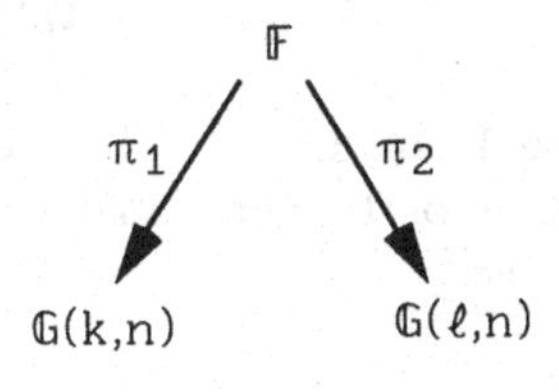

equivalently, we could describe $\mathbb{F}$ as mapping via π_2 to $\mathbb{G}(l, n)$ with fibers $\mathbb{G}(l-k-1, l)$ and derive the same result.

Exercise 11.41. Find the dimension of the flag manifold $\mathbb{F}(a_1, \ldots, a_l, n)$ in general.

Example 11.42. (Some) Schubert Cycles

We can use the preceding example to find the dimension of some of the subvarieties of the Grassmannian described in Lecture 6. As in Lecture 6, let $\Lambda \subset \mathbb{P}^n$ be a fixed m-plane, and let

$$\Sigma = \Sigma_k(\Lambda) = \{\Gamma: \dim(\Gamma \cap \Lambda) \geq k\} \subset \mathbb{G}(l, n).$$

We assume here that $m + l - n < k$, so that $\Sigma \neq \mathbb{G}(l, n)$.

To find the dimension of Σ, we introduce the incidence correspondence

$$\Psi = \{(\Theta, \Gamma): \Theta \subset \Gamma\} \subset \mathbb{G}(k, \Lambda) \times \mathbb{G}(l, n).$$

The projection on the first factor expresses Ψ as simply the inverse image of $\mathbb{G}(k, \Lambda) \subset \mathbb{G}(k, n)$ in the flag manifold $\mathbb{F}(k, l, n)$ discussed earlier, and from the same argument as there we conclude that Ψ is irreducible of dimension $(k+1)(m-k) + (l-k)(n-l)$; since the projection map $\pi_2: \Psi \to \Sigma$ is generally one to one, we deduce the same thing about Σ.

It may be more suggestive to express this by saying that the codimension of Ψ in $\mathbb{G}(k, n)$ is $(k+1)\cdot(k-(m+l-n))$; the first factor is the dimension of intersection of the vector spaces corresponding to Γ and Λ, the second the difference between k and the "expected" dimension $m+l-n$ of intersection of an m-plane and an l-plane in $\mathbb{P}^n$.

Exercise 11.43. Find the dimension of the more general incidence correspondence

$$\Psi = \{(\Gamma, \Lambda): \dim(\Gamma \cap \Lambda) \geq m\} \subset \mathbb{G}(k, n) \times \mathbb{G}(l, n).$$

Let $X \subset \mathbb{P}^n$ be an irreducible k-dimensional variety. In Lecture 4, we defined the universal hyperplane section $\Omega_X \subset \mathbb{P}^{n*} \times X$ and showed that it was irreducible. As another consequence of Theorem 11.14, we can verify this and extend the statement substantially. To start with, we make the following more general definition.

Definition. Let $X \subset \mathbb{P}^n$ by any variety. We define the *universal* $(n - l)$*-plane section* (or *universal l-fold hyperplane section*) $\Omega^{(l)}(X)$ of X to be the subvariety of the product $\mathbb{G}(n - l, n) \times X$ defined by

$$\Omega^{(l)}(X) = \{(\Lambda, p): p \in \Lambda\}.$$

Exercise 11.44. Let $X \subset \mathbb{P}^n$ be an irreducible k-dimensional variety. Find the dimension of the universal $(n - l)$-plane section $\Omega^{(l)}(X)$ and show that it is irreducible.

LECTURE 12

More Dimension Computations

Example 12.1. Determinantal Varieties

As in Lecture 9, let M be the projective space of nonzero $m \times n$ matrices up to scalars and $M_k \subset M$ the variety of matrices of rank k or less. To find the dimension of M_k, we introduce another incidence correspondence. We define

$$\Psi \subset M \times G(n-k, n)$$

by

$$\Psi = \{(A, \Lambda): \Lambda \subset \operatorname{Ker}(A)\}$$

(here, of course, we are viewing an $m \times n$ matrix A as a linear map $K^n \to K^m$). The point is that if we fix Λ the space of maps $A: K^n \to K^m$ such that $\Lambda \subset \operatorname{Ker}(A)$ is just the space $\operatorname{Hom}(K^n/\Lambda, K^m)$, so that fibers of the projection $\pi_2: \Psi \to G(n-k, n)$ are just projective spaces of dimension $km - 1$. We conclude that the variety Ψ is irreducible of dimension $\dim(G(n-k, n)) + km - 1 = k(m+n-k) - 1$; since the map $\pi_1: \Psi \to M$ is generically one to one onto M_k, we deduce that the same is true of M_k. One way to remember this is as the following proposition.

Proposition 12.2. *The variety $M_k \subset M$ of $m \times n$ matrices of rank at most k is irreducible of codimension $(m-k)(n-k)$ in M.*

Exercise 12.3. Show that equality holds in Exercise 11.29.

Exercise 12.4. Use an analogous construction to estimate the dimension of the spaces of (i) symmetric $n \times n$ matrices of rank at most k and (ii) skew-symmetric $n \times n$ matrices of rank at most $2k$.

Example 12.5. Fano Varieties

In Lecture 6 we described the Fano variety $F_k(X)$ parametrizing k-planes lying on a variety $X \subset \mathbb{P}^n$. Here we will estimate its dimension, at least in the case of X a general hypersurface of degree d in $\mathbb{P}^n$.

We do this by considering the space $\mathbb{P}^N$ parametrizing all hypersurfaces of degree d in $\mathbb{P}^n$ (N here is $\binom{n+d}{d} - 1$, though as we shall see that won't appear in the computation), and setting up an incidence correspondence between hypersurfaces and planes. Specifically, we define the variety $\Phi \subset \mathbb{P}^N \times \mathbb{G}(k, n)$ to be

$$\Phi = \{(X, \Lambda): \Lambda \subset X\}$$

so that the fiber of Φ over a point $X \in \mathbb{P}^n$ is just the Fano variety $F_k(X)$. If the image of the projection map $\pi_1: \Phi \to \mathbb{P}^N$ is of the maximal possible dimension, then we would expect the dimension of $F_k(X)$ for general X to be the dimension of Φ less N.

To find the dimension of Φ we simply consider the second projection map $\pi_2: \Phi \to \mathbb{G}(k, n)$. The fiber of this map over any point $\Lambda \in \mathbb{G}(k, n)$ is just the space of hypersurfaces of degree d in $\mathbb{P}^n$ containing Λ; since the restriction map

$$\{\text{polynomials of degree } d \text{ on } \mathbb{P}^n\} \to \{\text{polynomials of degree } d \text{ on } \Lambda \cong \mathbb{P}^k\}$$

is a surjective linear map, this is just a subspace of codimension $M = \binom{k+d}{d}$ in the space $\mathbb{P}^N$ of all hypersurfaces of degree d in $\mathbb{P}^n$. It follows that Φ is irreducible of dimension $(k+1)(n-k) + N - M$, and hence that the expected dimension on the general fiber of Φ over $\mathbb{P}^N$ is

$$\dim(F_k(X)) = (k+1)(n-k) - \binom{k+d}{d},$$

a number we will call $\varphi(n, d, k)$. Thus, for example, the expected dimension of the Fano variety of lines on a general surface $S \subset \mathbb{P}^3$ of degree d is $3 - d$. In particular, we expect a general plane in $\mathbb{P}^3$ to contain a 2-parameter family of lines (this is of course true); we expect a general quadric surface to contain a one-parameter family of lines (we have seen that this is also true); we expect a general cubic surface to contain a finite number of lines (this is also true, and will be verified in Exercise 12.7), and a general surface of degree $d \geq 4$ to contain no lines at all.

Note that we have to keep using the qualifier "expect": the preceding argument does not actually prove that the dimension of the Fano variety of a general hypersurface is given by the preceding formula. To be precise, in case $\varphi(n, d, k) < 0$ this argument does show that $F_k(X) = \emptyset$ for a general X; but in cases where $\varphi(n, d, k) \geq 0$ we don't a priori know whether Φ surjects onto $\mathbb{P}^N$ as expected or maps to a proper subvariety (so that the general X will contain no k-planes, while some special X will contain more than a $\varphi(n, d, k)$-dimensional family). Indeed, this does occur: for example, consider the case $n = 4$, $d = 2$ and $k = 2$, that is, 2-planes on quadrics in $\mathbb{P}^4$. Here the dimension of Φ equals the dimension $N = 14$ of the space of quadric hypersurfaces in $\mathbb{P}^4$; but, as we will see, the general quadric hypersurface $Q \subset \mathbb{P}^4$ does not in fact contain any 2-planes. (What happens in this case is that cones over quadrics in $\mathbb{P}^3$, which form a codimension 1 subvariety of the space of quadrics in $\mathbb{P}^4$, generically contain a one-parameter family of 2-planes. In fact, this observation, together with a dimension count and the irreducibility of Φ, implies that a general quadric in $\mathbb{P}^4$ contains no 2-planes.)

As it turns out, though, the only exceptional cases involve hypersurfaces of degree 2; when $d \geq 3$ the count is accurate. We will verify this explicitly in the case $n = 3$, $d = 3$, and $k = 1$ (i.e., lines on a cubic surface) and leave the general case, which is only notationally more difficult, as an exercise.

In general, to show that Φ surjects onto $\mathbb{P}^N$ we have simply to exhibit a fiber of Φ having the expected dimension $\varphi(n, d, k)$. In fact, by Theorem 11.12 it is enough to exhibit a point $(X, L) \in \Phi$ such that the local dimension of the fiber $(\pi_1)^{-1}(\pi_1(X, L)) = F_k(X)$ at (X, L) is exactly $\varphi(n, d, k)$. Thus, for example, in the case $n = 3$, $d = 3$, $k = 1$, in order to show that the general cubic surface $S \subset \mathbb{P}^3$ contains a positive finite number of lines it suffices to exhibit a cubic surface S that contains an isolated line, i.e., such that $F_1(S)$ has an isolated point L_0. In this particular case, we may construct such a surface S and line L_0 synthetically. For example, let $X = X_{1,2}$ be the rational normal scroll constructed as in Lecture 8 as the union of lines joining corresponding points on a line L and a conic curve C lying in a complementary 2-plane in $\mathbb{P}^4$. It is not hard to see by the construction that the only lines on X are L and the lines joining L to C; in particular, L is an isolated line of X. Moreover, if $S = \pi(X)$ is the image of X under projection from a general point $p \in \mathbb{P}^4$, the image $L_0 = \pi(L)$ is still isolated; any line on S is either the image of a line on X or the image of a plane curve of higher degree on X mapping multiple to one onto a line, and since $\pi: X \to S$ is birational there can be at most finitely many of the latter.

Exercise 12.6. (i) Verify that the surface S of the preceding argument is indeed a cubic surface in $\mathbb{P}^3$. (ii) Verify in general the statement made there that a birational map $\pi: X \to S$ of surfaces can fail to be birational on only finitely many curves in X.

Alternatively, we can simply write down a particular cubic surface $S \subset \mathbb{P}^3$ and line $L \subset S$ and verify that L is an isolated point of $F_1(S)$ by writing down the local equations for $F_1(S)$ near L. For this purpose, we choose the line

$$L_0: Z_2 = Z_3 = 0$$

and the surface

$$S: Z_0^2 Z_2 + Z_1^2 Z_3 = 0.$$

In a neighborhood of L_0 in the Grassmannian $\mathbb{G} = \mathbb{G}(1, 3)$, we can write a general line as the row space of the matrix

$$\begin{pmatrix} 1 & 0 & a & b \\ 0 & 1 & c & d \end{pmatrix};$$

that is, given parametrically by

$$[U, V] \mapsto [U, V, aU + cV, bU + dV].$$

Restricting the defining polynomial of S to this line, we have the cubic

$$U^2(aU + cV) + V^2(bU + dV) = aU^3 + cU^2V + bUV^2 + dV^3.$$

The coefficients a, c, b, and d of this polynomial are then the defining equations of the Fano variety $F_1(S)$ in the open subset $U \subset \mathbb{G}$, from which we deduce that the point L_0 is indeed an isolated point of $F_1(S)$.

In fact, the surface S we have written down is the surface described in the first approach, that is, it is the projection of a rational normal scroll from $\mathbb{P}^4$ (though that was certainly not necessary). It is worth remarking that, in either approach, the surface S we used in fact had a one-dimensional Fano variety $F_1(S)$.

Exercise 12.7. Consider here general values of n, d, and k for which $\varphi(n, d, k) = 0$, i.e., we expect a general hypersurface of degree d in $\mathbb{P}^n$ to contain a finite number of k-planes. In cases with $d \geq 3$, find a hypersurface $X_0 \subset \mathbb{P}^n$ of degree d whose Fano variety $F_k(X)$ contains an isolated point Λ_0 and deduce that every such hypersurface does indeed contain a k-plane.

In the general case $\varphi(n, d, k) > 0$, the situation is as in the case of the exercise, namely, with the exception of the case $d = 2$ of quadric hypersurfaces and some values of k, Φ always surjects. In sum, then, we have the following theorem.

Theorem 12.8. *Set*

$$\varphi(n, d, k) = (k + 1)(n - k) - \binom{k + d}{d}.$$

Then for any n, k, and $d \geq 3$ the Fano variety $F_k(X)$ of k-planes on a general hypersurface of degree d in $\mathbb{P}^n$ is empty if $\varphi < 0$, and of dimension φ if $\varphi \geq 0$.

The proof of this theorem follows exactly the lines of the preceding exercise, but is probably best left until after our discussion of the tangent spaces to Fano varieties in Example 16.21. Of course, we will have more to say later about the geometry of $F_k(X)$ in the case $\varphi \geq 0$, and for that matter there is more to be said in case $\varphi < 0$ as well; for example, we expect the locus in $\mathbb{P}^N$ of hypersurfaces that do contain a k-plane to have codimension $-\varphi$, and it does. We can ask further questions about the geometry of this locus. (For example, in the projective space of quartic surfaces in $\mathbb{P}^3$, the subvariety of those that do contain lines is a hypersurface; its degree is 320.)

In those cases where $\varphi(n, d, k) = 0$ (and $d \geq 3$), the actual number of k-planes on a general hypersurface $X \subset \mathbb{P}^n$ of degree d can be worked out; for example, there are 27 lines on a general cubic surface, 2875 lines on a general quintic threefold in $\mathbb{P}^4$, and so on.

Parameter Spaces of Twisted Cubics

We will consider next the dimension of some parameter spaces of varieties. Classically, if the family of varieties X of a given type was parametrized by a variety $\mathscr{H}$ of dimension k, we would say that a variety X of the family "varies with k degrees of freedom" or "depends on k parameters"; thus, for example, a plane conic $C \subset \mathbb{P}^2$ moves with five degrees of freedom.

Of course, as we said before we will not actually describe the construction of parameter spaces or indicate how to prove their existence until Lecture 21; for the time being we will just assume that they behave in reasonable ways and try to determine their dimensions in some cases.

Example 12.9. Twisted Cubics

Consider for example the space $\mathscr{H}$ of twisted cubics. There are several ways of estimating the dimension of $\mathscr{H}$. For one, we can use a trick: as we saw in Lecture 1, through any six points in general position in $\mathbb{P}^3$ there passes a unique twisted cubic curve. Thus, we can let $U \subset (\mathbb{P}^3)^6$ be the open set consisting of six-tuples $(p_1, \ldots, p_6)$ with no three collinear, and let $\Sigma \subset U \times \mathscr{H}$ be the incidence correspondence

$$\Sigma = \{((p_1, \ldots, p_6), C) \colon p_1, \ldots, p_6 \in C \text{ distinct}\}.$$

To see that Σ is indeed a variety, observe that it may be realized as an open subset of the sixth fiber product

$$\Sigma = \mathscr{C} \times_{\mathscr{H}} \mathscr{C} \times_{\mathscr{H}} \cdots \times_{\mathscr{H}} \mathscr{C}$$

of the universal twisted cubic $\mathscr{C} \subset \mathscr{H} \times \mathbb{P}^3$. By Theorem 1.18, Σ maps one to one onto U, and so $\dim(\Sigma) = 18$. On the other hand, the fiber of Σ over any point $C \in \mathscr{H}$ is just the complement of the main diagonal Δ in C^6, and so has dimension 6; thus, the dimension of $\mathscr{H}$ must be 12.

Here is another way, taking advantage of the fact that all twisted cubic curves are congruent under the group $\mathrm{PGL}_4 K$. We simply fix a standard twisted cubic $C_0 \in \mathscr{H}$ and define a map $\mathrm{PGL}_4 K \to \mathscr{H}$ by sending $g \in \mathrm{PGL}_4 K$ to $g(C_0)$. The fiber of this map over any $C \in \mathscr{H}$ is the subgroup of $\mathrm{PGL}_4 K$ carrying C to itself, which, as we observed in Lecture 10, is the group $\mathrm{PGL}_2 K$. We thus have

$$\dim(\mathscr{H}) = \dim(\mathrm{PGL}_4 K) - \dim(\mathrm{PGL}_2 K) = 12.$$

This approach is essentially equivalent to the following. Let $\mathfrak{M}$ be the space of maps $v \colon \mathbb{P}^1 \to \mathbb{P}^3$ mapping $\mathbb{P}^1$ one to one onto a twisted cubic. $\mathfrak{M}$ is an open subset of the projective space $\mathbb{P}^{15}$ associated to the vector space of 4-tuples of homogeneous polynomials of degree 3 on $\mathbb{P}^1$, so that $\dim(\mathfrak{M}) = 15$; on the other hand, two maps v and v' will have the same image if and only if they differ by composition $v' = v \circ A$ with an automorphism $A \in \mathrm{PGL}_2 K$ of $\mathbb{P}^1$, so again $\dim(\mathscr{H}) = 15 - 3 = 12$.

Exercise 12.10. Find the dimension of the parameter space of rational normal curves $C \subset \mathbb{P}^n$ twice, by methods analogous to each of the preceding two approaches.

Here is another, basically different, approach to determining the dimension of the space $\mathscr{H}$ of twisted cubics in $\mathbb{P}^3$. We have observed already that a twisted cubic curve $C \subset \mathbb{P}^3$ lies on a family of quadric surfaces $\{Q_\lambda\}$ parametrized by $\lambda \in \mathbb{P}^2$ and that the intersection of any two distinct quadrics of this family consists of the union of C and a line L; conversely, given a line $L \subset \mathbb{P}^3$, if Q and Q' are general quadrics containing L then the intersection $Q \cap Q'$ will consist of the union of L and a twisted cubic curve. This suggests introducing yet another incidence correspondence: let $\mathbb{P}^9$ be the space of quadric surfaces in $\mathbb{P}^3$ and set

$$\Psi = \{(C, L, Q, Q'): Q \cap Q' = C \cup L\} \subset \mathscr{H} \times \mathbb{G}(1, 3) \times \mathbb{P}^9 \times \mathbb{P}^9.$$

By what was said earlier, the fiber of Ψ over $L \in \mathbb{G}(1, 3)$ is an open subset of the product $\mathbb{P}^6 \times \mathbb{P}^6$, from which we may deduce that Ψ has dimension

$$\dim(\Psi) = \dim(\mathbb{G}(1, 3)) + 2 \cdot 6 = 16.$$

On the other hand, the fiber of Ψ over $C \in \mathscr{H}$ is an open subset of the product $\mathbb{P}^2 \times \mathbb{P}^2$; hence

$$\dim(\mathscr{H}) = \dim(\Psi) - 4 = 12.$$

Exercise 12.11. Find again the dimension of the parameter space of rational normal curves $C \subset \mathbb{P}^n$ using the construction of a rational normal scroll given in Example 9.15 and a computation analogous to the one just given. (*)

Example 12.12. Twisted Cubics on a General Surface

Having determined that the space of twisted cubics is 12-dimensional, we may now mimic the computation of the dimension of the Fano variety of a general hypersurface to answer the question: when does the general surface of degree d contain a twisted cubic curve? To answer this, we let as before $\mathbb{P}^N$ be the projective space of surfaces of degree d in $\mathbb{P}^3$, and let Φ be the incidence correspondence

$$\Phi = \{(S, C): C \subset S\} \subset \mathbb{P}^N \times \mathscr{H}.$$

To calculate the dimension of Φ, we consider the projection map $\pi_2: \Phi \to \mathscr{H}$. The fiber of this map over a point $C \in \mathscr{H}$ is simply the subset of $\mathbb{P}^N$ consisting of surfaces of degree d containing C; since the restriction map

$$\{\text{polynomials of degree } d \text{ on } \mathbb{P}^3\} \to \{\text{polynomials of degree } 3d \text{ on } C \cong \mathbb{P}^1\}$$

is a surjective linear map, this is a linear subspace of $\mathbb{P}^N$ having dimension $N - (3d + 1)$. We thus have

$$\dim(\Phi) = N - (3d + 1) + 12 = N - 3d + 11,$$

and in particular we may conclude that Φ does not surject onto $\mathbb{P}^N$ unless $3d \leq 11$. In other words, a general surface $S \subset \mathbb{P}^3$ of degree $d \geq 4$ contains no twisted cubic curves.

We may suspect a good deal more from this calculation: that the family of twisted cubic curves on a surface of degree $d \leq 3$ has dimension $11 - 3d$, and that the codimension in $\mathbb{P}^N$ of the subvariety of surfaces of degree d containing a twisted cubic curve is $3d - 11$. In fact, both these statements are true for $d > 1$, though we do not have the technique to prove them without straining. It is worth mentioning, however, that one special case of this is of some significance: the fact that a general cubic surface $S \subset \mathbb{P}^3$ contains a twisted cubic curve may be used to prove that S is rational. Briefly, suppose S is a cubic surface and $C \subset S$ a twisted cubic curve. Let F_0, F_1, F_2 be a basis for the space of quadric polynomials vanishing on C and consider the rational map $\varphi: S \to \mathbb{P}^2$ given by

$$[Z] \mapsto [F_0(Z), F_1(Z), F_2(Z)].$$

If $p \in \mathbb{P}^2$ is a general point—say given by the vanishing of two linear forms

$$a_0 W_0 + a_1 W_1 + a_2 W_2 = b_0 W_0 + b_1 W_1 + b_2 W_2 = 0$$

on $\mathbb{P}^2$—then the inverse image of p in S will consist of the points of intersection with S of the corresponding quadric surfaces

$$A: a_0 F_0 + a_1 F_1 + a_2 F_2 = 0 \qquad \text{and} \qquad B: b_0 F_0 + b_1 F_1 + b_2 F_2 = 0$$

away from C. But now we have seen that the intersection of A and B will consist of the union of C and a line L meeting C at two points $q, r \in C$. The inverse image of p will thus consist of the third point of intersection of L with S, so that the map φ is generically one to one.

Exercise 12.13. Find the dimension of the space of plane conic curves $C \subset \mathbb{P}^3$ and show that for $d \geq 4$ not every surface of degree d in $\mathbb{P}^3$ contains a conic. What do you expect to happen in case $d = 3$? Show directly that it does. (*)

Example 12.14. Complete Intersections

It is not hard to describe in general the dimension of the space of complete intersections of hypersurfaces of given degrees in $\mathbb{P}^n$. As an example, we will consider the space $\mathcal{U}_{d,e}$ of curves $C \subset \mathbb{P}^3$ that are complete intersections of surfaces of degrees d and e.

Note. In what follows, we will be using the fact that if S and T are general surfaces of degrees d and e then the intersection $S \cap T$ is a complete intersection—that is, the defining polynomials F and G of S and T together generate a radical ideal in $K[Z_0, Z_1, Z_2, Z_3]$. If this is not familiar to you from commutative algebra, it will be established in Proposition 17.18.

Suppose first that $d < e$. Then if C is a complete intersection of surfaces S and T of degrees d and e, S is uniquely determined by $C \subset \mathbb{P}^3$ as the zero locus of the unique (up to scalars) polynomial F in the dth graded piece of the homogeneous ideal $I(C)_d$. T, by contrast, is only determined modulo surfaces of degree e vanishing on S, that is, as an element of the projective space associated to the space of polynomials of degree e modulo $I(S)_e$. Thus, we consider the map of $\mathscr{U}_{d,e}$ to the space $\mathbb{P}^N$ of surfaces of degree d given by associating to C the surface S.

Now, the space of polynomials of degree e has dimension $\binom{e+3}{3}$ and the subspace of those vanishing on S—that is, divisible by F—may be identified with the space of polynomials of degree $e - d$, and so has dimension $\binom{e-d+3}{3}$. It follows that the fiber of $\mathscr{U}_{d,e}$ over a general $S \in \mathbb{P}^N$ is an open subset of a projective space of dimension

$$\binom{e+3}{3} - \binom{e-d+3}{3} - 1 = \frac{d^3 - 3d^2e - 6d^2 + 3de^2 + 12de + 11d - 6}{6}$$

$$= \binom{d-1}{3} + e \cdot d \cdot \frac{e-d+4}{2}.$$

Thus, $\mathscr{U}_{d,e}$ will have dimension

$$\binom{d+3}{3} - 1 + \binom{d-1}{3} + e \cdot d \cdot \frac{e-d+4}{2}$$

$$= \frac{d^3 + 11d - 3}{3} + e \cdot d \cdot \frac{e-d+4}{2}.$$

Now, if $d = e$, we can realize the space $\mathscr{U}_{d,d}$ as an open subset of the Grassmannian of two-dimensional subspaces of the space of polynomials of degree d on $\mathbb{P}^3$; thus

$$\dim(\mathscr{U}_{d,d}) = 2 \cdot \binom{d+3}{3} - 4.$$

Note that this differs by exactly one from the estimate for $\mathscr{U}_{d,e}$ in the previous case, the difference being that S is not uniquely determined by C here. To see the difference clearly, set up an incidence correspondence of pairs (C, S) with $C \subset S$.

Example 12.15. Curves of Type (a, b) on a Quadric

One more example of a family of curves in $\mathbb{P}^3$ whose dimension we can compute is the family $\mathscr{V}_{a,b}$ of curves $C \subset \mathbb{P}^3$ that are of type (a, b) on some smooth quadric surface $Q \subset \mathbb{P}^3$. (By "type (a, b)" we mean that in terms of the isomorphism $Q \cong \mathbb{P}^1 \times \mathbb{P}^1$ C is given as the zero locus of a bihomogeneous polynomial of bidegree (a, b) with no multiple components.) Here, if we assume that either a or $b \geq 3$, then C will determine Q; Q will be just the union of the trisecant lines to C.

The space $\mathscr{V}_{a,b}$ will thus map to the open subset U of the space $\mathbb{P}^9$ of all quadric surfaces parameterizing smooth quadrics. The fiber over a given $Q \in U$, moreover, will be just an open subset of the union of the two projective spaces associated to the vector spaces of bihomogeneous polynomials of bidegree (a, b) or (b, a) (given just Q, we don't have any canonical way of distinguishing one ruling from the other). This is a union of two projective spaces of dimension $(a + 1)(b + 1) - 1 = ab + a + b$, and so we have

$$\dim(\mathscr{V}_{a,b}) = ab + a + b + 9.$$

Note that it is not clear from this description whether $\mathscr{V}_{a,b}$ has one irreducible component or two; in fact it is irreducible.

Exercise 12.16. In case both a and b are ≤ 2, set up an incidence correspondence between curves and quadrics to compute the dimension of $\mathscr{V}_{a,b}$. In particular, check that you get the same answer as before for the space $\mathscr{V}_{2,1}$ of twisted cubics.

Exercise 12.17. Observe that a curve of type $(1, d - 1)$ on a smooth quadric $Q \subset \mathbb{P}^3$ is a rational curve of degree d—that is, the image of $\mathbb{P}^1$ under a map to $\mathbb{P}^3$ given by a fourtuple of homogeneous polynomials of degree d on $\mathbb{P}^1$, or equivalently a projection of the rational normal curve in $\mathbb{P}^d$. Use a count of parameters to show that for $d \geq 5$ the general rational curve of degree d is not a curve of type $(1, d - 1)$ on a quadric.

Example 12.18. Determinantal Varieties

Sometimes we can prove something about the general member of a family of varieties simply by counting dimensions. For example, recall that in Exercise 9.18 we stated that for sufficiently large d a general surface of degree d in $\mathbb{P}^3$ could not be expressed as a linear determinantal variety, i.e., was not the zero locus of any $d \times d$ matrix of linear forms on $\mathbb{P}^3$. In fact, as we can now establish, this is true for all $d \geq 4$. To prove this, we will count parameters, or in other words, estimate the dimension of the space of surfaces that are so expressible.

To this end, let U be the space of $d \times d$ matrices of linear forms on $\mathbb{P}^3$ whose determinants are not identically zero (as polynomials of degree d on $\mathbb{P}^3$), up to scalars. U is an open subset of a projective space of dimension $4d^2 - 1$, mapping to the space $\mathbb{P}^N$ of all surfaces of degree d in $\mathbb{P}^3$. Now, for any matrix $M = (L_{i,j})$ of linear forms, we can multiply M on either the left or the right by a $d \times d$ matrix of scalars without changing the zero locus of its determinant—in other words, $\mathrm{PGL}_d K \times \mathrm{PGL}_d K$ acts of the fibers of U over $\mathbb{P}^N$. Moreover, we can check that for general M the stabilizer of this action is trivial; we conclude that the dimension of the subset of $\mathbb{P}^N$ of surfaces given as linear determinantal varieties is at most

$$4d^2 - 1 - 2(d^2 - 1) = 2d^2 + 1.$$

Comparing this with the dimension $N = \binom{d+3}{3} - 1$ we conclude that for $d \geq 4$, the general surface of degree d in $\mathbb{P}^3$ is not a linear determinantal variety.

In fact, the dimension of the set of determinantal surfaces of degree d is indeed exactly $2d^2 + 1$. In particular, when $d = 4$ we get a hypersurface in the space $\mathbb{P}^{34}$ of all quartic surfaces.

Exercise 12.19. Consider possible expressions of a surface $S \subset \mathbb{P}^3$ of degree d as a more general determinantal variety, for example, as the zero locus of the determinant of a 2×2 matrix of polynomials; in the space $\mathbb{P}^N$ of all surfaces of degree d, let $\Phi_{a,b}$ be the subset of those of the form

$$S = \left\{[Z]: \begin{vmatrix} F(Z) & G(Z) \\ H(Z) & K(Z) \end{vmatrix} = 0\right\}$$

where $\deg F = a$, $\deg G = b$, $\deg H = d - b$, and $\deg K = d - a$. Estimate the dimension of $\Phi_{a,b}$ and deduce again that a general surface of degree $d \geq 4$ does not lie in $\Phi_{a,b}$ for any a and b.

Note that in case $d = 4$, every admissible choice of a and b yields a hypersurface in the space $\mathbb{P}^{34}$ of quartics. Indeed, some of these are hypersurfaces we have run into before; for example, a surface S given by the determinant

$$\begin{vmatrix} L(Z) & M(Z) \\ H(Z) & K(Z) \end{vmatrix} = 0,$$

where L and M are linear forms, certainly contains the line Λ in $\mathbb{P}^3$ given by $L(Z) = M(Z) = 0$. Conversely, if S contains this line, then its defining polynomial is in the homogeneous ideal $I(\Lambda)$, and so may be expressed as a linear combination $K(Z) \cdot L(Z) - H(Z) \cdot M(Z)$. Thus the hypersurface $\Phi_{1,1}$ is just the locus of quartics containing a line. Similarly, an irreducible quartic S is in $\Phi_{1,2}$ if and only if it contains a plane conic and may be given as the zero locus of a determinant

$$\begin{vmatrix} L_{11}(Z) & L_{12}(Z) & L_{13}(Z) \\ L_{21}(Z) & L_{22}(Z) & L_{23}(Z) \\ Q_1(Z) & Q_2(Z) & Q_3(Z) \end{vmatrix} = 0,$$

with L_{ij} linear and Q_i quadratic, if and only if it is in the closure of the locus of quartics containing a twisted cubic curve.

Exercise 12.20. Verify the last statement.

Exercise 12.21. Verify the preceding computation (Example 12.15) of the dimension of the family of curves of type $(a, a - 1)$ on a quadric surface by realizing such a curve $C \subset \mathbb{P}^3$ as a determinantal variety

$$C = \left\{ [Z]: \operatorname{rank} \begin{pmatrix} L_{11}(Z) & L_{12}(Z) & F(Z) \\ L_{21}(Z) & L_{22}(Z) & G(Z) \end{pmatrix} \leq 1 \right\}$$

with L_{ij} linear and F and G of degree $a - 1$ (cf Exercise 9.17).

Group Actions

As a final example of dimension computations, we consider the action of an algebraic group G on a variety X. In this context, Theorem 11.12, applied to the map $G \to X$ defined by sending $g \in G$ to $g(x_0)$ for some fixed $x_0 \in X$, says simply that the dimension of G is equal to the dimension of the orbit $G \cdot x_0$ plus the dimension of the stabilizer of x_0 in G. We will invoke here for the most part only a very crude form of this, the observation that the orbits of G can have dimension at most $\dim(G)$. This says in particular that the action of G on X can be *quasi-homogeneous*, that is, have a dense orbit, only if $\dim(G) \geq \dim(X)$.

Example 12.22. $\mathrm{GL}(V)$ acts on $\mathrm{Sym}^d V$ and $\wedge^k V$

Consider first the action of the group $\mathrm{GL}_n K$ on the space of homogeneous polynomials of degree d in n variables. Comparing the dimensions n^2 of $\mathrm{GL}_n K$ and $\binom{d+n-1}{n-1}$ of $\mathrm{Sym}^d(K^n)$, we may conclude that the only cases in which there can be only a finite number of hypersurfaces of degree d in $\mathbb{P}^{n-1}$ up to projective equivalence are $d = 1$ or 2 in general and $n = 2$, $d = 3$. That there are indeed only a finite number in each of these cases is easy to check.

A similar pattern emerges when we look at the action of $\mathrm{GL}_n K$ on $\wedge^k(K^n)$ with $n \geq 2k$; since the dimension of $\wedge^k(K^n)$ is $\binom{n}{k}$, we see that the only cases in which there can be only a finite number of skew-symmetric k-linear forms on an n-dimensional vector space V up to automorphisms of V are $k = 2$ in general and $k = 3$ with $n = 6$, 7, or 8.

Exercise 12.23. Show that there are indeed only finitely many orbits of $\mathrm{GL}_6 K$ on $\wedge^3 K^6$ (there are four, other than $\{0\}$). Try to relate these to the geometry of the Grassmannian $G(3, 6) \subset \mathbb{P}(\wedge^3 K^6)$. (*)

Example 12.24. $\mathrm{PGL}_{n+1} K$ Acts on $(\mathbb{P}^n)^l$ and $\mathbb{G}(k, n)^l$

Comparing the dimensions $n^2 + 2n$ of $\mathrm{PGL}_n K$ and $l \cdot n$ of $(\mathbb{P}^n)^l$, we see that general l-tuples of points in $\mathbb{P}^n$ can be carried into one another only when $l \leq n + 2$, a very weak form of the standard fact (Exercise 10.5) that there is a unique automorphism of $\mathbb{P}^n$ carrying any $n + 2$ points in general position into any other. Note that the action of $PGL_{n+1}K$ on $(\mathbb{P}^n)^l$ for $4 \leq l \leq n + 2$ is an example of a group action that has a dense orbit, but not finitely many orbits.

Things get substantially more interesting when we look at more general configurations of linear spaces—for example, l-tuples of k-planes. Since $\dim(\mathbb{G}(k, n)^l) = l \cdot (k + 1) \cdot (n - k)$, we can of course conclude that general l-tuples of k-planes in $\mathbb{P}^n$ can be carried into each other only when $l \cdot (k + 1) \cdot (n - k) \leq n^2 - 1$. As the following exercise shows, however, the action does not always behave as we'd expect.

Exercise 12.25. Show that the stabilizer in $PGL_4 K$ of a general point in $\mathbb{G}(1, 3)^4$ is one-dimensional, despite the fact that $\dim(\mathbb{G}(1, 3)^4) > \dim(PGL_4 K)$. Can you describe explicitly the set of orbits on the open set $U \subset \mathbb{G}(1, 3)^4$ consisting of fourtuples of pairwise skew lines? (*)

Exercise 12.26. Show that general 4-tuples of lines in $\mathbb{P}^4$ can be carried into each other by an element of $PGL_5 K$, as a dimension count would suggest. Is the same true for 4-tuples of 2-planes in $\mathbb{P}^6$? In general, is there an example where $\dim(\mathbb{G}(k, n)^l) \geq n^2 + 2n$, but the action of $PGL_{n+1}K$ on $\mathbb{G}(k, n)^l$ does not have a dense orbit? (*)

Exercise 12.27. Show that if we generalize the statement of the last problem further, the answer becomes no: for example, show that the action of $PGL_3 K$ on the space $(\mathbb{P}^2)^2 \times (\mathbb{P}^{2*})^2$ does not have a dense orbit.

LECTURE 13

Hilbert Polynomials

Hilbert Functions and Polynomials

Given that a projective variety $X \subset \mathbb{P}^n$ is an intersection of hypersurfaces, one of the most basic problems we can pose in relation to X is to describe the hypersurfaces that contain it. In particular, we want to know how many hypersurfaces of each degree contain X—that is, for each value of m, to know the dimension of the vector space of homogeneous polynomials of degree m vanishing on X. To express this information, we define a function

$$h_X: \mathbb{N} \to \mathbb{N}$$

by letting $h_X(m)$ be the codimension, in the vector space of all homogeneous polynomials of degree m on $\mathbb{P}^n$, of the subspace of those vanishing on X; i.e.,

$$h_X(m) = \dim(S(X)_m)$$

where $S(X) = K[Z_0, \ldots, Z_n]/I(X)$ is the homogeneous coordinate ring and the subscript m denotes "mth graded piece." The function h_X is called the *Hilbert function* of $X \subset \mathbb{P}^n$.

For example, to start with the simplest case, suppose X consists of three points in the plane $\mathbb{P}^2$. Then the value $h_X(1)$ tells us exactly whether or not those three points are collinear: we have

$$h_X(1) = \begin{cases} 2 & \text{if the three points are collinear.} \\ 3 & \text{if not.} \end{cases}$$

On the other hand, we claim that $h_X(2) = 3$, whatever the position of the points. To see this, note that there exists a quadratic polynomial vanishing at any two of the points p_i and p_j, but not at the third point p_k: take a product of a linear form vanishing at p_i but not at p_k with one vanishing at p_j and nonzero at p_k. Thus the map from the space of quadratic polynomials to K^3 given by evaluation at the p_i is surjective[1], and the kernel will have dimension 3. Similarly, we have $h_X(m) = 3$ for all $m \geq 3$, independent of the position of the points.

Similarly, if $X \subset \mathbb{P}^2$ consists of four points, there are two possible Hilbert functions. We could have

$$h_X(m) = \begin{cases} 2 & \text{for } m = 1 \\ 3 & \text{for } m = 2 \\ 4 & \text{for } m \geq 3 \end{cases}$$

if the four points are collinear or

$$h_X(m) = \begin{cases} 3 & \text{for } m = 1 \\ 4 & \text{for } m \geq 2 \end{cases}$$

if not.

More generally, we see that whenever $X \subset \mathbb{P}^n$ is a finite set of points, the values $h_X(m)$ for small values of m will give us information about the position of the points—$h_X(1)$, for example, tells us the size of the linear subspace of $\mathbb{P}^n$ they span—but in general, we have the following.

Exercise 13.1. Let $X \subset \mathbb{P}^n$ be a set of d points. Show that for sufficiently large values of m relative to d (specifically, for $m \geq d - 1$), the Hilbert function $h_X(m) = d$.

To give an example involving a positive-dimensional variety, suppose $X \subset \mathbb{P}^2$ is a curve, say the zero locus of the polynomial $F(Z)$ of degree d. The mth graded piece $I(X)_m$ of the ideal of X then consists of polynomials of degree m divisible by F. We can thus identify $I(X)_m$ with the space of polynomials of degree $m - d$, so that

$$\dim(I(X)_m) = \binom{m-d+2}{2}$$

and, for $m \geq d$,

$$h_X(m) = \binom{m+2}{2} - \binom{m-d+2}{2} = d \cdot m - \frac{d(d-3)}{2}.$$

One thing we may observe in all the cases so far is that h_X is basically a nice function of m—in particular, for large values of m it is a polynomial in m. We will see now that this is indeed the general picture.

[1] Of course, homogeneous polynomials do not take values at points p in projective space. In general, what we will mean by "the map from the space of polynomials to K^m given by evaluation at the points p_i" will be the map given by evaluation at vectors $v_i \in K^{n+1}$ lying over the points p_i; as far as the rank of the map is concerned it doesn't matter what v_i we pick.

Proposition 13.2. *Let $X \subset \mathbb{P}^n$ be a variety, h_X its Hilbert function. Then there exists a polynomial p_X such that for all sufficiently large m, $h_X(m) = p_X(m)$; the degree of the polynomial p_X is the dimension of X.*

The polynomial p_X is called the *Hilbert polynomial* of X. From a strictly logical point of view, it might make sense to define the dimension of a variety to be the degree of its Hilbert polynomial and then to deduce the various other definitions given earlier; since the Hilbert polynomial definition is perhaps the least intuitively clear, however, we chose to go in the other direction.

PROOF. We will prove Proposition 13.2 modulo one assertion whose proof will be given as Exercise 17.19 (it is an immediate consequence of Bertini's theorem). To state this assertion, let $X \subset \mathbb{P}^n$ be any irreducible variety of dimension k, and let $\Lambda \cong \mathbb{P}^{n-k} \subset \mathbb{P}^n$ be a general linear space of complementary dimension, so that the intersection $Y = X \cap \Lambda$ consists of a finite collection of points. We claim then that the ideals of X and Λ together locally generate the ideal of Y; i.e., the saturation

$$\overline{(I(X), I(\Lambda))} = I(Y).$$

(For the reader who is familiar with the notions of smoothness and tangent spaces to varieties, we can express this condition a little more geometrically: the claim amounts to saying that X and Λ intersect transversely at smooth points of X.)

Assuming the claim, let Λ be a general $(n-k)$-plane in $\mathbb{P}^n$ and $L_1, \ldots, L_k$ linear forms on $\mathbb{P}^n$ generating the ideal of Λ. Introduce ideals

$$I(X) = I^{(0)} \subset I^{(1)} \subset \cdots \subset I^{(k)} \subset K[Z_0, \ldots, Z_n]$$

defined by

$$I^{(\alpha)} = (I(X), L_1, \ldots, L_\alpha)$$

so that in particular the claim amounts to saying that the saturation of $I^{(k)}$ is $I(Y)$. Let $S^{(\alpha)} = K[Z_0, \ldots, Z_n]/I^{(\alpha)}$ be the quotient ring, and define functions $h^{(\alpha)}$ by

$$h^{(\alpha)}(m) = \dim((S^{(\alpha)})_m)$$

so that $h^{(0)}(m) = h_X(m)$ and by the claim and Exercise 13.1, $h^{(k)}(m)$ is constant (equal to the number d of points of Y) for sufficiently large m.

Now, since the dimension of Y is k less than the dimension of X, it follows that the dimension of the variety defined by $I^{(\alpha)}$ must be exactly $k - \alpha$ for all α; in particular, the image of $L_{\alpha+1}$ in $S^{(\alpha)}$ is not a zero-divisor. Multiplication by $L_{\alpha+1}$ thus defines an inclusion $(S^{(\alpha)})_{m-1} \hookrightarrow (S^{(\alpha)})_m$, with quotient $(S^{(\alpha+1)})_m$; we deduce that

$$h^{(\alpha+1)}(m) = h^{(\alpha)}(m) - h^{(\alpha)}(m-1),$$

i.e., $h^{(\alpha+1)}$ is just the successive difference function of $h^{(\alpha)}$.

Putting this all together, we see that for large values of m the kth successive difference function of the Hilbert function $h_X(m) = h^{(0)}(m)$ will be a constant d; i.e., for large $m h_X(m)$ will be a polynomial of degree k in m. □

In Remark 13.10 we will give another argument for the fact that the degree of $h_X(m)$ for large m is the dimension of X, given that it is a polynomial in the first place.

Note that by this argument the leading term of the Hilbert polynomial $p_X(m)$ is $(d/k!) \cdot m^k$, where d is the number of points of intersection of X with a general $(n-k)$-plane Λ. The number d is called the *degree* of X; it will be defined and discussed further in Lecture 18.

Example 13.3. Hilbert Function of the Rational Normal Curve

This is easy. Under the map

$$v_d \colon \mathbb{P}^1 \to \mathbb{P}^d$$

$$: [X_0, X_1] \mapsto [X_0^d, X_0^{d-1}X_1, \ldots, X_1^d],$$

the homogeneous polynomials of degree m in the coordinates $[Z_0, \ldots, Z_d]$ on $\mathbb{P}^d$ pull back to give all homogeneous polynomials of degree $m \cdot d$ in X_0 and X_1. If $X \subset \mathbb{P}^d$ is the image of this map, then $S(X)_m$ is isomorphic to the space of homogeneous polynomials of degree md in two variables; so

$$h_X(m) = p_X(m) = d \cdot m + 1.$$

Example 13.4. Hilbert Function of the Veronese Variety

We can, in exactly analogous fashion, write down the Hilbert function of the Veronese variety $X = v_d(\mathbb{P}^n) \subset \mathbb{P}^N$. We observe that polynomials of degree m on $\mathbb{P}^N$ pull back via v_d to polynomials of degree $d \cdot m$ on $\mathbb{P}^n$. The dimension of the space $S(X)_m$ is the dimension of the space of polynomials of degree $d \cdot m$ on $\mathbb{P}^n$, and we have

$$h_X(m) = p_X(m) = \binom{m \cdot d + n}{n}.$$

Exercise 13.5. Find the Hilbert function of a hypersurface of degree d in $\mathbb{P}^n$ and verify that the dimension of this variety is indeed $n-1$.

Exercise 13.6. Find the Hilbert function of the Segre variety $\Sigma_{n,m} = \sigma(\mathbb{P}^n \times \mathbb{P}^m) \subset \mathbb{P}^{(n+1)(m+1)-1}$ and verify that the dimension of this variety is indeed $n + m$.

Example 13.7. Hilbert Polynomials of Curves (for those familiar with the Riemann-Roch theorem for curves)

To see an example of what sort of information may be conveyed by the Hilbert polynomial, consider the case of a curve $X \subset \mathbb{P}^n$; for the moment, suppose that $K = \mathbb{C}$ and that X is a complex submanifold of $\mathbb{P}^n$, i.e., a Riemann surface. Suppose that the hyperplane $(Z_0 = 0)$ intersects X transversely in d points $p_1, \ldots, p_d$. Then to a homogeneous polynomial $F(Z)$ of degree m we may associate the meromorphic function $F(Z)/Z_0^m$ on X; this gives us an inclusion of the space $S(X)_m$ of homogeneous polynomials modulo those vanishing on X in the space $\mathscr{L}(m \cdot p_1 + \cdots + m \cdot p_d)$ of meromorphic functions on X, holomorphic away from the points p_i and with a pole of order at most m at each p_i. Two things now happen for m large. First, the divisor $m \cdot p_1 + \cdots + m \cdot p_d$ on X becomes nonspecial, so that by the Riemann-Roch formula, $\dim(\mathscr{L}(m \cdot p_1 + \cdots + m \cdot p_d)) = md - g + 1$; and the map $S(X)_m \to \mathscr{L}(m \cdot p_1 + \cdots + m \cdot p_d)$ becomes an isomorphism. Thus,

$$p_X(m) = md - g + 1.$$

Note that the coefficient of the linear term is d, the degree of the curve, and the constant term is $1 - g$, which tells us the genus. Indeed, we can use this to extend the notion of genus to arbitrary fields K and to singular curves as well as smooth ones; for any curve $X \subset \mathbb{P}^n$ with Hilbert polynomial $p_X(m) = a \cdot m + b$, the quantity $1 - b$ is called the *arithmetic genus* of X.

Observe that this computation of the Hilbert function of a plane curve $X \subset \mathbb{P}^2$ of degree d shows that the genus of a smooth plane curve is $\binom{d-1}{2}$, and more generally that this is the arithmetic genus of any plane curve of degree d.

Exercise 13.8. Determine the arithmetic genus of (i) a pair of skew lines in $\mathbb{P}^3$; (ii) a pair of incident lines in either $\mathbb{P}^2$ or $\mathbb{P}^3$; (iii) three concurrent but noncoplanar lines in $\mathbb{P}^3$; and (iv) three concurrent coplanar lines in either $\mathbb{P}^2$ or $\mathbb{P}^3$. (*)

Exercise 13.9. Consider a plane curve $X \subset \mathbb{P}^2$ of degree d and its image $Y = v_2(X) \subset \mathbb{P}^5$ under the quadratic Veronese map $v_2 \colon \mathbb{P}^2 \to \mathbb{P}^5$. Compare the Hilbert polynomials of the two and observe in particular that the arithmetic genus is the same.

It is in fact the case, as this exercise may suggest, that the constant term $p_X(0)$ of the Hilbert polynomial of a projective variety $X \subset \mathbb{P}^n$ is an isomorphism invariant, i.e., does not depend on the choice of embedding of X in projective space, though we would need cohomology to prove this. (One case is elementary: it is easy to see that it is invariant under Veronese reembeddings.) In general, we define the *arithmetic genus* of a projective variety X of dimension k to be $(-1)^k(p_X(0) - 1)$.

Finally, one natural problem we may pose in regard to Hilbert functions and Hilbert polynomials is to give explicit estimates for how large m has to be to insure that $p_X(m) = h_X(m)$. Specifically, given a polynomial $p(m)$, we would like to find the smallest value of m_0 such that for any irreducible variety $X \subset \mathbb{P}^n$ with Hilbert polynomial $p_X = p$, we have

$$h_X(m) = p(m)$$

for all $m \geq m_0$. This is very difficult; it is already a major theorem that such an m_0 exists, and very little is known about the actual value of m_0 even in as simple a case as curves in $\mathbb{P}^3$. For example, though Castelnuovo showed that taking $m_0 = d - 2$ sufficed for irreducible curves $C \subset \mathbb{P}^3$ with Hilbert polynomial $p(m) = d \cdot m + c$

(this is the best possible estimate for m_0 in terms of the leading coefficient d alone), and this was generalized by Gruson, Lazarsfeld and Peskine [GLP] for curves and Pinkham [P] for surfaces, we are still miles away from understanding the general question. To see a situation where this is of crucial importance, look at the construction of the Hilbert variety parametrizing subvarieties of $\mathbb{P}^n$, described in Lecture 21.

Syzygies

There is another way to see the existence of the Hilbert polynomial of a variety $X \subset \mathbb{P}^n$, which involves introducing a finer invariant.

We start with some notation: first, to save ink, we will write simply S for the homogeneous coordinate ring $K[Z_0, \ldots, Z_n]$ of $\mathbb{P}^n$. Next, suppose that M is any homogeneous module over the ring S—that is, a module M with grading

$$M = \bigoplus M_k$$

such that $S_d \cdot M_k \subset M_{d+k}$. For any integer l we let $M(l)$ be the same module as M, but with grading "shifted" by l, that is, we set

$$M(l)_k = M_{k+l}$$

so that $M(l)$ will be isomorphic to M as S-module, but not in general as graded S-module; $M(l)$ is called a *twist* of M. We introduce this notation so as to be able to make maps homogeneous of degree 0 (we say in general that a map $\varphi\colon M \to N$ of graded S-modules is *homogeneous of degree* d if $\varphi(M_k) \subset N_{k+d}$). For example, the module homomorphism $\cdot F$ from S to S given by multiplication by a homogeneous polynomial $F(Z)$ of degree d is clearly not homogeneous of degree 0; but the same map, viewed as a map from $S(-d)$ to S, is.

With this said, first consider the ideal $I(X)$ of X. $I(X)$ is generated by polynomials F_α of degree d_α, $\alpha = 1, \ldots, r$. We can express this by saying that we have a surjective homogeneous map of graded S-modules

$$\bigoplus S(-d_\alpha) \to I(X)$$

or equivalently that we have an exact sequence of graded S-modules

$$\bigoplus_{\alpha=1}^{r} S(-d_\alpha) \xrightarrow{\varphi_1} S \to A(X) \to 0$$

where φ_1 is given by the vector $(\ldots, \cdot F_\alpha, \ldots)$.

Next, we want to describe the relations among the polynomials F_α. We therefore introduce the *module* M_1 *of relations*, defined to be simply the kernel of φ_1, or, explicitly, the module of r-tuples $(G_1, \ldots, G_r)$ such that $\sum G_\alpha \cdot F_\alpha = 0$. Note that M_1 is indeed a graded module and that, except in the case $r = 1$ (i.e., X a hypersurface) it will be nonempty, since it will contain the relations $F_\alpha \cdot F_\beta - F_\beta \cdot F_\alpha$—that is, the vectors $(0, \ldots, -F_\beta, \ldots, F_\alpha, \ldots, 0)$.

In any event, since M_1 is a submodule of a finitely generated module it is again finitely generated; we let $\{G_{\beta,1}, \ldots, G_{\beta,r}\}$, $\beta = 1, \ldots, s$, be a set of generators[2]. Note that for each β we have

$$\deg(G_{\beta,1}) + d_1 = \cdots = \deg(G_{\beta,r}) + d_r = e_\beta$$

for some e_β. We can express all of this by saying that the exact preceding sequence may be lengthened to an exact sequence

$$\bigoplus_{\beta=1}^{s} S(-e_\beta) \xrightarrow{\varphi_2} \bigoplus_{\alpha=1}^{r} S(-d_\alpha) \xrightarrow{\varphi_1} S \to A(X) \to 0$$

where φ_2 is given by the $r \times s$ matrix $(G_{\beta,\alpha})$.

We now just repeat this process over and over: we let M_2 be the kernel of φ_2, called the module of relations among the relations, which is again finitely generated. Choosing a set of generators we obtain a map φ_3 from another direct sum of twists of S to $\bigoplus S(-e_\beta)$ extending the sequence one more step, and so on. The basic fact about this process is the famous *Hilbert syzygy theorem.* This says that after a finite number of steps (at most the number of variables minus 1) the module M_k we obtain is free, so that after one more step we can terminate the sequence. We arrive then at an exact sequence

$$0 \to N_k \to N_{k-1} \to \cdots \to N_1 \to S \to S(X) \to 0$$

in which each term N_i is a direct sum of twists $S(-a_{i,j})$ of S. Such a sequence is called a *free resolution* of the module $S(X)$; if at each stage we choose a minimal set of generators for the module M_i (i.e., the kernel of the previous map), it is called *minimal.*

One consequence of the existence of such a resolution is that we can use it to describe the Hilbert function of X. Explicitly, we have

$$\dim(S(-a)_m) = \dim(S_{m-a}) = \binom{m-a+n}{n}_0$$

where for $n \in \mathbb{N}$ and $c \in \mathbb{Z}$ the binomial coefficient $\binom{c}{n}_0$ is defined to be

$$\binom{c}{n}_0 = \begin{cases} \dfrac{c\cdot(c-1)\cdot\ldots\cdot(c-n+1)}{n!} & \text{if } c \geq n \\ 0 & \text{if } c < n. \end{cases}$$

(Note that this differs from some definitions of the binomial coefficient when $c < 0$, which is why we append the subscript.) It follows that if the ith term N_i in the resolution of $S(X)$ is $\bigoplus S(-a_{i,j})$, then the dimension of $S(X)_m$ is given by

[2] Note that we want these generators to be simultaneously homogeneous elements of the module M_1 and relations among the F_α; in order to know that we can indeed find such generators we have to check that the homogeneous parts of a relation are again relations.

$$\dim(S(X)_m) = \binom{m+n}{n}_0 + \sum_{i,j} (-1)^i \binom{m - a_{i,j} + n}{n}_0.$$

In particular, we observe that since the binomial coefficient $\binom{c}{n}_0$ is a polynomial in c for $c \geq 0$, for $m \geq \max(a_{i,j}) - n$ the Hilbert function $h_X(m) = \dim(S(X)_m)$ is a polynomial in m.

Remark 13.10. It is worth pointing out that, given that the Hilbert function $h_X(m)$ of a variety $X \subset \mathbb{P}^n$ is equal to a polynomial $p_X(m)$ for large m, one can see directly that the degree of this polynomial must be the dimension k of X. To do this, make a linear change of variables so that the plane Λ given by $Z_0 = \cdots = Z_k = 0$ is disjoint from X, and consider the projection $\pi_\Lambda \colon X \to \mathbb{P}^k$ from Λ. Since this is surjective, it gives an inclusion of the homogeneous coordinate ring $S' = K[Z_0, \ldots, Z_k]$ of $\mathbb{P}^k$ in $S(X)$, respecting degrees; this shows that

$$h_X(m) \geq \binom{m+k}{k}$$

for all m.

On the other hand, we claim that the map $S' \to S(X)$ expresses $S(X)$ as a finitely generated S'-module. To see this, it is enough to project from one point at a time, say from $[0, \ldots, 0, 1]$ to the plane $Z_n = 0$. In this case we let X' be the image of X in $\mathbb{P}^{n-1}$ and write $S(X) = S(X')[Z_n]/I(X)$; we observe that any homogeneous polynomial $F \in I(X)$ not vanishing at p gives a monic relation for Z_n over $S(X')$.

Now, if the generators of $S(X)$ as S'-module have degree d_i, we have a surjection $\bigoplus S'(-d_i) \to S(X)$, and hence an inequality

$$h_X(m) \leq \sum \binom{m+k-d_i}{k}_0.$$

Given that $h_X(m)$ is polynomial for large m, these two inequalities imply that it must have degree k.

Example 13.11. Three Points in $\mathbb{P}^2$

Consider again the case of three points in the plane. First, if they are not collinear, we may take them to be the points $[0, 0, 1]$, $[0, 1, 0]$, and $[1, 0, 0]$, so that their ideal is generated by the quadratic monomials

$$F_1 = XY, \qquad F_2 = XZ, \qquad \text{and} \qquad F_3 = YZ.$$

The relations among these three are easy to find; they are just

$$Z \cdot F_1 = Y \cdot F_2 = X \cdot F_3.$$

The minimal free resolution of Γ thus looks like

$$0 \to S(-3)^2 \to S(-2)^3 \to S \to S(\Gamma) \to 0$$

and we see in particular that the Hilbert function of Γ is

$$h_\Gamma(m) = \binom{m+2}{2} - 3\binom{m}{2} + 2\binom{m-1}{2}_0,$$

which for $m \geq 1$ is

$$\frac{(m+2)(m+1) - 3m(m-1) + 2(m-1)(m-2)}{2} = 3.$$

If, on the other hand, the three points of Γ are collinear, then they are the complete intersection of a line and a cubic—for example, if the three points are $[0, 0, 1]$, $[0, 1, 0]$, and $[0, 1, 1]$, then $I(\Gamma)$ is generated by $F_1 = X$ and $F_2 = YZ(Y - Z)$ (this can be checked directly or by applying Proposition 17.18). There is exactly one relation between these, the "trivial" relation $YZ(Y - Z) \cdot F_1 = X \cdot F_2$; so the resolution looks like

$$0 \to S(-4) \to S(-1) \oplus S(-3) \to S \to S(\Gamma) \to 0.$$

Of course, as you can check, this will also have Hilbert polynomial $p_\Gamma(m) \equiv 3$; but here the Hilbert function will coincide with this polynomial only for $m \geq 2$.

Example 13.12. Four Points in $\mathbb{P}^2$

We can likewise describe the free resolution of any configuration Γ of four points in the plane. To begin with the simplest case, if no three of the points are collinear, then they are the complete intersection of two conic curves—explicitly, if the points are the three coordinate points $[0, 0, 1]$, $[0, 1, 0]$, and $[1, 0, 0]$ plus the point $[1, 1, 1]$ then the ideal of Γ will be generated by $F_1 = XY - XZ$ and $F_2 = XZ - YZ$ (again, this can be checked either directly or using Proposition 17.18). The only relation between these will be the obvious one, $(XZ - YZ) \cdot F_1 = (XY - XZ) \cdot F_2$; so the free resolution will look like

$$0 \to S(-4) \to S(-2)^2 \to S \to S(\Gamma) \to 0.$$

At the other extreme, if all four points are collinear, then Γ is again a complete intersection, this time of a line and a quartic; the resolution will be

$$0 \to S(-5) \to S(-1) \oplus S(-4) \to S \to S(\Gamma) \to 0.$$

Finally, suppose that exactly three of the points are collinear—for example, say they are the three coordinate points and $[0, 1, 1]$. Now the ideal of Γ is generated by the two conics $F_1 = XY$ and $F_2 = XZ$ and the cubic $F_3 = YZ(Y - Z)$. The relations among these are

$$Z \cdot F_1 = Y \cdot F_2 \quad \text{and} \quad YZ \cdot F_1 - YZ \cdot F_2 = X \cdot F_3$$

so we have a sequence

$$0 \to S(-3) \oplus S(-4) \to S(-2)^2 \oplus S(-3) \to S \to S(\Gamma) \to 0.$$

Note one aspect of this example. While the first and third cases have the same Hilbert function, the terms of their minimal free resolutions differ. In other words, specifying the terms of the minimal free resolution of a variety—or giving the data of the integers $a_{i,j}$ in the general sequence—represents in general strictly more data than the Hilbert function, which of course is itself a refinement of the Hilbert polynomial.

Exercise 13.13. Verify the assertions made in Example 13.12—specifically, show that there are no further generators or relations beyond those listed.

(*)

Exercise 13.14. Find the minimal free resolutions of $\Gamma \subset \mathbb{P}^3$ in case (i) Γ consists of the four coordinate points; (ii) Γ consists of five points, no four coplanar.

Exercise 13.15. Find the minimal free resolution of the twisted cubic curve.

Example 13.16. Complete Intersections: Koszul Complexes

The reader will have noticed that the free resolutions were of a relatively simple form in those cases that were complete intersections, involving only the automatic relations $F_i \cdot F_j = F_j \cdot F_i$. In fact, this is the general pattern, though where more than two polynomials are involved the sequence will necessarily be longer. We will give here a description of the resolution of a complete intersection in general, but will not prove it; for details, see [E].

To describe the situation in general, suppose $X \subset \mathbb{P}^n$ is the complete intersection of hypersurfaces defined by polynomial equations $F_1, \ldots, F_k$, with F_α of degree d_α. The first term in the minimal free resolution of X is then of course the direct sum $N_1 = \bigoplus S(-d_\alpha)$. Next, all the relations will be the tautological ones, so that the next term in the resolution will be

$$N_2 = \bigoplus S(-d_\alpha - d_\beta),$$

where the direct sum runs over all (α, β) with $1 \leq \alpha < \beta \leq k$ and the map to N_1 carries the generator $e_{\alpha,\beta}$ in the (α, β)th factor to $(\ldots, -F_\beta, \ldots, F_\alpha, \ldots)$.

Next, the relations among the generators of N_2 are similarly easy to write down: the kernel of the map $N_2 \to N_1$ is generated by elements of the form

$$F_\gamma \cdot e_{\alpha,\beta} - F_\beta \cdot e_{\alpha,\gamma} + F_\alpha \cdot e_{\beta,\gamma}$$

for each $\alpha < \beta < \gamma$. Thus, the third term

$$N_3 = \bigoplus S(-d_\alpha - d_\beta - d_\gamma).$$

The general statement is then what you'd expect on the basis of these first steps: the ith term N_i in the minimal free resolution of a complete intersection $X \subset \mathbb{P}^n$ with ideal $I(X) = (F_1, \ldots, F_k)$ is the module

$$N_i = \bigoplus S(-d_{\alpha_1} - d_{\alpha_2} - \cdots - d_{\alpha_i}),$$

the sum ranging over all $\alpha_1 < \cdots < \alpha_i$ and the map sending the generator $e_{\alpha_1,\ldots,\alpha_i}$ to the sum $\Sigma(-1)^j F_{\alpha_j} \cdot e_{\alpha_1,\ldots,\hat{\alpha}_j,\ldots,\alpha_i}$. This sequence of modules is called the *Koszul complex* of X.

We can give a more intrinsic description of the Koszul complex of a complete intersection in terms of the first module

$$M = \bigoplus S(-d_\alpha).$$

We observe that the k-tuple of polynomials $(F_1, \ldots, F_k)$ determines a map $M \to S$, and so an element of the dual module M^*. This in turn gives us a series of contraction mappings

$$\cdots \to \wedge^3 M \to \wedge^2 M \to M \to S$$

and this is the Koszul complex.

Exercise 13.17. Use the preceding and Exercise 13.1 to prove a weak form of the *Bézout theorem* in $\mathbb{P}^2$: if F and G are polynomials of degrees d and e on $\mathbb{P}^2$ without common factors such that F and G generate the ideal of their intersection, then that intersection consists of $d \cdot e$ points. Similarly, show that if $\Gamma \subset \mathbb{P}^3$ is a complete intersection of surfaces of degrees d, e, and f then Γ consists of $d \cdot e \cdot f$ points.

Exercise 13.18. Find the Hilbert polynomial of a complete intersection in $\mathbb{P}^3$ of surfaces of degrees d and e. In case this intersection is a smooth curve, use this to say (based on Example 13.7) what the genus of the corresponding Riemann surface is.

To summarize, in this lecture we have associated to a variety $X \subset \mathbb{P}^n$ three collections of data. The most detailed information is the collection of integers $\{a_{i,j}\}$; these are sometimes called the *Betti numbers* of the variety X. The Betti numbers determine the Hilbert function, and this in turn determines the Hilbert polynomial.

As you might expect, the finer the information the more difficult it is to obtain; in particular, while the Hilbert polynomial of most varieties is relatively accessible, there are relatively few varieties whose Betti numbers can be calculated. By way of an example, the Hilbert polynomial of an arbitrary collection of d points in $\mathbb{P}^n$ is of course trivial; the Hilbert function is easy to determine for d general points and quite difficult in general; we do not even know the Betti numbers of a collection of d general points in $\mathbb{P}^n$ for $d \geq 2n$. In particular, finding the resolution of the ideal of a variety X is not an effective way to find things out about X, except in the realm of machine calculation; for the most part it is more a source of interesting questions than of answers.

LECTURE 14

Smoothness and Tangent Spaces

The Zariski Tangent Space to a Variety

The basic definition of a smooth point of an algebraic variety is analogous to the corresponding one from differential geometry. We start with the affine case; suppose $X \subset \mathbb{A}^n$ is an affine variety of pure dimension k, with ideal $I(X) = (f_1, \ldots, f_l)$. Let M be the $l \times n$ matrix with entries $\partial f_i/\partial x_j$. Then it's not hard to see that the rank of M is at most $n - k$ at every point of X, and we say a point $p \in X$ is a *smooth point* of X if the rank of the matrix M, evaluated at the point p, is exactly $n - k$. Note that in case the ground field $K = \mathbb{C}$, this is equivalent to saying that X is a complex submanifold of $\mathbb{A}^n = \mathbb{C}^n$ in a neighborhood of p, or that X is a real submanifold of $\mathbb{C}^n$ near p. (It is not, however, equivalent, in the case of a variety X defined by polynomials f_α with real coefficients, to saying that the locus of the f_α in $\mathbb{R}^n$ is smooth; consider, for example, the origin $p = (0, 0)$ on the plane curve $x^3 + y^3 = 0$.)

If X is smooth at p, we call the kernel of the matrix M at p the tangent space to X at p, denoted $T_p(X)$ or just T_pX. To be fastidious, this should be viewed as a subspace of the tangent space to $\mathbb{A}^n$ at p, which is naturally identified with K^n. In other words, it is the space of directional derivatives $D_v = \sum a_i \cdot (\partial/\partial z_i)$ such that $(D_v f)(p) = 0$ for all $f \in I(X)$, or equivalently,

$$T_pX = \{v \in T_p(\mathbb{A}^n) = K^n \colon (df)(v) = 0 \ \forall f \in I(X)\}.$$

In general, whether or not X is smooth at p, we will call the subspace $T_pX \subset T_p(\mathbb{A}^n) = K^n$ defined in this way the *Zariski tangent space* to X at p; to rephrase our definition of smoothness, we have in general

$$\dim(T_pX) \geq \dim(X)$$

and we say that X is smooth at p if equality holds. Note also that if X is smooth at p then T_pX may also be realized as the set of tangent vectors $v'(0)$ at $v(0) = p$ to analytic arcs

$$t \mapsto v(t) = (v_1(t), \ldots, v_n(t)) \in X \subset \mathbb{A}^n.$$

There is of course some potential here for confusion, since we like to draw pictures like the one at right and call the line L in question "the tangent space to X at p." The line drawn is what we will call, when we are being careful, the *affine tangent space* to X at p; that is, the affine subspace of $\mathbb{A}^n$ through p parallel to $T_p(X)$.

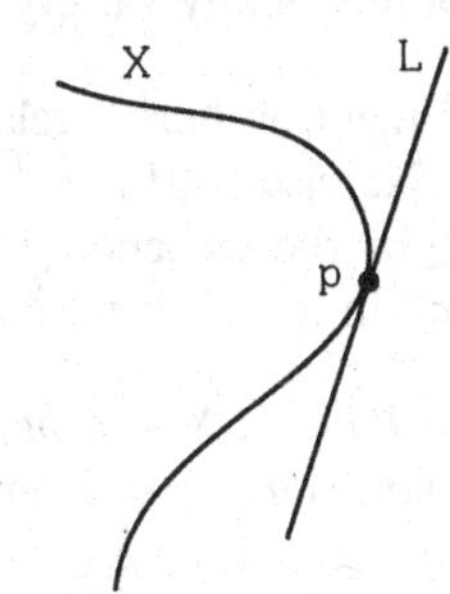

Exercise 14.1. Prove the statement that (in case $K = \mathbb{C}$) X is smooth at $p \Leftrightarrow X$ is a complex submanifold of $\mathbb{C}^n$ at $p \Leftrightarrow X$ is a real ($\mathscr{C}^\infty$) submanifold of $\mathbb{C}^n$ at p.

Exercise 14.2. Show that the dimension of the Zariski tangent space T_pX is an upper-semicontinuous function of p, in the Zariski topology on X.

Note that by this last exercise the locus of singular points of X is a subvariety of X; we will denote this subvariety X_{sing} and its complement X_{sm}. We will see in Exercise 14.3 that in fact X_{sing} is a proper subvariety, i.e., if X is any variety then X_{sm} is an open dense subset of X.

For those who want a more intrinsic definition of the Zariski tangent space (or just one that doesn't presuppose a definition of the tangent space to $\mathbb{A}^n$), we can give one as follows. First, we define the *Zariski cotangent space* to X at p, denoted $T_p^*(X)$, to be the vector space

$$T_p^*(X) = \mathfrak{m}/\mathfrak{m}^2$$

where $\mathfrak{m} \subset \mathcal{O}_{X,p}$ is the maximal ideal of functions vanishing at p. We then define the Zariski tangent space to be the dual vector space:

$$T_p(X) = (\mathfrak{m}/\mathfrak{m}^2)^*.$$

This description allows us to extend the definition of Zariski tangent space to arbitrary quasi-projective varieties.

Note that any regular map $f: X \to Y$ induces maps

$$f^*: \mathcal{O}_{Y,f(p)} \to \mathcal{O}_{X,p}$$

and, correspondingly, maps

$$df: T_p(X) \to T_{f(p)}(Y).$$

In particular, an embedding of a variety X into affine space $\mathbb{A}^n$ induces an inclusion of its Zariski tangent space at any point p in the tangent space to $\mathbb{A}^n$ at p; the image is the subspace of $T_p(\mathbb{A}^n)$ described earlier.

Exercise 14.3. (i) Show that if X is a hypersurface, then the locus of singular points of X is a proper subvariety of X. (ii) Use this, together with the fact that any variety is birational to a hypersurface (Exercise 11.23), to show that the singular points of any variety form a proper subvariety.

We should mention here a related result for maps in characteristic 0. Over the field $\mathbb{C}$, given Exercises 14.1 and 14.3, it is just Sard's theorem (and may be deduced from this case by the Lefschetz Principle 15.1). It is not so easy to prove algebraically, however; indeed, it is false over fields of characteristic $p > 0$.

Proposition 14.4. *Let $f: X \to Y$ be any surjective regular map of varieties defined over a field K of characteristic 0. Then there exists a nonempty open subset $U \subset Y$ such that for any smooth point $p \in f^{-1}(U) \cap X_{\text{sm}}$ in the inverse image of U the differential df_p is surjective.*

Exercise 14.5. In positive characteristic, even the weaker statement that there is a nonempty open subset of points $p \in X$ such that df_p is surjective may be false. To see a simple example, take $X \subset \mathbb{A}^2$ the plane curve $(y - x^p)$, $Y = \mathbb{A}^1$ the affine line with coordinate y, $f: X \to Y$ the projection $(x, y) \mapsto y$, and K any field of characteristic p. Show that in this case X and Y are smooth and f surjective, but that $df_q \equiv 0$ for all $q \in X$.

There is also a statement analogous to Exercise 14.2 for maps.

Exercise 14.6. Let $f: X \to Y$ be any regular map of affine varieties. Show that the dimension of the kernel of the differential df_p is an upper-semicontinuous function of p.

Note that this statement does not imply that the locus of smooth points of fibers of f—that is, the locus of $p \in X$ such that $X_{f(p)} = f^{-1}(f(p))$ is smooth at p—is open in X, or that the dimension of $T_p(X_{f(p)})$ is an upper-semicontinuous function of p, since the Zariski tangent space to $X_{f(p)}$ may be smaller than $\operatorname{Ker}(df_p)$. (All this is better behaved in the category of schemes; if by $X_{f(p)}$ we mean the scheme-theoretic fiber, then in fact the Zariski tangent space to $X_{f(p)}$ at p is $\operatorname{Ker}(df_p)$, and everything follows from this.)

Exercise 14.7. For an example of this phenomenon, let $X \subset \mathbb{A}^4$ be the hypersurface $(ax^2 + y^2 - b)$, $Y =$ the affine plane with coordinates (a, b) and f the projection; assume $\operatorname{char}(K) \neq 2$. Show that X is smooth of dimension 3, and that every fiber of f has dimension 1. Find the locus of points $p \in X$ such that df_p has rank <2, and verify Exercise 14.6 for this map. At the same time, find the locus of $p \in X$ such that the Zariski tangent space to $X_{f(p)}$ at p has dimension greater than one, and show that this is not closed.

The Zariski tangent space to a variety at a point p tells us more than just whether the variety is smooth at p or not; it may help to distinguish between varieties that may appear isomorphic. For example, consider the curves X and Y defined as follows. Let $X \subset \mathbb{A}^2$ be the cubic curve $xy(x-y)=0$ and $Y \subset \mathbb{A}^3$ be the union of the three coordinate axes, that is, the locus of the polynomials xy, yz, and xz. X and Y each consists of a union of three concurrent lines, and indeed the projection $\pi: \mathbb{A}^3 \to \mathbb{A}^2$ from the point at infinity on the line $\{(t, t, t)\}$ to the (x, y)-plane gives a bijection between them. The two varieties are not isomorphic, however; for one thing, the Zariski tangent space to X at its singular point $(0, 0)$ is two-dimensional, while the tangent space to Y at $(0, 0, 0)$ is three-dimensional. To express this differently (but equivalently), there is a regular function vanishing to order 2 on two of the three components of Y at the singular point of Y, but not on the third; on X no such function exists.

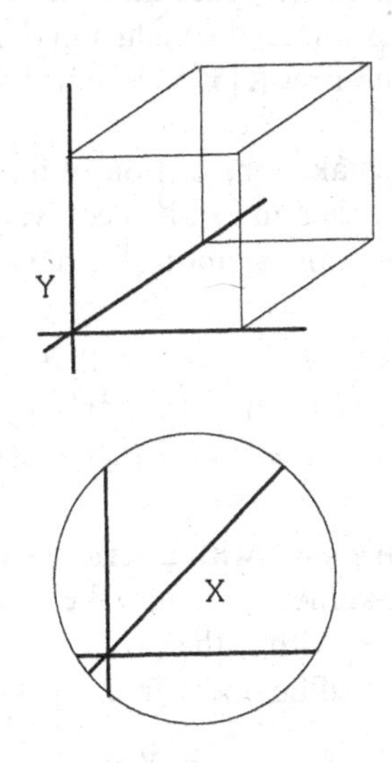

It may be worthwhile at this point to recall Exercise 13.8. In it, we computed the Hilbert polynomials of the closures $\overline{X} \subset \mathbb{P}^2$ and $\overline{Y} \subset \mathbb{P}^3$ of the curves X and Y and observed that the arithmetic genus of $\overline{X}$ (as defined in Example 13.7) is 1, while that of $\overline{Y}$ is 0. (Of course, to deduce from this that $\overline{X}$ and $\overline{Y}$ are not isomorphic we have to know that the constant term of the Hilbert polynomial is an isomorphism invariant of a projective variety, which is not proved in this text. On the other hand, as long as we are stating facts without proof, it is in fact the case that a projection $\pi_p: X \to Y$ of projective varieties from a point p not on X is an isomorphism if and only if the Hilbert polynomials p_X and p_Y coincide.)

A Local Criterion for Isomorphism

When we first introduced the notion of isomorphism, we said that a regular map $\pi: X \to Y$ between two varieties was an isomorphism if there existed an inverse regular map $\eta: Y \to X$. This is a relatively cumbersome thing to verify, especially given the awkwardness of writing down regular maps on projective varieties in coordinates. In categories such as $\mathscr{C}^\infty$ manifolds or complex manifolds, there is a much easier criterion: we can invoke the inverse function theorem in either context to show that a map $\pi: X \to Y$ is an isomorphism if it is bijective and has nonvanishing Jacobian determinant everywhere. In fact, we do have an analogous result in algebraic geometry and not just for smooth varieties.

In order to state this result in its appropriate generality, we need to introduce one further definition. We say that a map $\pi: X \to Y$ is *finite* if for every point $q \in Y$

there is an affine neighborhood U of q in Y such that $V = \pi^{-1}(U) \subset X$ is affine, and the pullback map

$$\pi^*: A(U) \to A(V)$$

expresses $A(V)$ as a finitely-generated module over $A(U)$. Note that this condition is local in Y; it implies that the fibers of π are finite but is certainly not implied by this (for example, the inclusion of $\mathbb{A}^1 - \{0\}$ in $\mathbb{A}^1$ certainly has finite fibers, but the coordinate ring $K[x, x^{-1}]$ of $\mathbb{A}^1 - \{0\}$ is not a finitely generated $K[x]$-module.)

What makes the notion of finite map useful in practice is that if X and Y are projective then in fact it is equivalent to the condition that the fibers of π are finite. In fact, we can say more: we have the following Lemma.

Lemma 14.8. *Let $\pi_0: X_0 \to Y_0$ be a regular map of projective varieties. Let $Y \subset Y_0$ be any open subset, $X = (\pi_0)^{-1}(Y)$ its inverse image, and π the restriction of π_0 to X. If the fibers of π are finite then it is a finite map.*

PROOF. We start with a series of reductions. First, since the question is local in Y, we can assume Y is affine. We can also assume that X is a closed subset of $Y \times \mathbb{P}^n$ for some n, with π the projection on the first factor. Next, (restricting if necessary to smaller affine opens in Y) we can factor π into a series of maps

$$X \subset Y \times \mathbb{P}^n \to Y \times \mathbb{P}^{n-1} \to \cdots \to Y \times \mathbb{P}^1 \to Y$$

where each map is given by projection from a point in $\mathbb{P}^k$; since a composition of finite maps is finite it is enough to prove it for each of these maps. Finally, using the birational isomorphism of $\mathbb{P}^k$ with $\mathbb{P}^{k-1} \times \mathbb{P}^1$ and again replacing the image variety by affine open subsets, we can take each of these maps to be of the form

$$X \subset Y \times \mathbb{P}^1 \to Y;$$

we will prove the assertion for such maps.

Now let $q \in Y$ be any point. The hypothesis that $\pi: X \to Y$ has finite fibers means simply that X does not contain the fiber $\{q\} \times \mathbb{P}^1$ over q; choose $\lambda \in \mathbb{P}^1$ such that $(q, \lambda) \notin X$. Restricting to an affine open contained in the complement of $\pi(X \cap (Y \times \{\lambda\}))$, we can assume that $X \cap (Y \times \{\lambda\}) = \emptyset$. Let Z and W be homogeneous coordinates on $\mathbb{P}^1$ with λ the point $(W = 0)$, and let $z = Z/W$ be the corresponding affine coordinate on $\mathbb{P}^1 - \{\lambda\} \cong \mathbb{A}^1$. In some affine open subset of Y, the ideal of

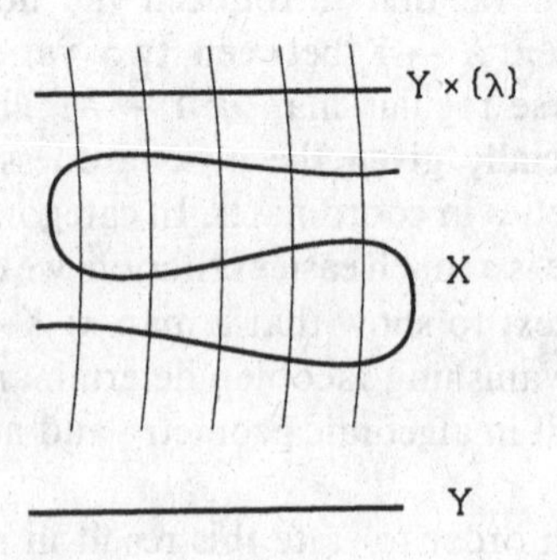

X in $Y \times \mathbb{P}^1$ will be generated by polynomials F of the form

$$F(Z, W) = a_0 Z^n + a_1 Z^{n-1} W + \cdots + a_{n-1} Z W^{n-1} + a_n W^n$$

with $a_i \in A(Y)$ in the coordinate ring of Y. Now, since $(q, \lambda) \notin X$, there must be one such F with $a_0(q) \neq 0$. Restricting (for the last time!) to the open neighborhood $(a_0 \neq 0)$ of q, we see that the coordinate ring

$$A(X) = A(Y)[z]$$

where z satisfies a monic polynomial with coefficients in $A(Y)$; it follows that $A(X)$ is a finitely generated $A(Y)$-module. □

We can now state and prove our "inverse function theorem" for varieties.

Theorem 14.9. *Let π: $X \to Y$ be a finite map of varieties. Then π is an isomorphism if and only if it is bijective and the map*

$$d\pi \colon T_p(X) \to T_{\pi(p)}(Y)$$

is an injection for all $p \in X$.

This theorem will be applied for the most part in conjunction with Lemma 14.8, in other words as the following Corollary.

Corollary 14.10. *Let X_0 be a projective variety and π_0: $X_0 \to \mathbb{P}^n$ any map. Let $U \subset \mathbb{P}^n$ be an open subset, $X = (\pi_0)^{-1}(U)$ its inverse image in X_0 and π the restriction of π_0 to X. If π is one-to-one and $d\pi_p$: $T_p X \to T_{\pi(p)} \mathbb{P}^n$ is an injection for all $p \in X$, then π is an isomorphism of X with its image. In particular, a bijection π: $X \to Y$ of projective varieties with injective differential everywhere is an isomorphism.*

Proof of Theorem 14.9. The proof involves somewhat more commutative algebra than we have been using; all of this may be found in [AM] or [E]. The key ingredient is just Nakayama's lemma.

First, we may reduce to the case where X and Y are affine, so that the map π is given by a ring homomorphism π^*: $A = A(Y) \to B = A(X)$. Note that since the map π is surjective, the map π^* in an injection (so that for the future we will consider A as a subring of B); we want, of course, to say that it is an isomorphism.

By the Nullstellensatz (Theorem 5.1), what the bijectivity of π tells us is this: we have a bijection between maximal ideals $\mathfrak{m} \subset A$ and $\mathfrak{n} \subset B$, given in one direction by sending the ideal $\mathfrak{n} \subset B$ to its intersection $\mathfrak{m} = \mathfrak{n} \cap A \subset A$ and in the other direction by sending a maximal ideal $\mathfrak{m} \subset A$ to the unique maximal ideal $\mathfrak{n} \subset B$ containing the ideal $\mathfrak{m}B$ generated by $\mathfrak{m}$ in B. Indeed, to show that π is an isomorphism, it is enough to show it induces an isomorphism between the localizations $A_{\mathfrak{m}}$ and $B_{\mathfrak{n}}$ for all $\mathfrak{m} \subset A$ and $\mathfrak{n} \supset \mathfrak{m}B$; so we can assume A and B are local rings with maximal ideals $\mathfrak{m}$ and $\mathfrak{n}$.

Note that the hypothesis of bijectivity does not insure that $\mathfrak{m}B = \mathfrak{n}$ for such a pair (consider, for example, the projection map described on page 177 between the coordinate axes in $\mathbb{A}^3$ and three concurrent lines in $\mathbb{A}^2$, or the map from the affine line to the cuspidal plane cubic given by $t \mapsto (t^2, t^3)$). The hypothesis that the differential of π is injective, on the other hand, does imply this: it says that the map

$$\mathfrak{m}/\mathfrak{m}^2 \to \mathfrak{n}/\mathfrak{n}^2$$

induced by the inclusion $\mathfrak{m} \hookrightarrow \mathfrak{n}$ is surjective, so that

$$\mathfrak{m}B + \mathfrak{n}^2 = \mathfrak{n}$$

as B-modules. By Nakayama's lemma applied to the B-module $\mathfrak{n}/\mathfrak{m}B$ it follows that $\mathfrak{m}B = \mathfrak{n}$. But now we have

$$B/A \otimes A/\mathfrak{m} = B/(\mathfrak{m}B + A) = B/(\mathfrak{n} + A) = 0$$

and applying Nakayama's lemma to the finitely-generated A-module B/A we deduce that $B/A = 0$. □

To see why some condition like the finiteness of π or the projectivity of X is necessary (apart from its use in the last sentence of the proof), consider the following example. Take X the disjoint union of the lines $(x = z = 0)$, $(y = z = 0)$ and the complement of the point $(0, 0, 1)$ in the line $(y - x = z - 1 = 0)$ in $\mathbb{A}^3$, Y the union of the lines $(x = 0)$, $(y = 0)$ and $(y - x = 0)$ in $\mathbb{A}^2$, and π the projection map $(x, y, z) \mapsto (x, y)$.

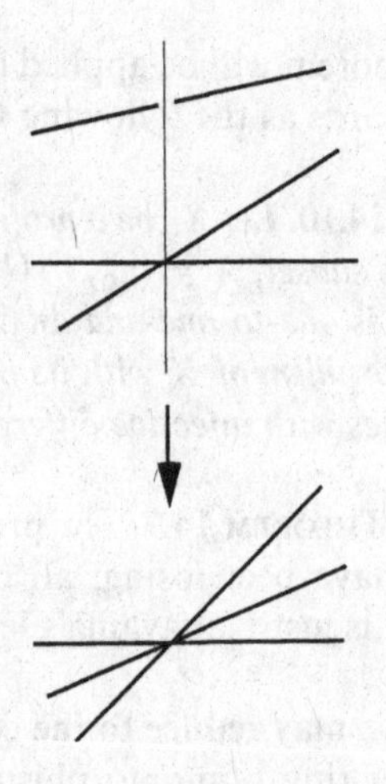

Finally, we should remark that the subtlety of Theorem 14.8—and the need for the commutative algebra in its proof—comes from the singular case. If X and Y are smooth, the result is intuitively clear, but in general it is far less so. For example, consider the three curves X, Y, and Z where $X = \mathbb{A}^1$ with coordinate t; Y is the plane cubic $y^2 = x^3$; and Z is the plane quartic curve $y^3 = x^4$. We have maps $f: X \to Y$ and $g: X \to Z$ defined by

$$t \mapsto (t^2, t^3)$$

and

$$t \mapsto (t^3, t^4)$$

respectively; these are clearly not isomorphisms because X is smooth, while the Zariski tangent spaces to Y and Z at the origin are two-dimensional. But now observe that the map g actually factors through f to give a regular map $h: Y \to Z$, which is a bijection such that the Zariski tangent spaces to Y and Z at corresponding points $p \in Y$ and $q \in Z$ have the same dimension; this map is also not an isomorphism, but it is less clear from the geometry of the curves why it cannot induce an isomorphism on tangent spaces at the origin.

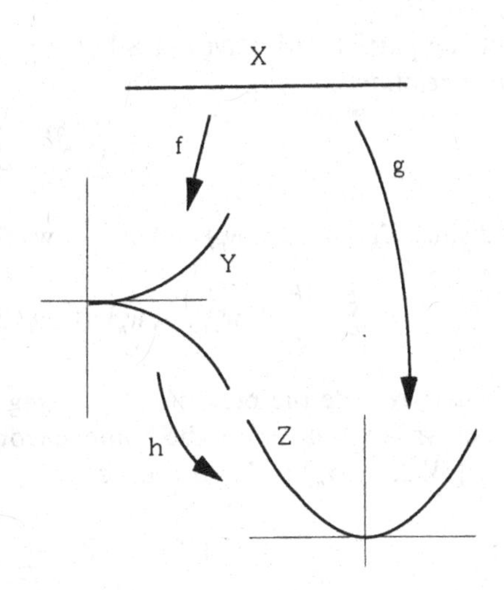

Exercise 14.11. (i) Verify directly that the map h does not induce an isomorphism between the tangent spaces to Y and Z at the origin (and in particular is not an isomorphism). (ii) Show that in fact Y and Z are not isomorphic. (*)

Projective Tangent Spaces

We have associated to a point p on an affine variety $X \subset \mathbb{A}^n$ both a Zariski tangent space T_pX, which is an abstract vector space, and an affine tangent space, which is an affine linear subspace of the ambient $\mathbb{A}^n$. Consider now a projective variety $X \subset \mathbb{P}^n$. We may also associate to it a projective tangent space at each point $p \in X$, denoted $\mathbb{T}_p(X)$, which is actually a projective subspace of the ambient $\mathbb{P}^n$. One way to describe this is to choose an affine open subset $U \cong \mathbb{A}^n \subset \mathbb{P}^n$ containing p and define the projective tangent space to X at p to be the closure in $\mathbb{P}^n$ of the affine tangent space at p of the affine variety $X \cap U \subset U \cong \mathbb{A}^n$.

This may seem a fairly cumbersome description, and it is; fortunately, there is a more direct one. To begin with, suppose that X is the hypersurface given by the homogeneous polynomial $F(Z)$; let $U \cong \mathbb{A}^n$ be the affine open $(Z_0 \neq 0)$ with Euclidean coordinates $z_i = Z_i/Z_0$ and set

$$f(z_1, \ldots, z_n) = F(1, z_1, \ldots, z_n)$$

so that $X \cap U$ is the zero locus of f. Now, at a point $p \in X$ with coordinates $(w_1, \ldots, w_n)$, the affine tangent space is given as the locus

$$\left\{(z_1, \ldots, z_n): \sum_{i=1}^{n} \frac{\partial f}{\partial z_i}(p) \cdot (z_i - w_i) = 0\right\}.$$

By our definition, then, the projective tangent space is given as the locus of homogeneous vectors

$$\left\{[Z_0, \ldots, Z_n]: \sum_{i=1}^{n} \frac{\partial F}{\partial Z_i}(1, w_1, \ldots, w_n) \cdot (Z_i - w_i \cdot Z_0) = 0\right\}.$$

But the partial derivatives of a homogeneous polynomial of degree d satisfy the *Euler relation*

$$\sum_{i=0}^{n} \frac{\partial F}{\partial Z_i} \cdot Z_i = d \cdot F$$

and since $F(1, w_1, \ldots, w_n) = 0$, it follows that

$$\sum_{i=0}^{n} \frac{\partial F}{\partial Z_i}(1, w_1, \ldots, w_n) \cdot (-w_i \cdot Z_0) = \frac{\partial F}{\partial Z_0}(1, w_1, \ldots, w_n) \cdot Z_0.$$

We may rewrite the preceding by saying that the projective tangent space to the hypersurface X given by the homogeneous polynomial $F(Z_0, \ldots, Z_n)$ at the point $P = [W_0, \ldots, W_n] \in X$ is the locus

$$\left\{[Z_0, \ldots, Z_n]: \sum_{i=1}^{n} \frac{\partial F}{\partial Z_i}(P) \cdot Z_i = 0\right\}.$$

In other words, if all the partial derivatives of F vanish at P, then X is singular at P with projective tangent space all of $\mathbb{P}^n$; if not, then X is smooth at P and the projective tangent space is the hyperplane corresponding to the point of $\mathbb{P}^{n*}$ with coordinates the partial derivatives of F at P. Note in particular that, in view of the Euler relation, if the characteristic of the field K is 0 (or just prime to d) then the singular locus of the hypersurface X is exactly the common zero locus in $\mathbb{P}^n$ of the partial derivatives of F.

Having thus described the projective tangent space to a hypersurface in $\mathbb{P}^n$, we may now describe the projective tangent space to an arbitrary subvariety $X \subset \mathbb{P}^n$ as the intersection of the projective tangent spaces to all the hypersurfaces containing it. In particular, if $\{F_1, \ldots, F_m\}$ is a collection of homogeneous polynomials generating the ideal of X then the projective tangent space will be just the subspace of $\mathbb{P}^n$ given by the kernel of the matrix $(\partial F_i/\partial Z_j)$, viewed as a map from K^{n+1} onto K^m.

In case X is smooth at p and our ground field $K = \mathbb{C}$, we may also describe the projective tangent space $\mathbb{T}_p X$ to a variety $X \subset \mathbb{P}^n$ in terms of a parametric representation of a neighborhood of p in X (in the analytic topology). Simply, say that an analytic neighborhood U of p is given as the image of a map

$$\varphi: \Delta \to \mathbb{P}^n$$

defined on an open set $\Delta \subset \mathbb{C}^k$ by a vector-valued function $v(z)$:

$$\varphi: (z_1, \ldots, z_k) \mapsto [v(z)] = [v_0(z), \ldots, v_n(z)].$$

Then the projective tangent space to X at $p = [v(0)]$ is given as the subspace $\mathbb{P}^k \subset \mathbb{P}^n$ associated to the multivector

$$\mathbb{T}_p X = \left[v(0) \wedge \frac{\partial v}{\partial z_1}(0) \wedge \cdots \wedge \frac{\partial v}{\partial z_k}(0) \right] \in \mathbb{G}(k, n) \subset \mathbb{P}(\wedge^{k+1}\mathbb{C}^{n+1});$$

that is, the row space of the matrix

$$\begin{bmatrix} v_0(0) & v_1(0) & \dots & v_n(0) \\ \frac{\partial v_0}{\partial z_1}(0) & \dots\dots\dots\dots & & \frac{\partial v_n}{\partial z_1}(0) \\ \frac{\partial v_0}{\partial z_k}(0) & \dots\dots\dots\dots & & \frac{\partial v_n}{\partial z_k}(0) \end{bmatrix}.$$

Exercise 14.12. Verify that the row space of this matrix is the projective tangent space to X at p.

It will often be helpful to describe projective tangent spaces in this way, especially in dealing with curves and surfaces. In particular, it allows us to describe the tangent space to a variety in terms of arcs; the projective tangent space $\mathbb{T}_p X$ to $X \subset \mathbb{P}^n$ at a smooth point $p \in X$ is the union of the lines $[v(0) \wedge v'(0)] \in \mathbb{G}(1, n)$ where $v(t)$ is any analytic arc on X with $v(0) = p$.

This may seem tied down to the analytic topology, but in fact the same idea can be carried out in a purely algebraic setting (at least in characteristic zero) by introducing and working with power series rings. One thing this approach is not (in general) useful for, however, is describing the Zariski tangent spaces to singular varieties.

Needless to say, there is an enormous potential for confusion between the Zariski tangent space to a variety $X \subset \mathbb{P}^n$ at a point p and the projective tangent space, the more so as each is referred to in different sources as "the tangent space to X at p" and denoted by the same symbol $T_p(X)$. Linguistically, we will try to deal with this by using the modifiers "Zariski" and "projective" whenever confusion seems likely. Notationally the problem is worse (we can't, for example, use the symbol $\mathbb{P}T_p(X)$ for the projective tangent space, since it a priori signifies the projective space of one-dimensional subspaces of the Zariski tangent space $T_p(X)$, a space we will want to use as well on occasion). For want of a better solution, we have chosen to denote the projective tangent space as $\mathbb{T}_p(X)$.

There is one more way to describe the projective tangent space to a variety $X \subset \mathbb{P}^n$ at a point $p \in X$. Let $\tilde{X} \subset K^{n+1}$ be the cone over X (i.e., the closure of the set of all $v \in K^{n+1}$ with $[v] \in X$, or equivalently the zero locus in K^{n+1} of the ideal $I(X) \subset K[Z_0, \dots, Z_n]$), and let $v \in \tilde{X}$ be any nonzero vector $v \in K^{n+1}$ lying over p. Then the projective tangent space $\mathbb{T}_p X$ is just the subspace of $\mathbb{P}^n$ corresponding to the Zariski tangent space $T_v \tilde{X} \subset T_v(K^{n+1}) = K^{n+1}$. This characterization is readily seen to be equivalent to each of the ones above, in particular to the description of $\mathbb{T}_p X$ in terms of the partials of homogeneous polynomials F_α generating $I(X)$.

This description also clarifies the relation between the projective and Zariski tangent spaces. Specifically, let $X \subset \mathbb{P}^n$ be any variety and $p \in X$ any point; and suppose that $\Lambda \subset K^{n+1}$ is the subspace corresponding to the projective tangent space $\mathbb{T}_p X$. Then by the above, Λ contains the one-dimensional subspace $K \cdot v$, and up to scalars we have a natural identification of the Zariski tangent space to X at p with $\Lambda/K \cdot v$. In particular, the normal space $N_p(X/\mathbb{P}^n)$ (or just $N_p(X)$) to X at p, defined a priori to be the quotient

$$N_p(X) = T_p(\mathbb{P}^n)/T_p(X),$$

may also be realized as the quotient space K^{n+1}/Λ.

(To be more precise, and to explain the "up to scalars," we will see in Lecture 16 that we have natural identifications

$$T_p(\mathbb{P}^n) = (K^{n+1}/K \cdot v) \otimes (K \cdot v)^*,$$

$$T_p(X) = (\Lambda/K \cdot v) \otimes (K \cdot v)^*,$$

and correspondingly

$$N_p(X) = (K^{n+1}/\Lambda) \otimes (K \cdot v)^*.)$$

In the following exercises, we verify the condition of smoothness (or singularity) for some of the varieties introduced in *I*.

Exercise 14.13. Use Corollary 14.10 to show that the Veronese and Segre varieties are smooth and verify this directly by looking at their defining equations. (Alternatively, given Exercise 14.3 this follows without calculation from their homogeneity.)

Exercise 14.14. Show that the rational normal scroll $X_{a_1,\ldots,a_k} \subset \mathbb{P}^n$ is smooth if and only if all $a_i > 0$.

Exercise 14.15. Let $X \subset \mathbb{P}^n$ be any irreducible nondegenerate variety i.e., such that X is not contained in any hyperplane in $\mathbb{P}^n$, and let $S(X)$ be its secant variety. Show that the projective tangent space to $S(X)$ at a point of X is all of $\mathbb{P}^n$ (and in particular $S(X)$ is singular along X if it is not equal to $\mathbb{P}^n$).

Example 14.16. Determinantal Varieties

Recall first the definition of the generic determinantal varieties M_k from Lecture 9: we let M denote the projective space $\mathbb{P}^{mn-1}$ of $m \times n$ matrices, up to scalars, and we let $M_k \subset M$ be the locus of matrices of rank k or less. In Example 12.1, we saw that M_k is an irreducible variety of codimension $(m-k)(n-k)$ in M; here we will determine its smooth and singular loci and describe its Zariski tangent space at both. We will do this explicitly, by introducing coordinates on M and examining the equations defining M_k; later, in Example 16.18, we will see a more intrinsic version.

Let us start by choosing a point $A \in M_k$ corresponding to a matrix of rank exactly k, i.e., $A \in M_k - M_{k-1}$. (Note that since the group $\mathrm{PGL}_m K \times \mathrm{PGL}_n K$ acts on M with orbits $M_l - M_{l-1}$, the smoothness or singularity of a point $A \in M_k$ can depend only on its rank.) We can choose bases for K^m and K^n so that A is represented by the matrix

$$\begin{bmatrix} 1 & 0 & 0 & \cdots\cdots & & 0 \\ 0 & 1 & 0 & \cdots\cdots & & 0 \\ \vdots & & & \cdots\cdots & & \\ 0 & \cdots & 1 & 0 & \cdots & 0 \\ 0 & \cdots & 0 & 0 & \cdots & 0 \\ \vdots & & & & & \\ 0 & \cdots\cdots & & & & 0 \end{bmatrix}$$

and then, in the affine neighborhood U of A given by $(X_{11} \neq 0)$, we may take as Euclidean coordinates the functions $x_{i,j} = X_{i,j}/X_{1,1}$; in terms of these we may write a general element of U as

$$\begin{bmatrix} 1 & x_{1,2} & x_{1,3} & \cdots\cdots & & & x_{1,m} \\ x_{2,1} & 1 + x_{2,2} & x_{2,3} & \cdots\cdots & & & x_{2,m} \\ \vdots & & & & & & \\ x_{l,1} & \cdots\cdots & & 1 + x_{l,l} & x_{l,l+1} & \cdots & x_{l,m} \\ x_{l+1,1} & \cdots\cdots & & x_{l+1,l} & x_{l+1,l+1} & \cdots & x_{l+1,m} \\ \vdots & & & & & & \\ x_{n,1} & \cdots\cdots & & & & & x_{n,m} \end{bmatrix}$$

with A of course corresponding to the origin in this coordinate system.

Consider first the case $l = k$. What are the linear terms of the $(k + 1) \times (k + 1)$ minors of this matrix? The answer is that the only such minors with nonzero differential at the origin A are those involving the first k rows and columns, so their linear terms are exactly the coordinates $x_{i,j}$ with $i, j > k$. Since there are exactly $(m - k)(n - k)$ of these, we may conclude that M_k is smooth at a point of $M_k - M_{k-1}$. Next, we see that if $l < k$, then in fact no $(k + 1) \times (k + 1)$ minors have any linear terms. Invoking the fact that these minors generate the ideal of M_k (stated but not proved earlier), we conclude that the projective tangent space to M_k at a point $A \in M_{k-1}$ is all of M.

We can give an intrinsic interpretation to the tangent space to M_k at a smooth point $A \in M_k - M_{k-1}$. To do this, note that we have, in terms of the coordinates $x_{i,j}$, identified the tangent space to M_k at A with the space of matrices whose lower right $(m - k) \times (n - k)$ submatrix is zero. But the bases $\{e_i\}$, $\{f_j\}$ for K^n and K^m are chosen so that the kernel of A is exactly the span of $e_{k+1}, \ldots, e_n$ and the image of A exactly the span of $f_1, \ldots, f_k$. We have thus

$$\mathbb{T}_A(M_k) = \mathbb{P}\{\varphi \in \operatorname{Hom}(K^n, K^m) : \varphi(\operatorname{Ker}(A)) \subset \operatorname{Im}(A)\};$$

i.e., we may say, without invoking coordinates, that the projective tangent space to M_k at a point A corresponding to a map $A: K^n \to K^m$ of rank exactly k is the linear space of maps $\varphi: K^n \to K^m$ carrying the kernel of A into the image of A.

LECTURE 15

Gauss Maps, Tangential and Dual Varieties

In the preceding lecture, we associated to each point of a projective variety $X \subset \mathbb{P}^n$ a linear subspace of $\mathbb{P}^n$. We investigate here how those planes vary on X, that is, the geometry of the *Gauss map*. Before we launch into this, however, we should take a moment to discuss a question that will be increasingly relevant to our analysis; the choice of our ground field K and in particular its characteristic.

A Note About Characteristic

Through Lecture 13, most of the statements we made were valid over an arbitrary (algebraically closed) ground field K. When we start talking about tangent spaces, however—and especially how the projective tangent spaces to a variety $X \subset \mathbb{P}^n$ vary—the situation depends very much on the characteristic of the field K. The reason is simple: in characteristic $p > 0$ a function can have all derivatives identically zero and not be constant. Thus, for example, the statement that not all tangent lines to a plane curve $C \subset \mathbb{P}^2$ can contain a given point $p \in \mathbb{P}^2$, self-evident if we trust our $\mathbb{R}$- or $\mathbb{C}$-based intuition and in any event easy to prove in characteristic 0, is in fact false in characteristic p. (Take, for example, C given by $XY - Z^2$, $p = [0, 0, 1]$ and $\operatorname{char}(K) = 2$.)

This is not to say that the subjects we will discuss in the next few lectures cannot be dealt with in finite characteristic. To do so, though, requires more advanced techniques than we have at our disposal. For our present purposes, then, we will go ahead and use the techniques of calculus, accepting the fact that this limits us, for the most part, to proofs valid only in characteristic zero. We will try to indicate in each statement whether it does in fact hold in arbitrary characteristic. (We should also say that starting in Lecture 18 we will go back to an essentially characteristic-free approach.)

There is a further restriction that will be extremely useful: if we assume that our ground field is actually $\mathbb{C}$, we can use the techniques of complex manifold theory and the theory of functions of several complex variables. We will not do this in any deep way, but it will be very handy, most notably in allowing us to parametrize varieties locally; if $X \subset \mathbb{P}^n$ is a k-dimensional variety and $p \in X$ a general point, we can describe a neighborhood U of p in X as the image of the analytic map

$$\varphi: \Delta \to X \subset \mathbb{P}^n$$
$$: (z_1, \ldots, z_k) \mapsto [f_0(z), \ldots, f_n(z)]$$

where $\Delta \subset \mathbb{C}^k$ is a polydisc and $f_0, \ldots, f_n$ are holomorphic functions. Clearly, no such parametrization exists in the category of algebraic varieties unless X is unirational.[3]

It may seem like a much more restrictive assumption to take $K = \mathbb{C}$ than merely to assume $\operatorname{char}(K) = 0$. This, as it turns out, is not the case. That it is not is known as the *Lefschetz principle*; it is, roughly, the following.

Lefschetz Principle 15.1. *A theorem in algebraic geometry involving a finite collection of algebraic varieties and maps that holds over $\mathbb{C}$ holds over an arbitrary algebraically closed field K of characteristic zero.*

To see why this is so, observe that any finite collection of varieties and maps is specified by a finite collection of polynomials; these are in turn given by their coefficients, which form a finite subset $\{c_\alpha\}$ of K. As far as the truth of a given statement goes, then, it makes no difference if we replace the field K with the algebraic closure $L = \overline{\mathbb{Q}(c_\alpha)} \subset K$ of the subfield $\mathbb{Q}(c_\alpha)$ generated by the c_α over $\mathbb{Q}$. At the same time, since the trancendence degree of $\mathbb{C}$ over $\mathbb{Q}$ is infinite, this subfield of K can also be embedded in $\mathbb{C}$; if we know the theorem to hold over $\mathbb{C}$, it follows that it holds over L and hence over K.

This argument should clarify what is meant by the phrase "theorem in algebraic geometry": essentially any theorem that asserts the existence of a solution of a system of polynomial equations or the existence of a variety or map with properties expressible in terms of solutions to systems of polynomial equations qualifies. This includes every theorem stated so far in this book; for example, Proposition 14.4, which as we observed is simply a case of Sard's theorem in case $K = \mathbb{C}$, may be deduced for arbitrary algebraically closed K of characteristic 0.

We have avoided using complex analytic techniques so far for two reasons: they were not necessary and they would require us to restrict to characteristic

[3] One way to make an analogous construction algebraically is to introduce the completion of the local ring of X at p.

0. Now that we are making the latter restriction anyway, however, we will invoke the complex numbers when it is useful to do so.

Example 15.2. Gauss Maps

The description of the projective tangent space $\mathbb{T}_p(X)$ of a variety $X \subset \mathbb{P}^n$ at a smooth point p as a linear subspace of $\mathbb{P}^n$ allows us to define the *Gauss map* associated to a smooth variety. If X is smooth of pure dimension k, this is just the map

$$\mathscr{G} = \mathscr{G}_X \colon X \to \mathbb{G}(k, n)$$

sending a point $p \in X$ to its tangent plane $\mathbb{T}_p(X)$. That this is a regular map is clear from the description of the projective tangent space as the kernel of the linear map given by the matrix of partial derivatives $(\partial F_\alpha/\partial Z_i)$, where $\{F_\alpha\}$ is a collection of generators of the ideal of X. Note that if X is singular then $\mathscr{G}$ is still defined and regular on the open set of smooth points of X (which is dense by Exercise 14.3) and so gives a rational map

$$\mathscr{G} \colon X \dashrightarrow \mathbb{G}(k, n).$$

In general, if we refer to the Gauss map of a variety $X \subset \mathbb{P}^n$ without specifying that X is smooth, we will be referring to this rational map. We will call the image $\mathscr{G}(X)$ the *variety of tangent planes* to X, or the *Gauss image* of X, and denote it $\mathscr{T}X$. Note that there is some potential for confusion here if X is singular, since the projective tangent spaces to X at singular points, being subspaces of dimension strictly greater than k, will not appear in $\mathscr{T}X$ and conversely $\mathscr{T}X$ will contain k-planes that are limits of tangent planes to X but not tangent planes themselves.

The simplest example of a Gauss map is the one associated to a hypersurface $X \subset \mathbb{P}^n$; if X is given by the homogeneous polynomial $F(Z_0, \ldots, Z_n)$, then the Gauss map $\mathscr{G}_X \colon X \to \mathbb{G}(n-1, n) = \mathbb{P}^{n*}$ is given simply by

$$\mathscr{G}_X(p) = \left[\frac{\partial F}{\partial Z_0}(p), \ldots, \frac{\partial F}{\partial Z_n}(p)\right].$$

Note that in case the degree of F is 2, this is a linear map. Indeed, as we have seen a smooth quadric hypersurface $X \subset \mathbb{P}V = \mathbb{P}^n$ may be given by a symmetric bilinear form

$$Q \colon V \times V \to K,$$

and the Gauss map is then just the restriction to X of the linear map $\mathbb{P}^n \to \mathbb{P}^{n*}$ associated to the induced isomorphism

$$\tilde{Q} \colon V \to V^*.$$

It is elementary to see from the preceding description that if $X \subset \mathbb{P}^n$ is a smooth hypersurface of degree $d \geq 2$, then the fibers of the Gauss map $\mathscr{G}_X \colon X \to \mathbb{P}^{n*}$ are finite: since the partial derivatives of F with respect to the Z_i do not vanish simultaneously at any point of X, the map $\mathscr{G}_X$ cannot be constant along a curve. It follows that the image $\mathcal{T}X$ of the Gauss map is again a hypersurface. It is a theorem of F. L. Zak, which we will not prove here, that for any smooth irreducible k-dimensional variety $X \subset \mathbb{P}^n$ other than a linear space, the Gauss map is finite, and in particular the dimension of the variety $\mathcal{T}X \subset \mathbb{G}(k, n)$ of tangent planes to X is again k ([FL]). This is in fact true in all characteristics. For curves, we have a much stronger statement, but one that holds in characteristic zero only.

Proposition 15.3. *Let $C \subset \mathbb{P}^n$ be an irreducible curve other than a line. Then the Gauss map $\mathcal{G}_C: C \to \mathbb{G}(1, n)$ is birational onto its image.*

PROOF. We will give a proof over the field $K = \mathbb{C}$ using the analytic topology.

Suppose that $\mathcal{G}_C$ is not birational. We may then find a positive-dimensional component Γ of the locus

$$\{(p, q): \mathbb{T}_p C = \mathbb{T}_q C\} \subset C_{sm} \times C_{sm}$$

other than the diagonal. Let (p, q) be a smooth point of Γ not on the diagonal, and let t be a local analytic parameter on Γ around (p, q). The points p and q on $C \subset \mathbb{P}^n$ may then be given by vector-valued functions $[v(t)]$, $[w(t)]$, respectively, with $[v(0)] \neq [w(0)]$. In these terms, the statements that $q \in \mathbb{T}_p C$ and $p \in \mathbb{T}_q C$ translate into the identities

$$v(t) \wedge w(t) \wedge v'(t) \equiv 0 \qquad \text{and} \qquad v(t) \wedge w(t) \wedge w'(t) \equiv 0.$$

(Indeed, any triple wedge product of the vectors $v(t)$, $w(t)$, $v'(t)$, and $w'(t)$ is zero.) Taking the derivatives of these equations, we have

$$v(t) \wedge w(t) \wedge v''(t) \equiv 0 \qquad \text{and} \qquad v(t) \wedge w(t) \wedge w''(t) \equiv 0.$$

Continuing in this way, we may deduce that all the derivatives $v^{(n)}(0)$ and $w^{(n)}(0)$ lie in the subspace of K^{n+1} spanned by $v(0)$ and $w(0)$, and hence that C is the line pq. □

We will see as a consequence of Theorem 15.24 that there are lots of hypersurfaces $X \subset \mathbb{P}^n$ of higher dimension (necessarily singular ones) whose Gauss images $\mathcal{T}X \subset \mathbb{P}^{n*}$ have lower dimension.

Example 15.4. Tangential Varieties

Let $X \subset \mathbb{P}^n$ be an irreducible variety of dimension k. The union

$$TX = \bigcup_{\Lambda \in \mathcal{T}X} \Lambda$$

of the k-planes corresponding to points of the image $\mathcal{T}(X)$ of the Gauss map is then also an algebraic subvariety, called the tangential variety of X. Of course, if X is smooth, this is just the union of its tangent planes; if X is singular, the description is not as clear (though we can describe it as the closure of the union of the tangent planes to X at smooth points).

What is the dimension of TX? We can answer this for the most part simply, as we did, for example, in the case of the chordal variety (Example 11.22). Precisely, we look at the incidence correspondence

$$\Sigma = \{(\Lambda, p): p \in \Lambda\} \subset \mathcal{T}X \times \mathbb{P}^n,$$

(or, alternatively, when X is smooth,

$$\Psi = \{(p, q): q \in \mathbb{T}_p(X)\} \subset X \times \mathbb{P}^n).$$

The image $\pi_2(\Sigma) \subset \mathbb{P}^n$ of Σ is the tangential variety TX. The projection map π_1 on the first factor is surjective, with all fibers irreducible of dimension k; thus Σ is irreducible of dimension at most $2k$. It follows then that the tangential variety TX of a k-dimensional variety has dimension at most $2k$, with equality holding if and only if a general point q on a general tangent plane $\mathbb{T}_p(X)$ lies on $\mathbb{T}_r(X)$ for only finitely many points $r \in X$.

Exercise 15.5. Describe the tangential surface to the twisted cubic curve $C \subset \mathbb{P}^3$. In particular, show that it is a quartic surface. What is its singular locus? (*)

Exercise 15.6. Let $X \subset \mathbb{P}^3$ be any irreducible nonplanar curve, and let Σ be the incidence correspondence introduced in Example 15.4. Show that the projection $\pi_2: \Sigma \to TX$ is birational (i.e., generically one to one). (*)

It may well happen that the tangential variety of a k-dimensional variety $X \subset \mathbb{P}^n$ has dimension strictly less than $\min(2k, n)$. The situation is very similar to that involving the secant variety, as described in Proposition 11.24 and the discussion following it: if we definite the *deficiency* of a variety $X \subset \mathbb{P}^n$ such that $TX \neq \mathbb{P}^n$ to be the difference $2 \cdot \dim(X) - \dim(TX)$, there are examples of varieties with arbitrarily large deficiency, though it is not known whether the deficiency of smooth varieties may be bounded above in general.

Example 15.7. The Variety of Tangent Lines

Having defined the projective tangent space to a variety $X \subset \mathbb{P}^n$, we can also talk about such things as tangent lines. If X is smooth, there is no ambiguity about how to define this notion; *tangent lines* are defined to be lines in $\mathbb{P}^n$ that contain a point $p \in X$ and lie in the corresponding tangent plane $\mathbb{T}_p(X)$. If X is singular, we will define the *variety of tangent lines to* X, denoted $\mathcal{T}_1(X)$, to be the closure, in the Grassmannian $\mathbb{G}(1, n)$, of the locus of lines containing a smooth point $p \in X$ and lying in $\mathbb{T}_p(X)$; equivalently, this is the set of lines L such that $p \in L \subset \Lambda$ for some pair (p, Λ) in the graph of the Gauss map $\mathcal{G}_X$.

(We should mention here that there are three other notions of a "tangent line" to a variety $X \subset \mathbb{P}^n$, all of which agree with the given one for X smooth and all of which may differ in general. We will discuss these other notions at the end of this section.)

Exercise 15.8. Show that if $X \subset \mathbb{P}^n$ is a smooth irreducible variety, then $\mathcal{T}_1(X) \subset \mathbb{G}(1, n)$ is indeed a variety.

The dimension of the locus of tangent lines to a k-dimensional variety $X \subset \mathbb{P}^n$ is fairly easy to find. We simply set up an incidence correspondence

$$\Sigma = \{(L, p): p \in L \subset \mathbb{T}_p(X)\} \subset \mathbb{G}(1, n) \times X_{sm}$$

and consider the fibers of Σ over X and over its image in $\mathbb{G}(1, n)$. The projection π_2 is surjective, with fibers isomorphic to $\mathbb{P}^{k-1}$, so that the dimension of Σ is $2k - 1$; unless X contains a general (and hence every) tangent line to itself—that is, unless X is a linear subspace of $\mathbb{P}^n$—the map π_1 will be finite to one. It follows that unless X is a linear space, the image $\pi_1(\Sigma)$, and correspondingly its closure $\mathcal{T}(X)$, has dimension $2k - 1$.

Having introduced the variety of tangent lines, we are now in a position to answer one of the questions raised in Lecture 8: what exactly is a secant line to a variety $X \subset \mathbb{P}^n$. We start with the relatively straightforward part, which we give as an exercise.

Exercise 15.9. Let $X \subset \mathbb{P}^n$ be any variety. Show that any tangent line to X is a secant line to X—that is, a point in the image $\mathcal{S}(X)$ of the rational map

$$s: X \times X \dashrightarrow \mathbb{G}(1, n)$$

sending $(p, q) \in X \times X$ to the line $\overline{pq} \in \mathbb{G}(1, n)$. (Note that it is enough to do this for a tangent line to X at a smooth point.)

In fact the converse of this exercise is also true, at least when X is smooth.

Proposition 15.10. *If $X \subset \mathbb{P}^n$ is a smooth variety, then every point of $\mathcal{S}(X)$ is either an honest secant line $\overline{pq}$ with $p \neq q \in X$ or a tangent line; in other words,*

$$\mathcal{S}(X) = s(X \times X - \Delta) \cup \mathcal{T}_1(x)$$

where $\Delta \subset X \times X$ is as usual the diagonal.

PROOF. To see this, we will actually write down a set of equations for $\mathcal{S}(X)$ as follows. First, choose $F_1, \dots, F_l \in K[Z_0, \dots, Z_n]$ a collection of polynomials generating the ideal $I(X)$ of X locally; we can take all the F_α to have the same degree d. Now, we can represent all lines $L \subset \mathbb{P}^n$ in an open subset U of the Grassmannian $\mathbb{G}(1, n)$ by matrices of the form

$$\begin{pmatrix} 1 & 0 & a_2 & a_3 & \dots & a_n \\ 0 & 1 & b_2 & b_3 & \dots & b_n \end{pmatrix}$$

or, parametrically, as

$$[S, T] \mapsto [S, T, a_2 S + b_2 T, a_3 S + b_3 T, \dots, a_n S + b_n T].$$

We can write the restriction of each polynomial F_α to the line $L_{a,b}$ as

$$F_\alpha|_{L_{a,b}} = p_{\alpha,d}(a, b)S^d + p_{\alpha,d-1}(a, b)S^{d-1}T + \cdots + p_{\alpha,0}(a, b)T^d,$$

where the coefficients $p_{\alpha,i}$ are polynomials in a and b. Now, let m be an arbitrary large integer, S_{m-d} the space of homogeneous polynomials of degree $m - d$ in $Z_0, \ldots, Z_n$ and V the space of polynomials of degree m in S and T. For any value of m and any a and b we can define a map

$$\varphi_{a,b}: (S_{m-d})^l \to V$$

by sending an l-tuple of homogeneous polynomials $G_1, \ldots, G_l$ of degree $m - d$ into the sum

$$(G_1, \ldots, G_l) \mapsto \sum (F_\alpha G_\alpha)|_{L_{a,b}}.$$

This is a linear map whose matrix entries are polynomials in the coordinates a and b on $U \subset \mathbb{G}(1, n)$. Moreover, if $L_{a,b}$ is an honest secant $\overline{pq}$ of the variety X, then since all the polynomials F_α have two common zeros $p, q \in L$, the image of this map will lie in the subspace of V consisting of polynomials vanishing at these two points. In particular, it follows that for (a, b) such that $L_{a,b} \in \mathcal{S}(X)$, the map $\varphi_{a,b}$ will have rank at most $m - 1$; or, in other words, the $m \times m$ minors of a matrix representative of $\varphi_{a,b}$, viewed as regular functions on the open subset $U \subset \mathbb{G}(1, n)$, all vanish on the variety $\mathcal{S}(X)$.

Conversely, let $\mathcal{S}'$ denote the locus of lines L such that the image of the map $\varphi_{a,b}$ has codimension 2 for all m. It is not hard to see that for $L_{a,b} \in \mathcal{S}'$ the image of $\varphi_{a,b}$ will consist either of all polynomials vanishing at two points $p, q \in L_{a,b}$ or of all polynomials vanishing to order 2 at some point $p \in L_{a,b}$. In the former case, $L = \overline{pq}$ is an honest secant of X; while in the latter case (as long as X is smooth) it follows that $L_{a,b} \subset \mathbb{T}_p X$, so that $L_{a,b} \in \mathcal{T}_1(X)$. We deduce that

$$\mathcal{S}' = s(X \times X - \Delta) \cup \mathcal{T}_1(X),$$

and by comparing dimensions we may deduce as well that $\mathcal{S}' = \mathcal{S}(X)$. □

Exercise 15.11. An alternative argument for the proposition is as follows: Show that the secant line map

$$s: X \times X \dashrightarrow \mathbb{G}(1, n)$$

extends to a regular map

$$\tilde{s}: \mathrm{Bl}_\Delta(X \times X) \to \mathbb{G}(1, n)$$

on the blow-up of $X \times X$ along the diagonal. Then show that the map $\tilde{s}$ carries the fiber of $\mathrm{Bl}_\Delta(X \times X) \to X \times X$ over a point $(p, p) \in \Delta$ to the projective space of lines through p in $\mathbb{T}_p X$.

Exercise 15.12. Show by example that Proposition 15.10 is false if we do not assume X smooth.

Note that since by Exercise 15.9 the tangential variety TX of a variety $X \subset \mathbb{P}^n$ is contained in its secant variety, the examples given in Exercises 11.26 and 11.27 of varieties with deficient secant varieties also furnish examples of varities with deficient tangential varieties.

The fact that the tangential variety TX of a variety $X \subset \mathbb{P}^n$ is contained in its secant variety has one other significant consequence: combining this with Theorem 14.8, we have a classical embedding theorem for smooth varieties $X \subset \mathbb{P}^n$. Specifically, we observe that the projection map $\pi_p\colon X \to \mathbb{P}^{n-1}$ of X from a point $p \in \mathbb{P}^n$ is one to one if and only if p lies on no proper chord to X (that is, a line $\overline{qr}$ with q, $r \in X$ distinct); and by Exercise 14.16 it induces injections from the Zariski tangent spaces to X to those of the image $\overline{X} = \pi_p(X) \subset \mathbb{P}^{n-1}$ if and only if p lies in no projective tangent plane to X. It follows then from Corollary 14.10 that $\pi_p\colon X \to \overline{X}$ is an isomorphism if and only if p does not lie on the chordal variety $S(X)$ of X. Since, finally, the dimension of $S(X)$ is at most twice the dimension of X plus one, we conclude that whenever $n > 2\cdot\dim(X) + 1$, X admits a biregular projection to $\mathbb{P}^{n-1}$. In particular, we have the following theorem.

Theorem 15.13. *Any smooth variety X of dimension k admits an embedding in projective space $\mathbb{P}^{2k+1}$.*

Note that this may be false if X is singular; the Zariski tangent spaces to a variety of dimension k may be of any dimension. For example, the Zariski tangent space to the curve $C \subset \mathbb{P}^n$ given as the image of the map

$$\rho\colon \mathbb{P}^1 \to \mathbb{P}^n$$

$$\colon t \mapsto [1, t^n, t^{n+1}, \ldots, t^{2n}]$$

at the point $\rho(0) = [1, 0, \ldots, 0]$ is all of $\mathbb{P}^n$, so that it cannot be projected isomorphically into any hyperplane. The point here is that for a singular variety $X \subset \mathbb{P}^n$ of dimension k, the projective tangent space to X at a singular point p may be much larger than the union of the planes $\Lambda \cong \mathbb{P}^k \subset \mathbb{P}^n$ such that (p, Λ) is in the graph of the Gauss map.

We should also point out that Theorem 15.13 is in general not at all useful; the varieties that come to us naturally embedded in large projective spaces (Veronese varieties, Segre varieties, Grassmannians, etc.) are almost always best left there, since once we project we lose the large groups of projective automorphisms that act on them.

Example 15.14. Joins of Intersecting Varieties

In Example 6.17 we defined the variety $\mathcal{J}(X, Y) \subset \mathbb{G}(1, n)$ of lines meeting each of two disjoint subvarieties X, $Y \subset \mathbb{P}^n$. In Example 8.1 we extended the definition to possibly intersecting pairs of varieties X, $Y \subset \mathbb{P}^n$, but did not say what additional lines, besides those containing distinct points $p \in X$, $q \in Y$, would be included in $\mathcal{J}(X, Y)$. We can now give a partial answer to this question.

Exercise 15.15. Let X and $Y \subset \mathbb{P}^n$ be subvarieties intersecting at a point p that is a smooth point of both X and Y with $\mathbb{T}_p X \cap \mathbb{T}_p Y = \{p\}$. Show that a line $L \subset \mathbb{P}^n$ through p that meets X and Y only at p will lie in the variety $\mathcal{J}(X, Y) \subset \mathbb{G}(1, n)$ of lines joining X to Y if and only if L lies in the span of $\mathbb{T}_p X$ and $\mathbb{T}_p Y$.

Exercise 15.16. Show by example that the conclusion of Exercise 15.15 is false if we do not assume $\mathbb{T}_p X \cap \mathbb{T}_p Y = \{p\}$.

We mentioned earlier that there are three other notions of "a tangent line to a variety X at a point p"; we will discuss these now. The first approach would be to take limits of secant lines $\overline{pq}$ to X as $q \in X$ approached p; specifically, we consider the map

$$s_p \colon X \dashrightarrow \mathbb{G}(1, n)$$

defined by sending $q \in X$ to $\overline{pq} \in \mathbb{G}(1, n)$ and take the set of tangent lines to X at p to be the image of $\{p\}$ under this map, that is, the image in $\mathbb{G}(1, n)$ of the fiber of the graph of s_p over p. In fact, we will see in Lecture 20 (Exercise 20.4 in particular) that the locus of tangent lines to X at p in this sense is what we call the *tangent cone* to X at p.

A second approach would be to define a tangent line to X at p to be any limiting position of a secant line $\overline{qr}$ as q and $r \in X$ approached p; that is, the image of the point (p, p) under the rational map

$$s \colon X \times X \dashrightarrow \mathbb{G}(1, n)$$

Finally, the third and most naive approach is simply to define a tangent line to X at p to be any line L containing p and contained in the projective tangent space $\mathbb{T}_p X$.

We will denote the union over all $p \in X$ of the set of tangent lines to X at p with respect to these three definitions as $\mathcal{T}'(X)$, $\mathcal{T}''(X)$, and $\mathcal{T}'''(X)$, respectively. The relationships among them are summed up in the following exercises.

Exercise 15.18. Show that $\mathcal{T}''(X)$, and $\mathcal{T}'''(X)$ are closed subvarieties of $\mathbb{G}(1, n)$, but that $\mathcal{T}'(X)$ need not be; and that they are equal for X smooth.

Exercise 15.19. Show that in general

$$\mathcal{T}'(X) \subset \mathcal{T}_1(X) \subset \mathcal{T}''(X) \subset \mathcal{T}'''(X).$$

Give examples where strict inclusion holds in each of these inclusions. (*)

One important instance where the second notion of tangent line is relevant is in a beautiful theorem of Fulton and Hansen. Let $\tan(X)$ denote the union of all tangent lines to X in the second sense, i.e.,

$$\tan(X) = \bigcup_{L \in \mathscr{T}''(X)} L$$

and let $S(X)$ be the secant variety to X, as before. Then we have the following theorem.

Theorem 15.20. *For any irreducible variety $X \subset \mathbb{P}^n$ of dimension k, either*

$$\text{(i)} \ \dim(S(X)) = 2k + 1 \quad \text{and} \quad \dim(\tan(X)) = 2k$$

or

$$\text{(ii)} \ S(X) = \tan(X).$$

For a discussion of this and related theorems, see [FL].

Example 15.21. The Locus of Bitangent Lines

In a similar fashion, we can introduce the variety of bitangent lines to a variety $X \subset \mathbb{P}^n$. To start, we set up an incidence correspondence

$$\Gamma = \{(L, p, q): p \neq q \text{ and } \overline{pq} = L \subset \mathbb{T}_p(X) \cap \mathbb{T}_q(X)\} \subset \mathbb{G}(1, n) \times X_{sm} \times X_{sm}.$$

This is just the double point locus associated to the projection $\pi_1 \colon \Sigma \to \mathbb{G}(1, n)$, where

$$\Sigma = \{(L, p): p \in L \subset \mathbb{T}_p X\} \subset \mathbb{G}(1, n) \times X_{sm}$$

is the incidence correspondence introduced in Example 15.7, that is, the union of the irreducible components of the fiber product $\Sigma \times_{\mathbb{G}} \Sigma$ other than the diagonal. We then define the *locus of bitangent lines to* X, denoted $\mathscr{B}(X)$, to be the closure of the image of Γ in $\mathbb{G}(1, n)$.

From the preceding example and the basic Theorem 11.12 (as applied, for example, in Exercise 11.15), we may conclude that the dimension of any component of the incidence correspondence Γ whose general member (L, p, q) does not satisfy $\overline{pq} \subset X$ is at least $4k - 2n$; correspondingly that any component of the locus $\mathscr{B}(X)$ of bitangent lines to X not consisting of lines lying on X has dimension at least $4k - 2n$. $\mathscr{B}(X)$ may indeed have smaller dimension if every bitangent line to X is contained in X, as will be the case, for example, when X is a quadric or cubic surface in $\mathbb{P}^3$ (in the former case, $\dim(\Gamma) = 3$ and $\dim(\mathscr{B}(X)) = 1$; in the latter, $\dim(\Gamma) = 2$ while $\dim(\mathscr{B}(X)) = 0$).

We should remark that just as the definition of double point locus given in Example 5.16 was not optimal, so the current definition of bitangent lines is not: it should more properly be defined as the image in $\mathbb{G}(1, n)$ of the double point locus of the projection $\Sigma \to \mathbb{G}(1, n)$, where double point locus is as defined in [F1] or [K] rather than as it is defined here. One problem with our definition is that it does not behave well with respect to families. For example, observe that by our definition the line $(Y = 0) \subset \mathbb{P}^2$ is not a bitangent line to the plane curve C_0 given by

$(Y^4 + YZ^3 = X^4)$, even though it is a bitangent line to the curves C_λ given by $(Y^4 + YZ^3 = (X^2 - \lambda Z^2)^2)$ for $\lambda \neq 0$. Thus, if we let $\mathbb{P}^{14}$ be the projective space parametrizing plane quartics, $U \subset \mathbb{P}^{14}$ the open subset corresponding to smooth quartics and

$$\Phi = \{(C, L): L \in \mathscr{B}(C)\} \subset U \times \mathbb{P}^{2*}$$

the incidence correspondence whose fiber over $C \in U$ is thc set of bitangent lines to the curve C, under the terms of our definition Φ will not be closed in $U \times \mathbb{P}^{2*}$; with the correct definition it will be.

Example 15.22. Dual Varieties

We can similarly define the notion of *tangent hyperplane* to a variety $X \subset \mathbb{P}^n$. As in the case of tangent lines, there is no ambiguity about what this should be if X is smooth: in this case a hyperplane $H \subset \mathbb{P}^n$ is called a tangent hyperplane if it contains a tangent plane to X. Again, as before, in case X is singular, we will simply define the locus of tangent hyperplanes to X to be the closure, in the dual space $\mathbb{P}^{n*}$, of the locus of hyperplanes containing the tangent plane to X at a smooth point. This is called the *dual variety* of X and is usually denoted $X^* \subset \mathbb{P}^{n*}$.

There is another characterization of the dual variety: in most cases if X is smooth we may say that it is the set of hyperplanes $H \subset \mathbb{P}^n$ such that the intersection $H \cap X$ is singular. The reason we cannot always say that X^* is the set of hyperplanes such that $X \cap H$ is singular is that, while the intersection of X with a hyperplane H having an isolated point p of tangency with X is necessarily singular at p, some hyperplanes may be tangent to X everywhere along their (smooth) intersection. For example, the hyperplane sections of the Veronese surface $S \cong \mathbb{P}^2 \subset \mathbb{P}^5$ correspond to conic plane curves; the hyperplane corresponding to a double line will be of this type. (In the language of schemes, it would be completely accurate to characterize the dual variety of a smooth variety X by the singularity of $X \cap H$). What is true is that such hyperplanes never form an irreducible component of the locus of tangent hyperplanes, so that for X smooth we can describe X^* as the closure of the locus of H such that $H \cap X$ is singular.

What is the dimension of the dual variety X^*? As usual, we may try to estimate this first by introducing an incidence correspondence, in this case the closure Φ of the set $\tilde{\Phi}$ of pairs of point and tangent hyperplanes:

$$\tilde{\Phi} = \{(p, H): p \in X_{sm} \text{ and } H \supset \mathbb{T}_p(X)\}.$$

The dimension of Φ is readily calculated from the projection map $\pi_1: \Phi \to X$ on the first factor. The fiber of this map over a smooth point $p \in X$ is just the subspace $\text{Ann}(\mathbb{T}_p(X)) \cong \mathbb{P}^{n-k-1} \subset \mathbb{P}^{n*}$ of hyperplanes containing the k-plane $\mathbb{T}_p(X)$. Thus the inverse image in Φ of the smooth locus of X is irreducible of dimension $n - 1$; and it follows that Φ is irreducible of dimension $n - 1$ in general. (We leave it as an exercise to check that the image $\pi_2(\Phi) \subset \mathbb{P}^{n*}$ is indeed the dual variety X^*.)

We conclude from this, of course, that the dual variety X^* is irreducible of dimension at most $n - 1$ and will in fact have dimension $n - 1$—that is, be a hypersurface—exactly when the general tangent hyperplane to X is tangent at only a finite number of points. We might naively expect that for most varieties a general tangent hyperplane is tangent at exactly one point, and indeed this is the case in a large number of examples; for example, the general hypersurface or complete intersection will have this property. This is reflected in the terminology: if the dual variety of a variety X fails to be a hypersurface, we say that it is *deficient*. There are, however, a number of examples where the dual variety fails to be a hypersurface (and, indeed, Theorem 15.24 will say that in some sense there are at least as many such examples as varieties whose duals are hypersurfaces).

Of course, one way in which the dual of a variety X may fail to be a hypersurface is if the tangent planes to X are constant along subvarieties of X—for example, a cone will have this property (as can easily be checked, the dual of the cone $\overline{p,Y} \subset \mathbb{P}^n$ with vertex p over a variety $Y \subset \mathbb{P}^{n-1}$ is just the dual of Y, lying in the hyperplane $p^* \subset \mathbb{P}^n$ of hyperplanes through p). This does not have to be the case, however; there are also examples of smooth varieties $X \subset \mathbb{P}^n$ whose Gauss maps are one to one, but whose dual varieties are deficient.

Probably the simplest examples of this are the rational normal scrolls $X = X_{a_1,\ldots,a_k}$ of dimension $k \geq 3$. To see this, recall from the definition in Lecture 8 that the scroll X is swept out by a one-parameter family (a "ruling") of $(k-1)$-planes. The tangent plane to X at any point $p \in X$ will contain the $(k-1)$-plane $\Lambda \subset X$ of the ruling through p; so that if a hyperplane section $Y = H \cap X$ is singular at p it must contain Λ. But now H must still meet the general plane of the ruling in a $(k-2)$-plane, and it follows from this that the closure Y_0 of $H \cap X - \Lambda$ will intersect Λ in a $(k-2)$-plane Γ. This means that Y will be singular along Γ, so that the general fiber of the projection map $\pi_2 \colon \Phi \to \mathbb{P}^{n*}$ will have dimension $k - 2$, and the image X^* correspondingly dimension $n - k + 1 < n - 1$.

There is another way to see that the dual of a scroll is deficient in the special case of the variety $X_{1,\ldots,1}$. This comes from the alternate description of $X_{1,\ldots,1}$ as the Segre variety $\sigma(\mathbb{P}^1 \times \mathbb{P}^{k-1}) \subset \mathbb{P}^{2k-1}$, or, in intrinsic terms, the locus

$$\mathbb{P}V \times \mathbb{P}W \subset \mathbb{P}(V \otimes W)$$

of reducible (i.e., rank 1) tensors $v \otimes w$ in the projective space associated to the tensor product of a two-dimensional vector space V and a k-dimensional vector space W. Now, the group $\mathrm{PGL}(V) \times \mathrm{PGL}(W)$ acts on the space $\mathbb{P}(V \otimes W)$, preserving the scroll $X = \mathbb{P}V \times \mathbb{P}W$, and likewise acts on the dual space preserving the dual X^*. But by easy linear algebra, there are only two orbits of the action of $\mathrm{PGL}(V) \times \mathrm{PGL}(W)$ on $\mathbb{P}(V \otimes W)^* = \mathbb{P}(V^* \otimes W^*)$: the tensors of rank 1 and the rest. It follows from this alone that the dual variety X^* can only be the locus of reducible tensors; in particular, it is isomorphic to X itself.

Note that we can see directly in this example what we claimed in the discussion of general scrolls: the singular hyperplane sections of the Segre variety $X = \mathbb{P}V \times \mathbb{P}W$ come from decomposable tensors $l \otimes m \in \mathbb{P}(V^* \otimes W^*)$, and they correspondingly consist simply of the union $(l = 0) \times \mathbb{P}W \cup \mathbb{P}V \times (m = 0)$ of the inverse images of hyperplanes in $\mathbb{P}V$ and $\mathbb{P}W$; in particular, they will be singular exactly along a $(k-2)$-plane.

Exercise 15.23. Identify the dual of the Segre variety $\sigma(\mathbb{P}^n \times \mathbb{P}^m) \subset \mathbb{P}^{nm+n+m}$ in general; in particular, show that it is deficient except in the case $m = n$.

Before moving on, we should mention two theorems about dual varieties. One is completely fundamental and elementary (though it involves one idea that will not be introduced until the following lecture, where its proof will be given). It is simply stated as follows.

Theorem 15.24. *Let $X \subset \mathbb{P}^n$ be an irreducible variety and $X^* \subset \mathbb{P}^{n*}$ its dual; let $\Phi_X \subset \mathbb{P}^n \times \mathbb{P}^{n*}$ and $\Phi_{X^*} \subset \mathbb{P}^{n*} \times \mathbb{P}^n = \mathbb{P}^n \times \mathbb{P}^{n*}$ be the incidence varieties associated to X and X^*. Then*

$$\Phi_X = \Phi_{X^*}.$$

In particular, the dual of the dual of a variety is the variety itself.

Note that, as promised, this theorem suggests that there are just about as many varieties with deficient dual as otherwise, since every variety is the dual of some variety.

Exercise 15.25. Prove directly that the dual of the dual of a plane curve $X \subset \mathbb{P}^2$ is X again by considering a parametric arc $\gamma: t \mapsto [v(t)] \in \mathbb{P}^2$, writing down the dual map γ^* from the disc to $\mathbb{P}^{2*}$ sending a point to its tangent line and iterating this process.

Classically, this case of Theorem 15.24 would have been argued just by drawing a picture to illustrate the statement "the limiting position of the intersection $\mathbb{T}_q(X) \cap \mathbb{T}_r(X)$ of the tangent lines to a smooth curve $X \subset \mathbb{P}^2$ at points q and r tending to a point $p \in X$ is p." More generally, Theorem 15.24 could be phrased, in classical language, as "the limit of the intersection of hyperplanes $H_1, \ldots, H_n$ tangent to a smooth variety $X \subset \mathbb{P}^n$ at points $q_1, \ldots, q_n$ tending to a point $p \in X$ contains p."

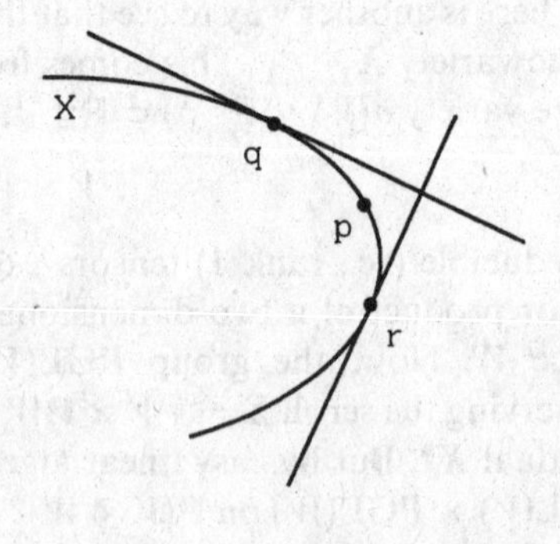

The other theorem is even more remarkable, and less elementary. It deals with smooth varieties $X \subset \mathbb{P}^n$ whose duals are deficient; for such a variety we may call the difference $n - 1 - \dim(X^*)$ the *deficiency of the dual* of X, called $\delta(X^*)$. We then have the following theorem.

Theorem 15.26. *Let $X \subset \mathbb{P}^n$ be any smooth variety with deficient dual. Then*

$$\delta(X^*) \equiv \dim(X) \qquad (\text{mod } 2).$$

This theorem is due to A. Landman; a second proof was given by L. Ein [Ein].

LECTURE 16

Tangent Spaces to Grassmannians

Example 16.1. Tangent Spaces to Grassmannians

We have seen that the Grassmannian $\mathbb{G}(k, n)$ is a smooth variety of dimension $(k+1)(n-k)$. This follows initially from our explicit description of the covering of $\mathbb{G}(k, n)$ by open sets $U_\Lambda \cong \mathbb{A}^{(k+1)(n-k)}$, though we could also deduce this from the fact that it is a homogeneous space for the algebraic group $\mathrm{PGL}_{n+1}K$. The Zariski tangent spaces to $\mathbb{G}$ are thus all vector spaces of this dimension. For many reasons, however, it is important to have a more intrinsic description of the space $T_\Lambda(\mathbb{G})$ in terms of the linear algebra of $\Lambda \subset K^{n+1}$. We will derive such an expression here and then use it to describe the tangent spaces of the various varieties constructed in Part I with the use of the Grassmannians.

To begin with, let us reexamine the basic open sets covering the Grassmannian. Recall from Lecture 6 that for any $(n-k)$-plane $\Gamma \subset K^{n+1}$ the open set $U_\Gamma \subset \mathbb{G}$ is defined to be the subset of planes $\Lambda \subset K^{n+1}$ complementary to Γ, i.e., such that $\Lambda \cap \Gamma = (0)$. These open sets were seen to be isomorphic to affine spaces $\mathbb{A}^{(k+1)(n-k)}$ as follows. Fixing any subspace $\Lambda \in U_\Gamma$, a subspace $\Lambda' \in U_\Gamma$ is the graph of a homomorphism $\varphi\colon \Lambda \to \Gamma$, so that

$$U_\Gamma = \mathrm{Hom}(\Lambda, \Gamma).$$

In particular, this isomorphism of spaces induces an isomorphism of tangent spaces

$$T_\Lambda(\mathbb{G}) = \mathrm{Hom}(\Lambda, \Gamma).$$

Now suppose we start with a subspace Λ and do not specify the subspace Γ. To what extent is this identification independent of the choice of Γ? The answer is straightforward: any subspace Γ complementary to Λ may naturally be identified with the quotient vector space K^{n+1}/Λ, and if we view the isomorphism of tangent spaces in this light, it is independent of Γ. We thus have a natural identification

$$T_\Lambda(\mathbb{G}) = \operatorname{Hom}(\Lambda, K^{n+1}/\Lambda).$$

In order to see this better, and because it will be extremely useful in applications, we will also describe this identification in terms of tangent vectors to arcs in $\mathbb{G}$. (As usual, when calculating with arcs, we will assume our ground field $K = \mathbb{C}$ and use the language of complex manifolds; but the techniques and results apply to maps of smooth curves over any field of characteristic 0.) Specifically, suppose that $\{\Lambda(t)\} \subset \mathbb{G}$ is a holomorphic arc in $\mathbb{G}$—that is, a family of planes parametrized by t in a disc (or in a smooth curve) with $\Lambda(0) = \Lambda$—and let $v \in \Lambda$ be any vector. We can measure the extent to which Λ is moving away from the vector v as follows. Choose any holomorphic arc $\{v(t)\} \in K^{n+1}$ with $v(t) \in \Lambda(t)$ for all t, and associate to v the vector

$$\varphi(v) = v'(0) = \left.\frac{d}{dt}\right|_{t=0} v(t).$$

This measures the movement of Λ away from v in the sense that the only thing preventing us from choosing $v(t) \equiv v$ is this movement. Of course, it is not unique; but if $w(t)$ is any other choice of holomorphic arc with $w(t) \in \Lambda(t)$ for all t and $w(0) = v$, then since $w(0) = v(0)$ we can write

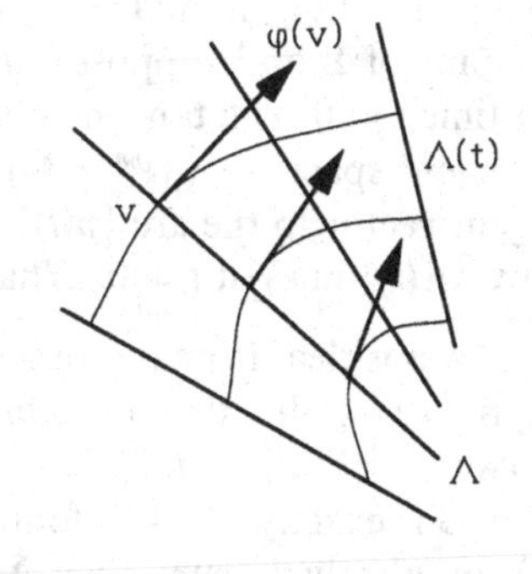

$$w(t) - v(t) = t \cdot u(t)$$

with $u(t) \in \Lambda(t)$ for all t. It follows that $\varphi(v)$ is well-defined as an element of the quotient K^{n+1}/Λ. The arc $\{\Lambda(t)\}$ thus determines a linear map $\varphi\colon \Lambda \to K^{n+1}/\Lambda$, which is just its tangent vector in terms of the preceding identification.

There are numerous other ways to view the identification. For example, the Grassmannian $\mathbb{G}(k, n) = G(k + 1, n + 1)$ is a homogeneous space for the group $\mathrm{GL}_{n+1}K$, so that its tangent space at a point $\Lambda \in G$ will be naturally identified with the quotient of the Lie algebra of $\mathrm{GL}_{n+1}K$ by the Lie algebra of the stabilizer of Λ. The first of these is just the vector space of endomorphisms of K^{n+1}, the second just the subspace of endomorphisms carrying Λ into itself, so that we have

$$\begin{aligned} T_\Lambda G &= \operatorname{Hom}(K^{n+1}, K^{n+1})/\{\varphi\colon \varphi(\Lambda) \subset \Lambda\} \\ &= \operatorname{Hom}(\Lambda, K^{n+1}/\Lambda). \end{aligned}$$

As we said earlier, however, the identification via tangent vectors to arcs is the one we will find most useful for our purposes.

Before going on, we should raise one point of notation. We would like, for many reasons, to put Grassmannians and projective spaces on the same footing, at least notationally. For example, we have been using the symbol Λ to denote, simultaneously, a vector subspace K^{k+1} of K^{n+1} and a projective subspace $\mathbb{P}^k \subset \mathbb{P}^n$. To be consistent, then, we should also use the same symbol p that we have been using to denote a point in projective space to denote a line in K^{n+1} at the same time. This leads to some funny-looking statements—e.g., $T_p(\mathbb{P}^n) = \mathrm{Hom}(p, K^{n+1}/p)$—but will be easier in the long run than keeping separate systems of symbols for $\mathbb{P}^n$ and $\mathbb{G}(k, n)$, as the next series of examples will bear out.

Example 16.2. Tangent Spaces to Incidence Correspondences

We consider here the basic incidence correspondence introduced in Lecture 6, that is, we let $\mathbb{G} = \mathbb{G}(k, n)$ and set

$$\Sigma = \{(p, \Lambda): p \in \Lambda\} \subset \mathbb{P}^n \times \mathbb{G}.$$

It is not hard to see that Σ is smooth, either by homogeneity or by direct examination of its equations. We want to identify the tangent space to Σ as a subspace of the tangent space to the product $\mathbb{P}^n \times \mathbb{G}$. To do this, let $(p, \Lambda) \in \Sigma$ be any point of Σ and suppose $\sigma(t) = (p(t), \Lambda(t))$ is an arc in Σ passing through (p, Λ) at time $t = 0$. The tangent vector to the arc σ at $t = 0$, viewed as an element of the tangent space $T_{(p,\Lambda)}(\mathbb{P}^n \times \mathbb{G}) = T_p(\mathbb{P}^n) \times T_\Lambda(\mathbb{G})$, is (η, φ) where $\eta \in T_p(\mathbb{P}^n)$ is the tangent vector to the arc $\{p(t)\}$ in $\mathbb{P}^n$ at $t = 0$ and $\varphi \in T_\Lambda$ is the tangent vector to the arc $\{\Lambda(t)\}$ in $\mathbb{G}$ at $t = 0$. What can we say about the pair (η, φ)?

The answer is clear from our description of the tangent space to $\mathbb{G}$: if $p(t) = [v(t)]$ with $v(t) \in \Lambda(t)$ for all t, then by definition $\varphi(v) = \eta(v)$ modulo Λ. Since this relation represents $n - k$ linear conditions on the pair (η, φ) and since the codimension of Σ in $\mathbb{P}^n \times \mathbb{G}$ is exactly $n - k$, it follows that this determines the tangent space to Σ at (p, Λ) completely. In sum, then, Σ is smooth and the tangent space to Σ at (p, Λ) is the space of pairs

$$T_{(p,\Lambda)}(\Sigma) = \left\{(\eta, \varphi): \begin{matrix} \eta \in \mathrm{Hom}(p, K^{n+1}/p) \\ \varphi \in \mathrm{Hom}(\Lambda, K^{n+1}/\Lambda) \end{matrix} \text{ and } \varphi|_p \equiv \eta \pmod{\Lambda}\right\}.$$

Exercise 16.3. Consider one generalization of this. For each $k < l$ we consider the *flag manifold* $\mathbb{F}(k, l, n)$ of Example 11.40, i.e., the incidence correspondence

$$\Omega = \{(\Lambda, \Gamma): \Lambda \subset \Gamma\} \subset \mathbb{G}(k, n) \times \mathbb{G}(l, n).$$

Show that Ω is smooth and that at a pair (Λ, Γ) the tangent space is given by

$$T_{(\Lambda,\Gamma)}(\Omega) = \left\{(\eta, \varphi): \begin{matrix} \eta \in \mathrm{Hom}(\Lambda, K^{n+1}/\Lambda) \\ \varphi \in \mathrm{Hom}(\Gamma, K^{n+1}/\Gamma) \end{matrix} \text{ and } \varphi|_\Lambda \equiv \eta \pmod{\Gamma}\right\}.$$

Exercise 16.4. Consider another generalization of the basic example: the incidence correspondence of pairs of incident planes, given by

$$\Omega = \{(\Lambda, \Gamma): \Lambda \cap \Gamma \neq \emptyset\} \subset \mathbb{G}(k, n) \times \mathbb{G}(l, n).$$

(We assume here that $k + l < n$ so that $\Omega \neq \mathbb{G}(k, n) \times \mathbb{G}(l, n)$.) Show that Ω is smooth at a point (Λ, Γ) if and only if Λ and Γ intersect in exactly one point; and that at such a pair the tangent space is given by

$$T_{(\Lambda,\Gamma)}(\Omega) = \left\{(\eta, \varphi): \begin{array}{l}\eta \in \mathrm{Hom}(\Lambda, K^{n+1}/\Lambda) \\ \varphi \in \mathrm{Hom}(\Gamma, K^{n+1}/\Gamma)\end{array} \text{ and } \varphi|_{\Lambda\cap\Gamma} \equiv \eta|_{\Lambda\cap\Gamma} \ (\mathrm{mod}\ \Lambda + \Gamma)\right\}.$$

Exercise 16.5. Generalize the preceding two exercises to the incidence correspondence Ξ of pairs (Λ, Γ) meeting in a subspace of dimension at least m.

Example 16.6. Varieties of Incident Planes

We will now identify the tangent spaces to the variety $\mathscr{C}_k(X) \subset \mathbb{G}(k, n)$ of k-planes meeting a variety $X \subset \mathbb{P}^n$, introduced in Example 6.14. Recall from Example 11.18 that if X has dimension m, then $\mathscr{C}_k(X)$ is a subvariety of codimension $n - m - k$ in $\mathbb{G}(k, n)$ (we assume throughout that $k < n - m$, so that $\mathscr{C}_k(X) \subsetneqq \mathbb{G}(k, n)$).

Suppose now that $\Lambda \in \mathscr{C}_k(X)$ is a plane meeting X at exactly one point p, that p is a smooth point of X, and that $\Lambda \cap \mathbb{T}_p(X) = \{p\}$. We claim first that under these conditions, $\mathscr{C}_k(X)$ is smooth at Λ. To see this, we go back to the description of $\mathscr{C}_k(X)$ that was useful before: we have

$$\mathscr{C}_k(X) = \pi_2(\Sigma \cap \pi_1^{-1}(X))$$

where $\Sigma \subset \mathbb{P}^n \times \mathbb{G}(k, n)$ is the incidence correspondence described in Example 16.2. Now, we have already shown that Σ is smooth at a point (p, Λ), with tangent space

$$T_{(p,\Lambda)}(\Sigma) = \left\{(\eta, \varphi): \begin{array}{l}\eta \in \mathrm{Hom}(p, K^{n+1}/p) \\ \varphi \in \mathrm{Hom}(\Lambda, K^{n+1}/\Lambda)\end{array} \text{ and } \varphi|_p \equiv \eta \ (\mathrm{mod}\ \Lambda)\right\}$$

and if X is smooth at p then of course $\pi_1^{-1}(X)$ is smooth at (p, Λ) with tangent space $\{(\eta, \varphi): \eta(p) \subset \mathbb{T}_p(X)\}$. If Λ has no nontrivial intersection with $\mathbb{T}_p(X)$, then these two tangent spaces intersect transversely, and so $\Sigma \cap \pi_1^{-1}(X)$ will be smooth at (p, Λ). Note moreover that in this case the tangent space to $\Sigma \cap \pi_1^{-1}(X)$ will not intersect the tangent space to the fibers of $\mathbb{P}^n \times \mathbb{G}(k, n)$ over $\mathbb{G}(k, n)$. If we assume in addition that Λ meets X at only the one point p—i.e., π_2: $\Sigma \cap \pi_1^{-1}(X) \to \mathscr{C}_k(X)$ is one to one over Λ—then it follows (using Corollary 14.10) that $\mathscr{C}_k(X)$ is smooth at Λ with tangent space

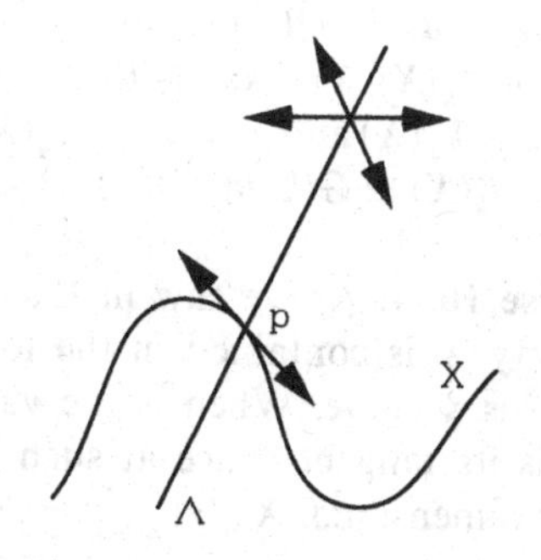

$$T_\Lambda(\mathscr{C}_k(X)) = \{\varphi \in \mathrm{Hom}(\Lambda, K^{n+1}/\Lambda): \varphi(p) \subset \mathbb{T}_p(X) + \Lambda\}.$$

Exercise 16.7. Let $\Lambda \in \mathscr{C}_k(X)$ and assume that $\Lambda \cap X$ is finite; assume further that $k + \dim(X) < n$ so that $\mathscr{C}_k(X) \neq \mathbb{G}(k, n)$. Show that $\mathscr{C}_k(X)$ is smooth at Λ if and only if $\Lambda \cap X = \{p\}$, $p \in X_{sm}$ and $\Lambda \cap \mathbb{T}_p X = \{p\}$. Note: you may need to use the following weak form of what is called Zariski's Main Theorem.

Proposition 16.8. *Let $f: X \to Y$ be a regular birational map of projective varieties, $q \in Y$ a point. If the fiber $f^{-1}(q)$ is disconnected, then q is a singular point of Y.*

This is immediate over $\mathbb{C}$ (and hence over any field of characteristic 0 by the Lefschetz Principle 15.1) but is tricky to prove algebraically; we won't do it here.

Exercise 16.9. (a) Let X_1, X_2, and $X_3 \subset \mathbb{P}^3$ be smooth curves, and let $l \subset \mathbb{P}^3$ be a line meeting each curve X_i in a unique point p_i; assume the p_i are distinct, that $l \neq \mathbb{T}_{p_i}(X_i)$ for each i, and that the lines $\mathbb{T}_{p_i}(X_i)$ are not all coplanar. Show that the subvarieties $\mathscr{C}_1(X_i)$ intersect transversely at the point $l \in \mathbb{G}(1, 3)$. (b) Now say we have four curves $X_i \subset \mathbb{P}^3$ and a line l meeting each X_i in a single point p_i, with the same hypotheses as in the first part. Let $l' \subset \mathbb{P}^3$ be a line skew to l, and set $q_i = l' \cap \overline{l, \mathbb{T}_{p_i}(X)}$. Assuming that the points $\{p_i\}$ and $\{q_i\}$ are distinct, show that the cycles $\mathscr{C}_1(X_i)$ intersect transversely at l if and only if the cross-ratio of the points $p_i \in l$ is not equal to the cross-ratio of the points $q_i \in l'$.

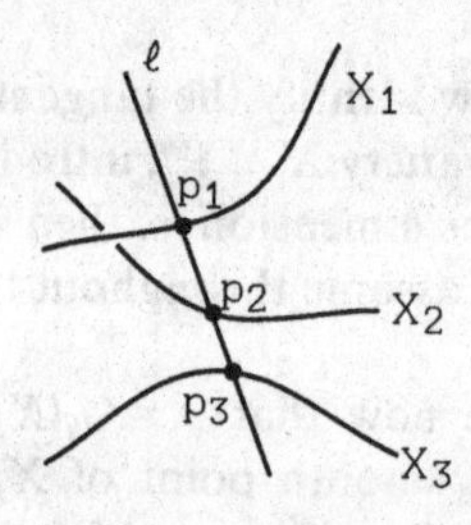

Example 16.10. The Variety of Secant Lines

In exactly the same way as earlier, we can describe the tangent space to the variety $\mathscr{S}(X)$ of secant lines to a variety $X \subset \mathbb{P}^n$ at a point corresponding to a line $l = \overline{pq}$ meeting X at exactly two distinct smooth points, with l not contained in either $\mathbb{T}_p(X)$ or $\mathbb{T}_q(X)$; it is exactly the space of homomorphisms from l to K^{n+1}/l carrying p into $l + \mathbb{T}_p(X)$ and q into $l + \mathbb{T}_q(X)$. An analogous statement can be made for the variety $\mathscr{S}_k(X) \subset \mathbb{G}(k, n)$ of $(k + 1)$-secant k-planes to X.

Exercise 16.11. As we saw in Exercise 15.9, the locus $\mathscr{T}_1(X)$ of tangent lines to a variety X is contained in the locus $\mathscr{S}(X)$ of secant lines to X. Suppose now that X is a curve. When is the variety $\mathscr{S}(X)$ singular at a point of $\mathscr{T}_1(X)$, and what is its tangent space at such a point in general? What is the situation for higher-dimensional X? (*)

Example 16.12. Varieties Swept out by Linear Spaces

We have looked at some examples of subvarieties of the Grassmannian. We now turn our attention to some varieties expressed as the unions of families of linear spaces.

We start with the general question: suppose $\Gamma \subset \mathbb{G}(k, n)$ is any subvariety, and we form the union

$$X = \bigcup_{\Lambda \in \Gamma} \Lambda.$$

When is X smooth and what is its tangent space?

We will restrict ourselves here to the "general" case, where the dimension of X is the expected $\dim(\Gamma) + k$. We will also assume that the variety X is swept out just once by the family of planes, i.e., that a general $p \in X$ lies on a unique $\Lambda \in \Gamma$. To put it differently, let $\Sigma \subset \mathbb{P}^n \times \mathbb{G}(k, n)$ be the incidence correspondence discussed in Example 16.2, and let $\Psi = \pi_2^{-1}(\Gamma) \subset \Sigma$. Then our two assumptions together are equivalent to the statement that the projection map $\pi_1: \Psi \to X$ is generically one to one, i.e., birational.

In this case, we simply have to identify the tangent space to the incidence correspondence Ψ and say when it intersects the tangent space to the fiber of $\mathbb{P}^n \times \mathbb{G}(k, n)$ over $\mathbb{P}^n$. Since we have already identified the tangent space to Σ, this is easy; the tangent space to Ψ at a point (p, Λ) is given by

$$T_{(p,\Lambda)}(\Phi) = \left\{(\eta, \varphi): \begin{matrix} \eta \in \operatorname{Hom}(p, K^{n+1}/p) \\ \varphi \in T_\Lambda(\Gamma) \subset \operatorname{Hom}(\Lambda, K^{n+1}/\Lambda) \end{matrix} \text{ and } \varphi|_p \equiv \eta \pmod{\Lambda}\right\}.$$

When does this contain a nonzero pair (η, φ) with $\eta = 0$? Exactly when $p \in \operatorname{Ker}(\varphi)$ for some nonzero tangent vector $\varphi \in T_\Lambda(\Gamma)$; if this is not the case then the image of this tangent space in $T_p(\mathbb{P}^n)$ will be just the span of the images $\varphi(p)$ over all $\varphi \in T_\Lambda(\Gamma)$.

To put it another way, we may view the inclusion

$$i: T_\Lambda(\Gamma) \to T_\Lambda(\mathbb{G}(k, n)) = \operatorname{Hom}(\Lambda, K^{n+1}/\Lambda)$$

as a linear map

$$i: T_\Lambda(\Gamma) \otimes \Lambda \to K^{n+1}/\Lambda,$$

or, equivalently, as a map

$$j: \Lambda \to \operatorname{Hom}(T_\Lambda(\Gamma), K^{n+1}/\Lambda).$$

In these terms, we have the following theorem.

Theorem 16.13. *Let $\Gamma \in \mathbb{G}(k, n)$ be any subvariety of dimension l, let $X \subset \mathbb{P}^n$ be the union of all $\Lambda \in \Gamma$, and let $p \in X$ be a point. Suppose that p lies on a unique plane $\Lambda \in \Gamma$, that Γ is smooth at Λ, and that the map*

$$j(p) \in \operatorname{Hom}(T_\Lambda(\Gamma), K^{n+1}/\Lambda)$$

is injective. Then X is smooth of dimension $k + l$ at p, with tangent space $\Lambda + \operatorname{Im}(j(p))$.

Exercise 16.14. Let $\mathbb{P}^k$ and $\mathbb{P}^l \subset \mathbb{P}^n$ be two disjoint linear subspaces of $\mathbb{P}^n$, and $X \subset \mathbb{P}^k$ and $Y \subset \mathbb{P}^l$ any smooth subvarieties other than linear subspaces. Let

$J = J(X, Y)$ be the join of X and Y, as described in Examples 6.17, 8.1, and 11.36. Show that J is singular exactly along $X \cup Y$, and that at a point $r \in \overline{pq} \subset J$ other than p and q the tangent space $\mathbb{T}_r(J)$ is the span of the spaces $\mathbb{T}_p(X)$ and $\mathbb{T}_q(Y)$.

Exercise 16.15. Let $\mathbb{P}^k$, $\mathbb{P}^l$, X and Y be as above, but say X and Y are curves, and let $\varphi\colon X \to Y$ be an isomorphism. Let $K = K(\varphi)$ be the surface formed by the joins of corresponding points, as in Example 8.14. Show that K is smooth and that the tangent spaces to K at points $r \in \overline{pq}$ are exactly the 2-planes containing the line $\overline{pq}$ and lying in the 3-plane spanned by $\mathbb{T}_p(X)$ and $\mathbb{T}_q(Y)$.

Finally, we can use the preceding description of tangent spaces to varieties swept out by linear spaces to prove an assertion made in connection with Example 11.30: the secant variety $S_l(X)$ of a curve $X \subset \mathbb{P}^n$ has dimension $2l + 1$ whenever $n \geq 2l + 1$. We do this in two steps.

Exercise 16.16. Let $X \subset \mathbb{P}^n$ be an irreducible nondegenerate curve and $p_1, \ldots, p_{l+1} \in X$ general points. Show that if $2l + 1 \leq n$, the tangent lines $\mathbb{T}_{p_i}X$ span a linear space $\mathbb{P}^{2l+1}$, while if $2l + 1 \geq n$, they span all of $\mathbb{P}^n$. (Note: this is false in characteristic $p > 0$.) (*)

Exercise 16.17. Let $X \subset \mathbb{P}^n$ be any irreducible variety and suppose that $p_1, \ldots, p_{l+1} \in X$ are general points. Show that the dimension of the secant variety $S_l(X)$ is exactly the dimension of the span of the tangent planes $\mathbb{T}_{p_i}X$. Show also that if this span is a plane Λ of dimension $(l + 1)\cdot\dim(X) + l$, and $q \in S_l(X)$ is a point lying on the span of the p_i and on no other secant l-plane, then $S_l(X)$ is smooth at q with tangent plane Λ.

Example 16.18. The Resolution of the Generic Determinantal Variety

As promised, we will now go back and discuss tangent spaces to determinantal varieties, this time in the spirit of the preceding examples. To begin with, let $M_k \subset \mathbb{P}^{mn-1}$ be the variety of $m \times n$ matrices of rank k or less, in the projective space $\mathbb{P}^{mn-1}$ of all nonzero $m \times n$ matrices modulo scalars. Recall the definition of the fundamental variety Ψ associated to M_k in Example 12.1; this is the incidence correspondence consisting of pairs (A, Λ) where A is an $m \times n$ matrix up to scalars and $\Lambda \cong K^{n-k}$ is a plane contained in the kernel of A, i.e.,

$$\Psi = \{(A, \Lambda)\colon A|_\Lambda \equiv 0\} \subset M \times G(n - k, n).$$

As we saw in the computation of the dimension of M_k, it is relatively easy to describe Ψ in terms of the projection on the second factor $G(n - k, n)$: if we fix Λ the space of maps $A\colon K^n \to K^m$ such that $\Lambda \subset \mathrm{Ker}(A)$ is just the space $\mathrm{Hom}(K^n/\Lambda, K^m)$, so that the fibers of the projection $\pi_2\colon \Psi \to G(n - k, n)$ are just projective spaces of dimension $km - 1$. We concluded in Example 12.1 that the variety Ψ is irreducible of dimension $\dim(G(n - k, n)) + km - 1 = k(m + n - k) - 1$ (and since the map $\pi_2\colon \Psi \to M$ is generically one to one onto M_k, the same is true of M_k); we may now deduce as well that Ψ is smooth and will here determine its tangent space.

We do this in much the same way as Example 16.2. Let $(A, \Lambda) \in \Psi$ be any point of Ψ, and suppose $\psi(t) = (A(t), \Lambda(t))$ is an arc in Ψ passing through (A, Λ) at time $t = 0$. We will write the tangent vector to the arc ψ at $t = 0$ as (α, φ) where

$$\alpha \in T_p(M) = \mathrm{Hom}(K^n, K^m)/K \cdot A^{4}$$

is the tangent vector to the arc $\{A(t)\}$ at $t = 0$ and

$$\varphi \in T_\Lambda(G(n-k, n)) = \mathrm{Hom}(\Lambda, K^n/\Lambda)$$

is the tangent vector to the arc $\{\Lambda(t)\}$ in $G(n-k, n)$ at $t = 0$. As before, we ask what we can say about the pair (α, φ) on the basis of the relation $A(t)|_{\Lambda(t)} \equiv 0 \; \forall t$.

The answer is clear from the product rule. If $\{v(t)\}$ is any arc with $v(t) \in \Lambda(t)$ for all t, then we have

$$A(t) \cdot v(t) \equiv 0;$$

taking the derivative with respect to t and setting $t = 0$ we have

$$A'(0) \cdot v(0) + A(0) \cdot v'(0) = 0,$$

i.e.,

$$\alpha(v(0)) + A(\varphi(v(0))) = 0.$$

(Note that this is well-defined, since α is well-defined modulo multiples of A, which kills $v(0) \in \Lambda$, and $\varphi(v(0))$ is well-defined modulo $\Lambda \subset \mathrm{Ker}(A)$.) Since $v(0)$ is an arbitrary element of Λ, we may write this condition in general as

$$\alpha|_\Lambda + A \circ \varphi = 0 \qquad \text{in } \mathrm{Hom}(\Lambda, K^m).^5$$

Now, this represents exactly $\dim(\mathrm{Hom}(\Lambda, K^m)) = m(n-k)$ linear conditions on the pair $(\alpha, \varphi) \in \mathrm{Hom}(K^n, K^m)/K \cdot A \times \mathrm{Hom}(\Lambda, K^n/\Lambda)$. Since this is, as observed, the codimension of $\Psi \subset M \times G(n-k, n)$, to say that any tangent vector $(\alpha, \varphi) \in T_{(A,\Lambda)}(\Psi)$ satisfies this condition immediately implies that Ψ is smooth at (A, Λ), with tangent space given by

$$T_{(A,\Lambda)}(\Psi) = \left\{(\alpha, \varphi)\colon \begin{matrix} \alpha \in \mathrm{Hom}(K^n, K^m)/K \cdot A) \\ \varphi \in \mathrm{Hom}(\Lambda, K^n/\Lambda) \end{matrix} \text{ and } \alpha|_\Lambda = -A \circ \varphi\colon \Lambda \to K^m\right\}.$$

As promised, this statement also yields the description of the tangent spaces to M_k found earlier in Example 14.16. At a point $A \in M_k$ corresponding to a map $A\colon K^n \to K^m$ of rank exactly k, the map $\pi_1\colon \Psi \to M_k$ is one to one (the unique point of Ψ over A is (A, Λ) with $\Lambda = \mathrm{Ker}(A)$), and the tangent space $T_{(A,\Lambda)}(\Psi)$ does not

[4] To be precise, $T_p(M)$ is identified with $\mathrm{Hom}(K \cdot A, \mathrm{Hom}(K^n, K^m)/K \cdot A)$. In the interests of reasonable notation we will work for the most part modulo scalars, identify $T_p(M)$ with $\mathrm{Hom}(K^n, K^m)/K \cdot A$, and indicate the precise form of the final results (see, for example, the next footnote).

[5] More precisely, if $\alpha \in \mathrm{Hom}(K \cdot A, \mathrm{Hom}(K^n, K^m)/K \cdot A)$, the condition is: $\alpha(A)|_\Lambda + A \circ \varphi = 0$.

meet the tangent space to the fiber of π_1. Thus, M_k is smooth at such a point; the tangent space $T_A(M_k) = (\pi_1)_*(T_{(A,\Lambda)}(\Psi))$ is the space of maps $\alpha: K^n \to K^m$ such that $\alpha|_\Lambda = -A \circ \varphi$ for some $\varphi \in \mathrm{Hom}(\Lambda, K^n/\Lambda)$—i.e., the space of maps α carrying Λ into $\mathrm{Im}(A)$.

Exercise 16.19. The entire analysis of the determinantal variety M_k carried out in the preceding example could be done by introducing not the auxiliary variety Ψ given but the analogously defined

$$\Xi = \{(A, \Gamma): \mathrm{Im}(A) \subset \Gamma\} \subset M \times G(k, m).$$

Do this. Also, describe the fiber product $\Psi \times_{M_k} \Xi$—in particular, is this smooth?

Example 16.20. Tangent Spaces to Dual Varieties

Of course, the identification of the tangent space to the dual variety X^* of a variety $X \subset \mathbb{P}^n$ is the content of Theorem 15.24, which we will now prove (assuming as before that $\mathrm{char}(K) = 0$).

We do this by first identifying the tangent space to the incidence correspondence Φ introduced in Example 15.22. Recall that Φ is defined to be the closure of the set of pairs

$$\tilde{\Phi} = \{(p, H): p \in X_{sm} \text{ and } H \supset \mathbb{T}_p(X)\},$$

and that $\tilde{\Phi}$ is irreducible of dimension $n - 1$. Suppose now that X is of dimension k and that the dual variety X^*, to which Φ maps, is of dimension l; let $p \in X$ and $H \in X^*$ be smooth points such that (p, H) is a general point of Φ (so that in particular by Proposition 14.4 the tangent space $T_{(p,H)}(\Phi)$ surjects onto $T_p(X)$ and $T_H(X^*)$) and let $\alpha \in T_p(X)$ and $\beta \in T_H(X^*)$ be tangent vectors. To describe the tangent space to Φ at the point (p, H), note first that since Φ is contained in both $X \times X^*$ and the incidence correspondence

$$\Sigma = \{(p, H): p \in H\}$$

we have from Example 16.2

$$T_{(p;q)}(\Phi) \subset \left\{(\alpha, \beta): \begin{array}{c}\alpha \in T_p(X)\\ \beta \in T_H(X^*)\end{array} \text{ and } \beta|_p \equiv \alpha \ (\mathrm{mod}\ H)\right\},$$

where we view α as a homomorphism from p to K^{n+1}/p and $\beta \in \mathrm{Hom}(H, K^{n+1}/H)$. If we think of $\mathbb{P}^{n*}$ as one-dimensional subspaces q of the dual space $(K^{n+1})^*$ instead of n-dimensional subspaces H of K^{n+1} and the tangent vector β correspondingly as a homomorphism from q to $(K^{n+1})^*/q$, then we can write this more symmetrically as

$$T_{(p,q)}(\Phi) \subset \left\{(\alpha, \beta): \begin{array}{c}\alpha \in T_p(X)\\ \beta \in T_q(X^*)\end{array} \text{ and } \langle\alpha(v), w\rangle + \langle v, \beta(w)\rangle = 0 \text{ for } v \in p, w \in q\right\}.$$

Now use the fact that $(p, q) \in \Phi$, i.e., that the hyperplane q contains the tangent plane $\mathbb{T}_p(X)$. This amounts to the statement that

$$\langle \alpha(v), w \rangle = 0 \qquad \text{for all } \alpha \in T_p(X), \qquad v \in p, \qquad w \in q.$$

The symmetry in the statement of Theorem 15.4 now emerges. We have already established that $\langle \alpha(v), w \rangle = -\langle v, \beta(w) \rangle$ for $(\alpha, \beta) \in T_{(p,q)}(\Phi)$, and since by hypothesis $T_{(p,q)}(\Phi)$ surjects onto $T_q(X^*)$, the condition $\langle \alpha(v), w \rangle = 0$ implies the opposite condition

$$\langle v, \beta(w) \rangle = 0 \qquad \text{for all } \beta \in T_q(X^*), \qquad v \in p, \qquad w \in q.$$

This condition in turn says that p, viewed as a hyperplane in $\mathbb{P}^{n*}$, contains the tangent space to X^*. Thus an open subset of, and hence all of Φ is contained in the incidence correspondence Φ_{X^*}; since both are irreducible of dimension $n - 1$, we deduce that

$$\Phi_X = \Phi_{X^*}. \qquad \square$$

Example 16.21. Tangent Spaces to Fano Varieties

Consider finally the Fano variety $F_k(X)$ of k-planes lying on a variety $X \subset \mathbb{P}^n$. In example 12.5 we estimated the "expected" dimension of the variety $F_k(X)$ associated to a general hypersurface $X \subset \mathbb{P}^n$ of degree d; here we will describe the tangent space to $F_k(X)$ and, in so doing, complete the proof of Theorem 12.8.

This is actually completely straightforward. Suppose first that X is the hypersurface given by the single homogeneous polynomial $F = F(Z_0, \ldots, Z_n)$, and let $\{\Lambda(t)\}$ be an arc in the Fano variety $F_k(X)$, that is, a family of k-planes lying on X. For each point $p \in \Lambda_0$, then, we can draw an arc $\{p(t) = [Z_0(t), \ldots, Z_n(t)]\}$ with $p(t) \in \Lambda(t)$ for all t; since $\Lambda(t) \subset X$ for all t we have

$$F(p(t)) \equiv 0.$$

Taking the derivative with respect to t, we find that

$$\sum \frac{\partial F}{\partial Z_i}(p_0) \cdot Z_i'(0) = 0,$$

or, in other words, the tangent vector to the arc $\{p(t)\}$ lies in the tangent plane to X at p_0. It follows that the tangent plane to the Fano variety $F_k(X)$ at $\Lambda = \Lambda_0$ is contained in the space $\mathscr{H}$ of homomorphisms $\varphi\colon \Lambda \to K^{n+1}/\Lambda$ defined by

$$\mathscr{H} = \{\varphi\colon \varphi(p) \in \mathbb{T}_p(X)/\Lambda \text{ for all } p \in \Lambda\}.$$

Since the Fano variety of an arbitrary variety $X \subset \mathbb{P}^n$ is just the intersection of the Fano varieties of the hypersurfaces containing X, it follows that the same is true in general. In particular, when the subspace $\mathscr{H} \subset \operatorname{Hom}(\Lambda, K^{n+1}/\Lambda)$ determined by this condition has dimension equal to the dimension of $F_k(X)$, we may deduce that $F_k(X)$ is smooth at Λ, with tangent space $\mathscr{H}$.

In fact, this condition, morally speaking, determines the tangent space to the Fano variety, in a sense made precise in Exercise 16.23. Note in particular that the "expected" dimension of the space $\mathscr{H}$ coincides, in the case of a hypersurface

$X \subset \mathbb{P}^n$, with the expected dimension $\varphi(n, d, k)$ of the Fano variety $F_k(X)$ as described in Example 12.5: given X: $(F = 0)$ and a plane $\Lambda \subset X$, to any homomorphism $\varphi: \Lambda \to K^{n+1}/\Lambda$ we may associate a polynomial

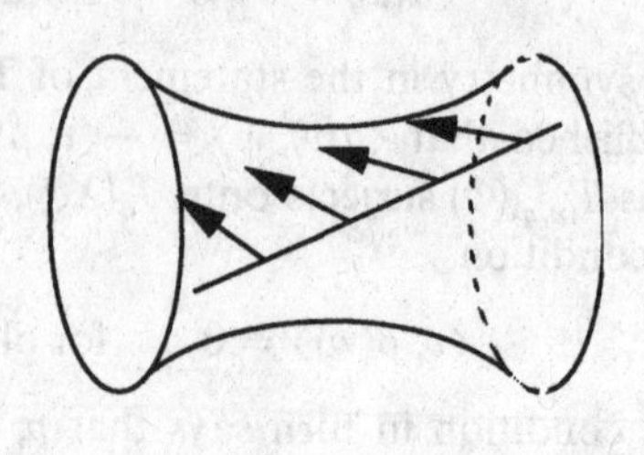

$$G_\varphi(p) = \left(\frac{\partial F}{\partial Z_0}(p), \ldots, \frac{\partial F}{\partial Z_n}(p)\right) \cdot \varphi(p).$$

Note that this is well-defined—$\varphi(p)$ is only defined modulo Λ, but Λ is in the kernel of the linear form given by the partials of F—and defines a linear map from $\mathrm{Hom}(\Lambda, K^{n+1}/\Lambda)$ to the space of polynomials of degree d on Λ whose kernel is exactly $\mathscr{H}$. We may thus expect that the codimension of $\mathscr{H}$ in $\mathrm{Hom}(\Lambda, K^{n+1}/\Lambda)$ is $\binom{k+d}{d}$ and hence that the dimension of $\mathscr{H}$ is $(k+1)(n-k) - \binom{k+d}{d} = \varphi(n, d, k)$.

Exercise 16.22. For each d, n, and k such that the expected dimentsion $\varphi(n, d, k)$ of the Fano variety $F_k(X)$ of k-planes on a hypersurface $X \subset \mathbb{P}^n$ of degree d (as determined in Example 12.5) is nonnegative, exhibit a hypersurface $X \subset \mathbb{P}^n$ of degree d in $\mathbb{P}^n$ and a k-plane $\Lambda \subset X$ such that the tangent space to $F_k(X)$ at Λ has dimension at most $\varphi(n, d, k)$, and deduce the statement of Theorem 12.8.

Exercise 16.23. The word "morally" in the preceding paragraph means, as usual, "scheme-theoretically": the tangent space to the Fano *scheme*—defined to be the subscheme of the Grassmannian $\mathbb{G}(k, n)$ determined by the equations described in Example 6.19—is always equal to the space $\{\varphi: \varphi(p) \in \mathbb{T}_p(X)/\Lambda\ \forall p\})$. Verify this by explicit calculations, i.e., show that the gradients of the functions introduced in Example 6.19 describe exactly this space. (Suggestion: do this for a hypersurface X and the case $k = 1$.)

LECTURE 17

Further Topics Involving Smoothness and Tangent Spaces

Example 17.1. Gauss Maps on Curves

Let $X \subset \mathbb{P}^n$ be a smooth curve. In Example 15.2 we introduced the Gauss map

$$\mathcal{G}: X \to \mathbb{G}(1, n)$$

sending a point $p \in X$ to the tangent line $\mathbb{T}_p(X)$. We will now investigate this map further, and in particular characterize, among all maps of curves to Grassmannians $\mathbb{G}(1, n)$, those that arise in this way.

To begin with, we will restrict to the case $\mathrm{char}(K) = 0$, and will work with a parametric form of X; that is, we will consider an arc

$$\gamma: t \mapsto [v(t)] = [v_0(t), \ldots, v_n(t)] \in \mathbb{P}^n.$$

We will assume that $v(t_0) \neq 0$ (we can always divide v by a power of $(t - t_0)$ if $v(t_0) = 0$).

Now let $\mathcal{G}: X \to \mathbb{G}(1, n)$ be the Gauss map associated to X and let $\varphi(t): \mathcal{G}(t) \to K^{n+1}/\mathcal{G}(t)$ be a tangent vector to $\mathcal{G}(X)$ at the point $\mathcal{G}(t)$. Then $\mathcal{G}(t)$ is spanned by $v(t)$ and $v'(t)$, and by definition

$$\varphi(t)(v(t)) = v'(t) \equiv 0 \qquad (\mathrm{mod}\ \mathcal{G}(t))$$

and

$$\varphi(t)(v'(t)) = v''(t) \qquad (\mathrm{mod}\ \mathcal{G}(t)).$$

In particular, $\varphi(t)$ has rank at most 1; what we are saying in effect is that, to first order, the motion of the tangent line $T_p(X)$ to a curve is the same as a rotation around the point p. Note also that $\varphi(t)$ is zero exactly when the wedge product $v(t) \wedge v'(t) \wedge v''(t) = 0$ (such a point is called a *flex point* of the curve; we will see in Example 18.4 that—except in case C is a line—not every point can be a flex). Note that the kernel of $\varphi(t)$ is the vector $v(t)$ itself, while the image is the plane spanned by $v(t)$, $v'(t)$, and $v''(t)$ (modulo $\mathscr{G}(t)$, of course). In fact, we have the following converse.

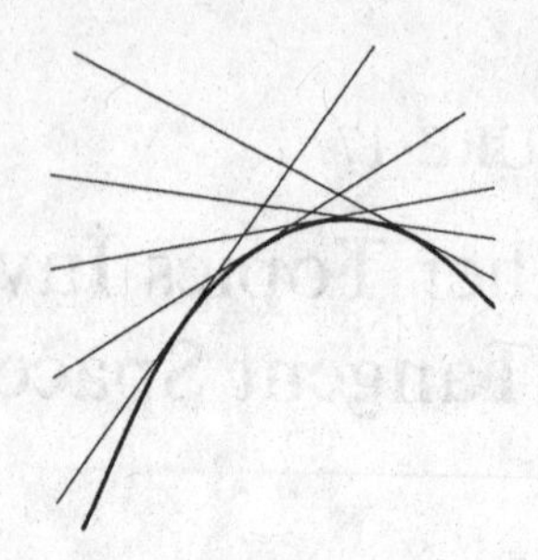

Theorem 17.2. *Let $\mathscr{G}: X \to \mathbb{G}(1, n)$ be any map of a smooth curve X to a Grassmannian of lines and suppose that the tangent vectors $\varphi(t)$ to $\mathscr{G}(X)$ have rank at most 1 for all t. Then either $\mathscr{G}$ is the Gauss map associated to some map $v: X \to \mathbb{P}^n$ or all the lines $\mathscr{G}(t)$ pass through a fixed point $p \in \mathbb{P}^n$.*

PROOF. The way to prove this is to recover the map v from the map $\mathscr{G}$, and the way to do this was suggested earlier: $v(t)$ should be the kernel of the tangent vector to $\mathscr{G}(X)$ at $\mathscr{G}(t)$. So assume that $\mathscr{G}: X \to \mathbb{G}(1, n)$ is a map with tangent vectors $\varphi(t)$ of rank at most 1; since the rank cannot be zero everywhere (if it is, $\mathscr{G}$ is constant, and so is the Gauss map associated to a map v of X to a line in $\mathbb{P}^n$) we may assume it has rank exactly 1 outside a finite set of points in X. Away from these points, define

$$v: X \to \mathbb{P}^n$$

by

$$v: t \mapsto \operatorname{Ker}(\varphi(t)).$$

This then extends to a regular map on all of X: if we write

$$v(t) = [v_0(t), \ldots, v_n(t)]$$

with $v_i(t)$ meromorphic functions, after multiplying through by a suitable power of t we may assume that the v_i are all holomorphic and not all $v_i(t)$ are 0.

We claim now that either v is constant or $\mathscr{G}(t)$ is spanned by $v(t)$ and $v'(t)$ for general t. In the former case, all the lines $\mathscr{G}(t)$ contain the point v. On the other hand, to say that $v(t) \in \operatorname{Ker}(\varphi(t))$ for all t is equivalent to saying that $v'(t) \in \mathscr{G}(t)$ for all t; since $v'(t)$ will be independent from $v(t)$ for all but a finite set of values of t, it follows that $\mathscr{G}(t)$ coincides with the span of $v(t)$ and $v'(t)$ for all but a finite set S of t. Thus $\mathscr{G} = \mathscr{G}_v$ for all $t \notin S$, and hence for all t. □

Exercise 17.3. Let $X \subset \mathbb{P}^n$ be a smooth curve and

$$TX = \bigcup_{p \in X} \mathbb{T}_p(X)$$

be its tangential surface. Let $q \in TX$ be a point lying on a unique tangent line $\mathbb{T}_p(X)$ to X, $q \neq p$. Show that q is a smooth point of the surface TX if and only if p is not a flex point of X, and that in this case the tangent plane to TX at q is constant along the lines $\mathbb{T}_p(X) \subset TX$ (the tangent plane to TX along this line is the *osculating* plane to X at p—see Example 17.5).

Exercise 17.4. Observe that by the preceding exercise if $S \subset \mathbb{P}^n$ is the tangential surface of a curve then the image of the Gauss map $\mathcal{G}_S: S \to \mathbb{G}(2, n)$ will be one-dimensional. Classify all ruled surfaces S with the property that $\mathcal{G}_S$ is constant along the lines of the ruling; if you're feeling reasonably energetic after that, classify all surfaces S such that the variety $\mathcal{T}(S)$ of tangent planes to S has dimension 1. (*)

Example 17.5. Osculating Planes and Associated Maps

As is suggested by the preceding, we can generalize the definition of the tangent line to a curve by introducing its osculating planes. To do this, suppose first that $C \subset \mathbb{P}^n$ is a smooth curve, given parametrically as the image of a vector-valued function

$$\gamma: t \mapsto [v(t)] = [v_0(t), \ldots, v_n(t)] \in \mathbb{P}^n$$

(if C is singular, we can still by Theorem 17.23 (proved in Lecture 20 for curves) view C as the image of a smooth curve under a generically one-to-one map, and so proceed). As we have seen, the Gauss map $\mathcal{G}$ on C may be described by

$$\mathcal{G}: t \mapsto [v(t) \wedge v'(t)],$$

which cannot vanish identically. More generally, if C is irreducible and nondegenerate, we see that the multivector $v(t) \wedge \cdots \wedge v^{(k)}(t)$ cannot vanish identically for any $k < n$: if it did, we could choose k to be the smallest such integer and take the derivative with respect to t to obtain

$$\begin{aligned} 0 &\equiv \frac{d}{dt}(v(t) \wedge \cdots \wedge v^{(k)}(t)) \\ &= v(t) \wedge \cdots \wedge v^{(k-1)}(t) \wedge v^{(k+1)}(t), \end{aligned}$$

i.e., at all times t the $(k + 1)$st derivative of v will be linearly dependent on $v(t)$ and the first k derivatives $v'(t), \ldots, v^{(k)}(t)$. Differentiating again, we see that the same is true of $v^{(k+2)}$ and so on, so that the entire curve C must lie in the k-plane spanned by $v(t), \ldots, v^{(k)}(t)$.

Since the vectors $v(t), \ldots, v^{(k)}(t)$ are not everywhere dependent, we can for each $k = 1, \ldots, n$ define a rational map

$$\begin{aligned} \mathcal{G}^{(k)}: C &\dashrightarrow \mathbb{G}(k, n) \\ : p &\mapsto [v(t) \wedge \cdots \wedge v^{(k)}(t)]. \end{aligned}$$

As in the proof of Theorem 17.2, this will extend to a regular map on all of C: writing

$$\mathcal{G}^{(k)}(t) = [w_0(t), \ldots, w_N(t)]$$

for some meromorphic functions $w_0, \ldots, w_N$, we can multiply through by a power of t to make all w_i holomorphic, not all $w_i(t) = 0$. For $k = 1$, this is of course the

Gauss map; in general, we call $\mathscr{G}^{(k)}$ the kth *associated map*, its image the kth *associated curve*, and its image point $\Lambda = \mathscr{G}^{(k)}(p)$ at $p \in C$ the *osculating k-plane* to C at p.

Exercise 17.6. Show that the kth associated map as defined earlier is independent of the parametrization of C chosen.

Exercise 17.7. There is another way to characterize the osculating k-plane to a curve at a point. We say that a hyperplane $H \subset \mathbb{P}^n$ has *contact of order m* with a smooth curve $C \subset \mathbb{P}^n$ at a point p if the restriction to C of a linear form vanishing on H vanishes to order m at p (again, we can still make sense of this if C is singular by parametrizing it by a smooth curve). More generally, we define the order of contact of a linear space $\Lambda \subset \mathbb{P}^n$ with C at P to be the minimum order of contact with C at p of any hyperplane containing Λ. In these terms, show that the osculating k-plane to C at p is simply the (unique) k-plane having maximal order of contact with C at p. (*)

In general, we say a point p on a curve $C \subset \mathbb{P}^n$ is an *inflectionary point* if for some $k < n$ the osculating k-plane to C at p has contact of order strictly greater than $k + 1$ (equivalently, if the osculating hyperplane has contact of order greater than n). This is equivalent to saying that for some parametrization of C by a vector valued function $v(t)$ with $p = [v(0)]$, the wedge product $v(0) \wedge v'(0) \wedge \cdots \wedge v^{(n)}(0) = 0$. By the analysis above, a curve can have at most finitely many inflectionary points; on the other hand, it is a fact (which we will not prove here) that the only curve with no inflectionary points is the rational normal curve.

By analogy with the preceding description of the tangent lines to the Gauss image $\mathscr{G}(C) \subset \mathbb{G}(1, n)$ of a curve $C \subset \mathbb{P}^n$, we can say what the tangent line to the kth associated curve of C is at a general point $\Lambda = \mathscr{G}^{(k)}(p)$: it is the homomorphism of Λ into K^{n+1}/Λ, unique up to scalars, with kernel containing $\mathscr{G}^{(k-1)}(p)$ and image contained in $\mathscr{G}^{(k+1)}(p)$. We likewise have a converse, analogous to Theorem 17.2.

Theorem 17.8. *Let $\mathscr{G}: X \to \mathbb{G}(k, n)$ be any map of a smooth curve X to a Grassmannian and suppose that the tangent vectors $\varphi(t)$ to $\mathscr{G}(X)$ have rank at most 1 for all t. Then either $\mathscr{G}$ is the* kth *associated map of some curve $C \subset \mathbb{P}^n$ or all the planes $\mathscr{G}(t)$ pass through a fixed point $p \in \mathbb{P}^n$.*

Exercise 17.9. Prove this.

Exercise 17.10. Just as we can form the tangential surface of a curve, we can more generally define the *osculating* $(k + 1)$*-fold* $T^{(k)}(C)$ of a smooth curve $C \subset \mathbb{P}^n$ to be the union of the osculating k-planes to C. Say when a point of $T^{(k)}(C)$ is smooth, and describe its tangent space at such a point. (*)

Example 17.11. The Second Fundamental Form

While we will not be able to go into it as deeply as in the case of curves, we will consider here the differential of the Gauss map on a smooth variety $X \subset \mathbb{P}^n$ of general dimension k. Such a Gauss map $\mathcal{G}$ induces a differential map on tangent spaces: for any $p \in X$ with tangent space $\mathbb{T}_p(X) \subset \mathbb{P}^n$, we have

$$(d\mathcal{G})_p\colon T_p(X) \to \operatorname{Hom}(\mathbb{T}_p X, K^{n+1}/\mathbb{T}_p X).$$

The first thing to observe is that, just as in the curve case, every homomorphism $\varphi\colon \mathbb{T}_p X \to K^{n+1}/\mathbb{T}_p X$ in the image of $(d\mathcal{G})_p$ has p in its kernel. $(d\mathcal{G})_p$ thus induces a map

$$(d\mathcal{G})_p\colon T_p(X) \to \operatorname{Hom}(\mathbb{T}_p X/p, K^{n+1}/\mathbb{T}_p X),$$

and using the identifications of the tangent space

$$T_p(X) = \operatorname{Hom}(p, \mathbb{T}_p X/p)$$

and normal space

$$N_p(X) = T_p(\mathbb{P}^n)/T_p(X) = \operatorname{Hom}(p, K^{n+1}/\mathbb{T}_p(X)),$$

we see that this is equivalent to a map

$$(d\mathcal{G})_p\colon T_p(X) \to \operatorname{Hom}(T_p(X), N_p(X)).$$

Equivalently, we can view this as a map

$$(d\mathcal{G})_p\colon T_p(X) \otimes T_p(X) \to N_p(X).$$

In these terms, the basic fact about $(d\mathcal{G})_p$ is that it is symmetric in its two arguments; i.e., it factors through the symmetric square of $T_p(X)$ to give a map

$$II_p\colon \operatorname{Sym}^2(T_p(X)) \to N_p(X),$$

or, dualizing, a map

$$II_p^*\colon N_p(X)^* \to \operatorname{Sym}^2(T_p^*(X)).$$

This map is called the *second fundamental form*. It is, unlike the first fundamental form of a submanifold of $\mathbb{P}^n$, invariant under the action of $PGL_{n+1}K$. For a further discussion, see [GH1].

Exercise 17.12. Verify that the map $(d\mathcal{G})_p$ defined earlier is symmetric. (Use a parametrization of X by a vector-valued function $v = (v_0, \ldots, v_n)$ of k variables $z_1, \ldots, z_k$.)

Exercise 17.13. Now let $X \subset \mathbb{P}^n$ be a smooth k-dimensional variety, TX its tangential variety, and $q \in TX$ a point lying on a unique tangent plane $\mathbb{T}_p(X)$. Under what circumstances is TX smooth at q, and what is its tangent plane?

Example 17.14. The Locus of Tangent Lines to a Variety

Let $X \subset \mathbb{P}^n$ be a smooth variety once more and let $\mathcal{T}_1(X) \subset \mathbb{G}(1, n)$ be the variety of tangent lines to X, as introduced in Example 15.7. As one application of the construction of the second fundamental form, we will give a condition for the smoothness of the variety $\mathcal{T}_1(X)$ at a point L.

To do this, we first set up the incidence correspondence $\Sigma \subset \mathbb{G}(1, n) \times X$ consisting of pairs (L, p) such that $p \in L \subset \mathbb{T}_p(X)$ and ask for the tangent space to Σ at (L, p). The answer is as follows: if (φ, v) is the tangent vector to an arc $\{(L(t), p(t))\}$ in Σ at $(L(0), p(0)) = (L, p)$, with $\varphi: L \to K^{n+1}/L$ and $v \in T_p(X)$, then

(i) the condition that $p(t) \in L(t)$ for all t says, as in Example 16.6, that $\varphi|_p = v \bmod L$ (where $v \in T_p(X) \subset T_p(\mathbb{P}^n) = \operatorname{Hom}(p, K^{n+1}/p)$); and
(ii) the condition that $L(t) \subset \mathbb{T}_{p(t)}(X)$ for all t says that the homomorphism

$$(d\mathcal{G})_p(v): \mathbb{T}_p(X) \to K^{n+1}/\mathbb{T}_p(X),$$

when restricted to L, coincides with the homomorphism φ modulo $\mathbb{T}_p(X)$ (of course, since both are zero on p, this need only be checked for one other point $q \neq p \in L$).

Now (i) represents $n - 1$ linear conditions on the pair (φ, v), while (ii) gives $n - k$, so that together they describe a subspace of codimension $2n - k - 1$ in $T_L(\mathbb{G}(1, n)) \times T_p(X)$, i.e., a space of dimension $2k - 1$. Since we have seen that Σ is of pure dimension $2k - 1$, it follows that for any L, Σ is smooth at (L, p) with tangent space specified by (i) and (ii).

Finally, we consider the projection of Σ to $\mathbb{G}(1, n)$ and ask when this is injective on tangent spaces. The answer is clear: if $(v, 0) \in T_{(L,p)}(\Sigma)$, then by condition (i) we must have $v \in L$, i.e., v belongs to the one-dimensional subspace of $T_p(X)$ corresponding to $L \subset \mathbb{T}_p(X)$, and by (ii) we must have $(d\mathcal{G})_p(v)(L) \equiv 0$, or in other words $II_p(v, v) = 0$ where I_p is the second fundamental form of X at p. In sum, then, the point $L \in \mathcal{T}_1(X)$ will be a smooth point if L is tangent to X at a unique point p and $II_p(v, v) \neq 0$, where $v \in T_p(X)$ is any nonzero vector in the direction of L.

Exercise 17.15. A line L tangent to a variety X at a smooth point p and such that $II_p(v, v) = 0$, where $v \in T_p(X)$ is in the direction of L, is called a *flex line* of X.

(a) Estimate the dimension of the variety of flex lines to a variety $X \subset \mathbb{P}^n$ of dimension k.

(b) Show that L is a flex line of X if and only if the polynomials $F \in I(X)$ have a common triple zero on L.

Bertini's Theorem

In closing this lecture, we should mention a few topics of interest in connection with the notion of smoothness and singularity. The first is a very fundamental and classical result.

Theorem 17.16. Bertini's Theorem. *If X is any quasi-projective variety, $f: X \to \mathbb{P}^n$ a regular map, $H \subset \mathbb{P}^n$ a general hyperplane, and $Y = f^{-1}(H)$, then*

$$Y_{\text{sing}} = X_{\text{sing}} \cap Y.$$

This theorem is usually expressed by saying that "the general member of a linear system on X is smooth outside the singular locus of X and the base locus of the linear system," meaning that if $f_0, \ldots, f_n$ are functions on X, then for a general choice of $\alpha = [\alpha_0, \ldots, \alpha_n]$, the locus $\sum \alpha_i f_i = 0$ is smooth away from X_{sing} and the locus $f_0 = \cdots = f_n = 0$.

On important warning: Bertini's theorem is false in general for varieties in characteristic p. For example, in characteristic 2 every tangent line to the hypersurface $X \subset \mathbb{P}^4$ given by $UV + WY + Z^2$ passes through the point $p = [0, 0, 0, 0, 1]$, and conversely every hyperplane through p is tangent to X somewhere, contradicting the statement of Bertini's theorem as applied to the projection map $f = \pi_p$ from p. Given that it's a theorem valid in characteristic zero only, it's not surprising that the proof invokes the complex numbers in an essential way, in the form of Sard's theorem.

PROOF. Consider the incidence correspondence

$$\Gamma = \{(p, H): f(p) \in H\} \subset X \times \mathbb{P}^{n*}.$$

If X is k-dimensional, Γ will have dimension $k + n - 1$. Now, let $p \in X$ be any smooth point and $H \subset \mathbb{P}^n$ a hyperplane containing $f(p)$; we can choose coordinates so that $f(p) = [0, \ldots, 0, 1]$ and H is the hyperplane $Z_0 = 0$. Then in a neighborhood U of $p \in X$ we can write the map f by

$$f(q) = [f_0(q), \ldots, f_{n-1}(q), 1];$$

if we let H_α be the hyperplane

$$\{[Z]: Z_0 + \alpha_1 Z_1 + \cdots + \alpha_n Z_n = 0\},$$

$\alpha_1, \ldots, \alpha_n$ will give Euclidean coordinates on $\mathbb{P}^{n*}$ in a neighborhood V of H. In terms of these coordinates, Γ will be given in $U \times V$ by the single equation

$$f_0 + \alpha_1 f_1 + \cdots + \alpha_{n-1} f_{n-1} + \alpha_n = 0.$$

Since the partial derivative of this with respect to α_n is nonzero, we conclude that

$$\dim(T_{(p,H)}(\Gamma)) = \dim(T_{(p,H)}(X \times \mathbb{P}^{n*})) - 1;$$

in particular, the singular locus of Γ is exactly the inverse image $\pi_1^{-1}(X_{\text{sing}})$ of the singular locus of X.

Now look at the restricted map $\tilde{\pi}_2: \Gamma_{sm} \to \mathbb{P}^{n*}$. Bertini's theorem follows from applying Proposition 14.4 to $\tilde{\pi}_2$; or, more directly, by Exercise 14.6 the locus $U \subset \mathbb{P}^{n*}$ of hyperplanes H such that the fiber $(\tilde{\pi}_1)^{-1}(H) = X_{sm} \cap H$ is smooth is either contained in a proper subvariety of $\mathbb{P}^{n*}$ or contains an open subset of $\mathbb{P}^{n*}$, and we can simply apply Sard's theorem to $\tilde{\pi}_2$ deduce that the former must be the case. □

Note that it is not always the case that $X_{\text{sing}} \cap H \subset (X \cap H)_{\text{sing}}$; for example, consider the cone $X \subset \mathbb{A}^3$ given by $(xy - z^2)$ and the plane H given by (x). (This is one more argument for the language of schemes, where this inclusion does hold in general.)

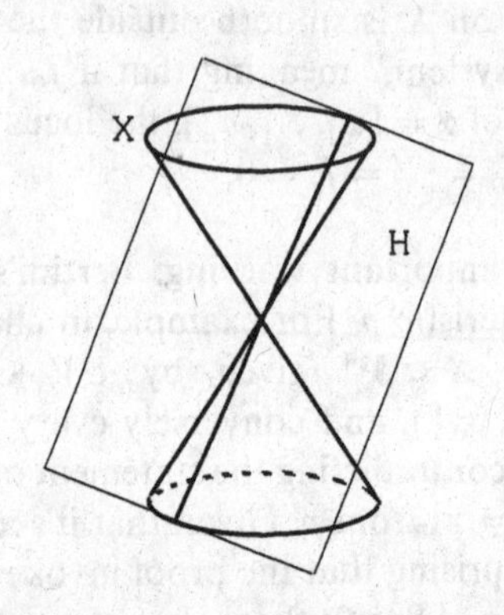

Exercise 17.17. Use Bertini's theorem to show that (a) the general hypersurface of degree d in $\mathbb{P}^n$ is smooth, and more generally that (b) for $k < n$, if $F_1, \ldots, F_k$ are general homogeneous polynomials of degrees $d_1, \ldots, d_k$ in $n + 1$ variables, the corresponding hypersurfaces in $\mathbb{P}^n$ intersect transversely in a smooth, $(n - k)$-dimensional variety $X \subset \mathbb{P}^n$.

Note that by part (b) of Exercise 17.17 the polynomials $F_1, \ldots, F_k$ generate the ideal of X locally, i.e., the ideal $(F_1, \ldots, F_k) \subset K[Z_0, \ldots, Z_n]$ is radical. It is also the case that it is saturated; this follows from the fact that $(F_1, \ldots, F_k)$ has no associated primes (see [E]). We conclude that they in fact generate the homogeneous ideal of X; we have thus the following.

Proposition 17.18. *For $k < n$, if $Y_1, \ldots, Y_k \subset \mathbb{P}^n$ are hypersurfaces intersecting transversely, then their intersection $X = Y_1 \cap \cdots \cap Y_k$ is a complete intersection in $\mathbb{P}^n$; in particular, the intersection of k general hypersurfaces in $\mathbb{P}^n$ is a complete intersection.*

Exercise 17.19. Let $X \subset \mathbb{P}^n$ be a k-dimensional variety, let $L_1, \ldots, L_k$ be general linear forms on $\mathbb{P}^n$, and let $\Lambda \cong \mathbb{P}^{n-k} \subset \mathbb{P}^n$ the linear subspace they define. Show that the ideal of X, together with the forms L_i, generate the ideal of the intersection $Y = X \cap \Lambda$ locally, thereby completing the proof of Proposition 13.2. (Note: the fact that the general $(n - k)$-plane Λ is transverse to X follows immediately from Bertini's theorem. It may also be seen in arbitrary characteristic by using the irreducibility of the universal $(n - k)$-plane section

$$\Omega^{(k)}(X) = \{(\Lambda, p) \colon p \in \Lambda\} \subset \mathbb{G}(n - k, n) \times X$$

(Exercise 11.44) and arguing that the locus of pairs (Λ, p) such that $\Lambda \cap \mathbb{T}_p X \supsetneqq \{p\}$ is a proper subvariety of $\Omega^{(k)}(X)$.)

One warning: Bertini's theorem tells us, for example, that if $X \subset \mathbb{P}^n$ is projective and smooth, and $V \subset K[Z_0, \ldots, Z_n]_d$ is any vector space of homogeneous polynomials of degree d, then the zero locus on X of the general $F \in V$ is smooth outside the common zero locus of all $F \in V$. It does not say, however, that this singular locus is constant. For example, we have the following.

Exercise 17.20. Let $\Lambda \subset \mathbb{P}^4$ be a 2-plane. Show that the general hypersurface $Y \subset \mathbb{P}^4$ of degree $m > 1$ containing Λ has singular locus a finite (nonempty) set of points in Λ, and that these points all vary with Y.

This exercise suggests (correctly) that the following one is nontrivial.

Exercise 17.21. Let $C \subset \mathbb{P}^3$ be a smooth curve. Show that for all sufficiently large m there exists a smooth surface $S \subset \mathbb{P}^3$ of degree m containing C. (*)

In fact, the techniques of Exercise 17.21 allow us to prove a stronger statement: if C is any curve such that for all points $p \in C$ the Zariski tangent space T_pC has dimension at most 2, then C lies on a smooth surface. It is not so clear whether any k-dimensional variety X with $\dim(T_pX) \leq k + 1$ $\forall p \in X$ lies on a smooth $(k + 1)$-dimensional variety (it is certainly not the case, given examples like that of Exercise 17.20, that any such k-dimensional $X \subset \mathbb{P}^{k+2}$ lies on a smooth hypersurface).

There is a nice generalization of Bertini's theorem, due to Kleiman ([K1]). It is as follows.

Theorem 17.22. *Let X be any variety on which an algebraic group G acts transitively, $Y \subset X$ any smooth subvariety, and $f: Z \to X$ any map. For any $g \in G$ let $W_g = (g \circ f)^{-1}(Y) \subset Z$—that is, the inverse image of the translate $g(Y)$. Then for general $g \in G$,*

$$(W_g)_{\text{sing}} = W_g \cap Z_{\text{sing}}.$$

In particular, if f is an inclusion, this says that a general translate of one smooth subvariety of a homogeneous space meets any other given smooth subvariety transversely.

Blow-ups, Nash Blow-ups, and the Resolution of Singularities

There are numerous respects in which it is easier to deal with smooth varieties than with singular ones; to mention just the most important, in analyzing smooth varieties in characteristic zero we can use the tools and techniques of complex manifold theory.

Clearly we cannot restrict our attention exclusively to smooth varieties, since singular ones arise all the time in the course of the standard constructions of algebraic geometry. To the extent that we are interested in properties of varieties that are invariant under birational isomorphism, however, this does raise a fundamentally important question: given an arbitrary variety X, does there exist a smooth variety Y birational to it? In practice, this is equivalent to the statement of the following theorem.

Theorem 17.23. Resolution of Singularities. *Let X be any variety. Then there exists a smooth variety Y and a regular birational map $\pi: Y \to X$.*

Such a map $\pi: Y \to X$ is called a *resolution of singularities* of X. This theorem was proved by Heisuke Hironaka [Hi] for varieties X in characteristic zero; in characteristic p it is known for curves and surfaces but is unknown in general. We will give a proof of this theorem for curves later.

In fact, the proof says something more specific than what was stated earlier: it says that the singularities of any variety may be resolved by a sequence of blow-ups—that is, given any X there is a sequence of varieties X_i, regular maps

$$Y \xrightarrow{\pi_n} X_n \to X_{n-1} \to \cdots \xrightarrow{\pi_2} X_2 \xrightarrow{\pi_1} X_1 = X$$

and subvarieties $Z_i \subset X_i$ for $i = 1, \ldots, n$ such that the map $\pi_i \colon X_{i+1} \to X_i$ is the blow-up of X_i at Z_i.

To see why it is reasonable to expect that blowing up might resolve singularities, consider the simplest of all singular varieties, the nodal cubic $C \subset \mathbb{P}^2$ given by $Y^2Z = X^3 + X^2Z$ (or, in Euclidean coordinates $x = X/Z$, $y = Y/Z$ on $\mathbb{A}^2 \subset \mathbb{P}^2$, by the equation $y^2 = x^2(x+1)$). The singularity of this curve at the point $p = [0, 0, 1]$ is called a *node*; in case our ground field $K = \mathbb{C}$ it may be characterized by saying that a neighborhood of p in C in the analytic topology consists of the union of two smooth arcs meeting with distinct tangents at p.

What happens if we blow up the curve C at the point p? To see this, we can first of all just write everything out in coordinates; the blow-up is the graph of the rational map

$$\pi \colon C \to \mathbb{P}^1$$

given by

$$[X, Y, Z] \mapsto [U, V] = [X, Y],$$

or, in terms of Euclidean coordinates (x, y) on $\tilde{C} = C \cap \mathbb{A}^2$ and $v = V/U$ on $\mathbb{A}^1 \subset \mathbb{P}^1$, by

$$(x, y) \mapsto y/x.$$

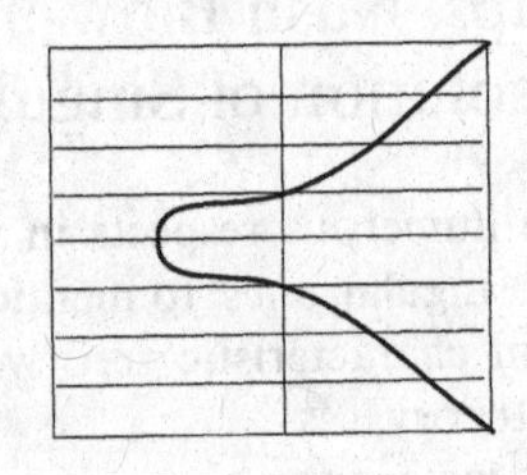

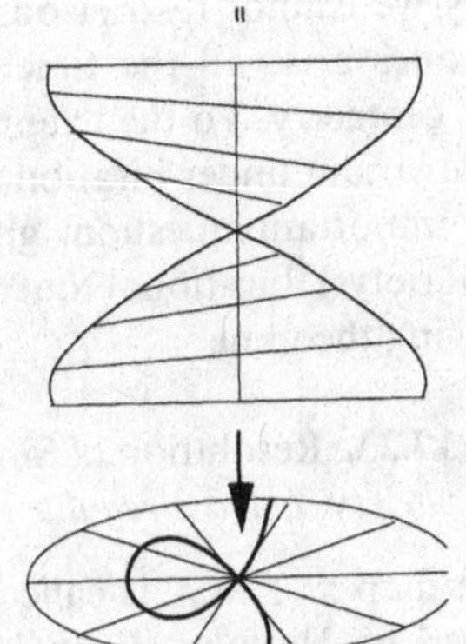

In the open subset of $\mathbb{A}^2 \times \mathbb{A}^1$ where $x \neq 0$, then, the graph is defined by the equations $y^2 = x^2(x+1)$ and $v = y/x$, or, equivalently, $v^2 = x + 1$ and $vx = y$. The portion $\tilde{\Gamma}$ of the graph lying in $\mathbb{A}^2 \times \mathbb{A}^1$ thus has local equation

$$\Gamma = \{(x, y, u) \colon v^2 - (x+1)$$
$$= vx - y = 0\};$$

in particular, we see that there are two points of Γ—$(0, 0, 1)$ and $(0, 0, -1)$—lying over the point $p \in C$, and that these are both smooth points of Γ.

It is perhaps easiest to visualize this blow-up by looking at the blow-up $\pi \colon \tilde{\mathbb{P}}^2 \to \mathbb{P}^2$ of $\mathbb{P}^2$ at p. Recall from Lecture 7 the basic picture of $\tilde{\mathbb{P}}^2$, which is the disjoint union of lines L_v, with L_v mapping to the line through p in $\mathbb{P}^2$ with slope v. (We may thus think of $\tilde{\mathbb{P}}^2$ as the lines in $\mathbb{P}^2$ through p made disjoint, and lying over the corresponding lines in $\mathbb{P}^2$ at "heights" corresponding to their slope.) Each point of $C - \{p\}$ is thus "lifted" to a height equal to the ratio $v = y/x$ of its coordinates; in particular, as we approach the point p the limiting value of this slope will be either 1 or -1 depending on which branch of C we are on.

A similar picture applies when we look at slightly more complicated singularities. For example, if a plane curve C has a *tacnode*—that is, a singular point p whose analytic neighborhood consists of two smooth arcs simply tangent to one another, such as the origin on the curve $y^2 = x^4 - x^5$—then the blow-up of C at p will not be smooth, but will have an ordinary node, so that a further blow-up will resolve its singularities. Likewise, if C has a *cusp* (like the curve $y^2 = x^3$ encountered in Lecture 1) the blow-up will be smooth.

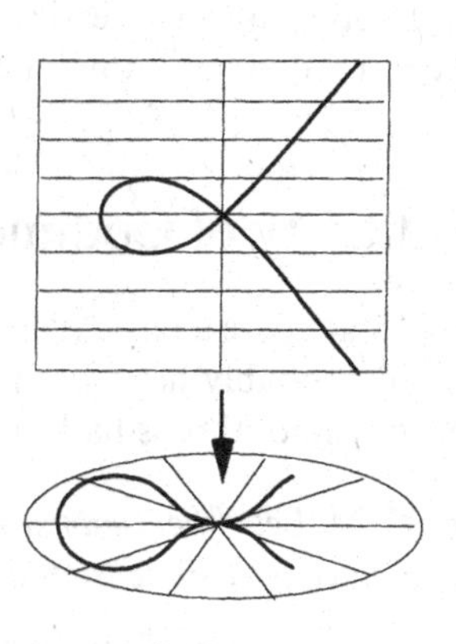

There is one problem with resolving singularities by blowing up. If the singularities of a variety X are sufficiently complicated—for example, if the singular locus is positive-dimensional, as indeed it may be after blowing up even a variety with isolated singularities—it is not clear what subvarieties of X we should blow up in what order to achieve the desired resolution. There is an alternate approach that is more canonical, though it is not known to work in every case; this is called the *Nash blow-up* of X.

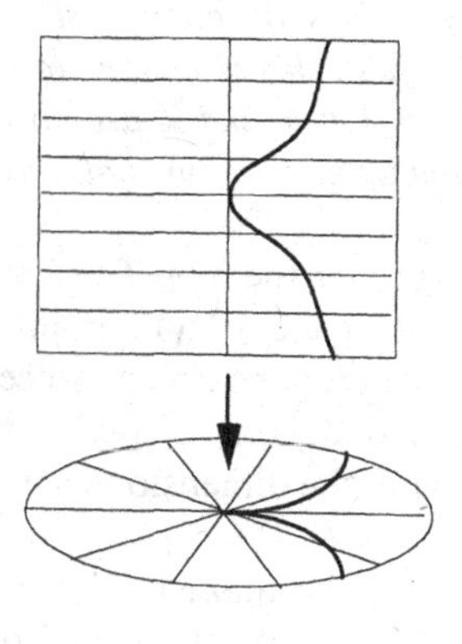

Very simply, the Nash blow-up of a k-dimensional variety $X \subset \mathbb{P}^n$ is the graph Γ of the Gauss map $\mathscr{G}_X: X \dashrightarrow \mathbb{G}(k, n)$. (In particular, it does not depend on any choice of subvariety of X.) Of course, the map $\mathscr{G}_X$ is regular on the smooth locus of X, so that the projection $\pi: \Gamma \to X$ will be an isomorphism over X_{sm}. At the same time, we may think of our inability to extend this map to a regular map on all of X as a measure of the singularity of X. For example, if $X = C$ is the nodal cubic given earlier, the map $\mathscr{G}_C$ will clearly not be well-defined at p; both the lines L and M will be limits of tangent lines to p, so that there will be two points (p, L) and $(p, M) \in \Gamma$ lying

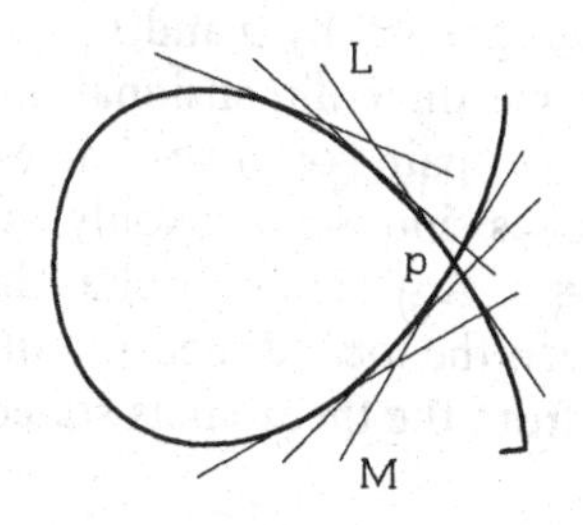

over $p \in C$. Even in cases where there is apparently a unique limiting position of the tangent plane, the Nash blow-up effects a change. For example, if X is the cuspidal curve above it is less obvious (but straightforward to check) that Γ is smooth; while if X has a tacnode, again it may be checked that Γ will have an ordinary node. It may thus be hoped that the singularities of an arbitrary variety could be resolved by a succession of Nash blow-ups; but while this is known to be true for curves, it is not known in general.

Subadditivity of Codimension of Intersections

We should state here a basic theorem on the dimension of intersections of varieties. This should probably have been stated in Lecture 11, but because of a necessary hypothesis of smoothness had to wait until now. It is as follows.

Theorem 17.24. *Let Y be a subvariety of a smooth variety X, $f: Z \to X$ a regular map. If Y and Z have pure dimension and Y and $f(Z)$ are not disjoint, then*

$$\dim(f^{-1}(Y)) \geq \dim(Z) + \dim(Y) - \dim(X);$$

or, in other words, the codimension of the inverse image is at most the codimension of Y in X (in fact, this is true of every component of $f^{-1}(Y)$). In particular, if f is an inclusion, so that Y and Z are both subvarieties of X, then the codimension of their intersection (if nonempty) is at most the sum of their codimensions.

Proof. Consider the map $f \times i: Z \times Y \to X \times X$, where $i: Y \to X$ is the inclusion. The inverse image $f^{-1}(Y)$ is just the inverse image $(f \times i)^{-1}(\Delta_X)$, where $\Delta_X \subset X \times X$ is the diagonal. Now, since X is smooth, Δ_X is a local complete intersection on $X \times X$, that is, it is locally the zero locus of $\dim(X)$ regular functions. By Exercise 11.6, the dimension estimate on $f^{-1}(Y)$ follows. □

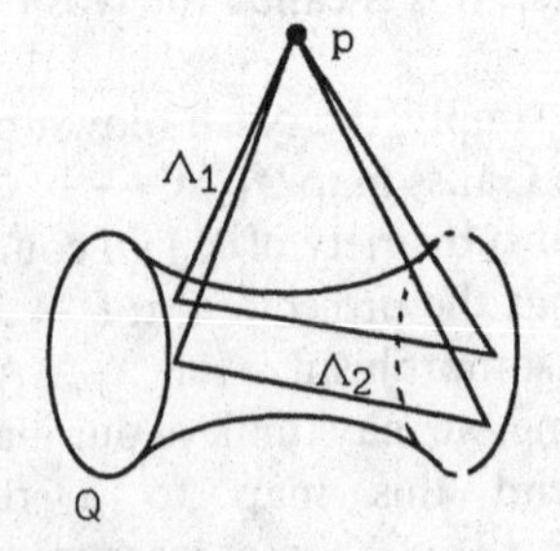

One remark to make here is that the statement is false if we allow X arbitrary; for example, take $Q \subset \mathbb{P}^3$ a smooth quadric surface, $X \subset \mathbb{P}^4$ the cone $\overline{pQ}$ over Q with vertex p, L_1 and L_2 lines of the same family on Q (so that in particular $L_1 \cap L_2 = \emptyset$), and $\Lambda_i = \overline{p, L_i}$ the plane spanned by p and L_i. Then X is of course three-dimensional and the $\Lambda_i \subset X$ a pair of two-dimensional subvarieties intersecting only at the point $p \in X$. A second remark is that the theorem is true locally; if $p \in f^{-1}(Y)$ is any point, then the local dimension $\dim_p(f^{-1}(Y)) \geq \dim(Z) + \dim(Y) - \dim(X)$. This follows from the theorem as stated simply by replacing Z by a neighborhood of p in Z.

The last remark is that we can combine this theorem with dimension calculations made in Lecture 11 to derive more general dimension estimates. For example, we have the following.

Proposition 17.25. *Let X be any variety, $(f_{i,j})$ an $n \times m$ matrix of regular functions on X, and $Z \subset X$ the locus defined by*

$$Z = \{p \in X \colon \operatorname{rank}(f_{i,j}(p)) \leq k\}.$$

Then either Z is empty or

$$\dim(Z) \geq \dim(X) - (m - k)(n - k).$$

PROOF. This is straightforward: the functions define a map f from X to the vector space $\tilde{M}$ of matrices, and Z is just the inverse image of the subvariety $\tilde{M}_k \subset \tilde{M}$ of matrices of rank at most k; combining Theorem 17.24 and Example 12.1 we arrive at the statement. □

Note that we could also apply this to a matrix of homogeneous polynomials on a projective variety $X \subset \mathbb{P}^l$, just applying the result as stated locally.

Given this statement, we may introduce a notion analogous to the notion of complete intersections: if $Z \subset \mathbb{P}^l$ is a subvariety whose ideal is generated by the $(k + 1) \times (k + 1)$ minors of an $m \times n$ matrix of homogeneous polynomials $F_{i,j}$, we will say that Z is a *proper determinantal variety* if

$$\dim(Z) = l - (m - k)(n - k).$$

Such proper determinantal varieties are like complete intersections in the sense that their numerical invariants (e.g., their Hilbert polynomials) may be completely described in terms of the degrees of the polynomials $F_{i,j}$. (In fact, complete intersections may be viewed as special cases of proper determinantal varieties; by definition, a complete intersection $X \subset \mathbb{P}^n$ of codimension l is the proper determinantal variety associated to a $1 \times l$ matrix). One word of warning: in some texts, the term "determinantal variety" is used to mean what we here call "proper determinantal variety."

LECTURE 18
Degree

Our next fundamental notion, that of the *degree* of a projective variety $X \subset \mathbb{P}^n$, may be defined in ways for the most part exactly analogous to the notion of dimension. As in the case of the various definitions of dimension, the fact that these definitions are equivalent (or, in some cases, that they are well-defined at all) will not be established until we have introduced them all.

To start with, we defined dimension by saying that the dimension of $\mathbb{P}^n$ was n. We then said that a projective variety $X \subset \mathbb{P}^n$ admits a finite surjective map to some projective space $\mathbb{P}^k$, which may be given as the projection π_Λ from a general linear subspace $\Lambda \cong \mathbb{P}^{n-k-1}$ disjoint from X, and we defined the dimension of X to be that k.

To define degree, we may again start with a case where we have an a priori notion of what degree should be: if $X \subset \mathbb{P}^{k+1}$ is a hypersurface, given as the zero locus of an irreducible polynomial $F(Z)$, then the degree of X is the degree of the polynomial F. Now start with an arbitrary irreducible $X \subset \mathbb{P}^n$. As we have seen, unless X is already a hypersurface, the projection map $\pi_p\colon X \to \mathbb{P}^{n-1}$ from a general point $p \in \mathbb{P}^n$ is birational onto its image; in fact, if we choose $q \in X$ any point, it's enough for p to lie outside the cone $\overline{q, X}$. Thus, projecting successively from general points, we see that the projection map $\pi_\Gamma\colon X \to \mathbb{P}^{k+1}$ from a general $(n-k-2)$-plane is birational onto its image $\overline{X}$, which is a hypersurface in $\mathbb{P}^{k+1}$; and we may take the degree of X to be the degree of this hypersurface. Of course, to define the degree of X to be the degree of the "general" projection $\overline{X}$ is illegal unless we know that in fact the function $d(\Gamma) = \deg(\pi_\Gamma(X))$ is constant on an open subset of the Grassmannian $\mathbb{G}(n-k-2, n)$; this will be established in Exercise 18.2.

Another way to express this is in terms of the projection $\pi_\Lambda\colon X \to \mathbb{P}^k$. This is just the composition of the projection $\pi_\Gamma\colon X \to \mathbb{P}^{k+1}$ with the projection map $\pi_p\colon \overline{X} \to \mathbb{P}^k$ from a general point $p \in \mathbb{P}^{k+1}$, so that a general fiber of this map consists of the intersection of $\overline{X}$ with a general line. But if $\overline{X}$ is the zero locus of a homogeneous polynomial of degree d in $\mathbb{P}^{k+1}$ a general line in $\mathbb{P}^{k+1}$ will meet it in d points; so we can also take the degree of X to be the degree of the map π_Λ as defined in Lecture 7, that is, the number of points in a general fiber. Equivalently, another characterization of the dimension of X is that number k such that the general $(n-k)$-plane $\Omega \subset \mathbb{P}^n$ meets X in a finite collection of points; since for $\Omega \supset \Lambda$ the intersection $\Omega \cap X$ is the fiber of the map $\pi_\Lambda: X \to \mathbb{P}^k$ over the point $\Omega \cap \mathbb{P}^k$, we see that the degree can further be defined to be the number of points of intersection of X with a general $(n-k)$-plane $\Omega \subset \mathbb{P}^n$. In sum, then, we have the following.

Definition 18.1. Let $X \subset \mathbb{P}^n$ be an irreducible k-dimensional variety. If Γ, Λ, and Ω are a general $(n-k-2)$-plane, $(n-k-1)$-plane, and $(n-k)$-plane, respectively, then the *degree* of X, denoted $\deg(X)$, is either

(i) the degree of the hypersurface $\overline{X} = \pi_\Gamma(X) \subset \mathbb{P}^{k+1}$;
(ii) the degree of the finite surjective map $\pi_\Lambda: X \to \mathbb{P}^k$; or
(iii) the number of points of intersection of Ω with X.

Exercise 18.2. Assuming $\mathrm{char}(K) = 0$, use Exercises 7.16 and 11.44 to show that part (iii) of Definition 18.1 is well-defined (part (iii) amounts to saying that the degree of X is the degree of the projection $\Omega^{(k)}(X) \to \mathbb{G}(n-k, n)$, where $\Omega^{(k)}(X) \subset \mathbb{G}(n-k, n) \times X$ is the universal k-fold hyperplane section of X), and use this to deduce that parts (i) and (ii) are also well-defined.

In fact, following Bézout's theorem we will be able to say exactly which $(n-k-2)$-planes, $(n-k-1)$-planes, and $(n-k)$-planes are general in the sense of giving the correct degree of X.

Of course, we can restate part (ii) of Definition 18.1 in terms of the function field $K(X)$ (though, since degree is, unlike dimension, very much dependent on choice of embedding, it cannot be in terms of $K(X)$ alone). Recall that the projection π_Λ induced an inclusion of function fields

$$K(\mathbb{P}^k) = K(z_1, \ldots, z_k) \hookrightarrow K(X),$$

expressing $K(X)$ as an algebraic extension of $K(z_1, \ldots, z_n)$; the degree of X will then be the degree of this field extension.

To complete the analogy between the definitions of dimension and degree, there is a definition of degree in terms of the Hilbert polynomial. We have already laid the groundwork for this: as we saw in Lecture 13, if h_X is the Hilbert function of a variety $X \subset \mathbb{P}^n$ of dimension k, then the kth difference function—that is, the function $h^{(k)}$ defined inductively by setting $h^{(0)} = h_X$ and

$$h^{(l+1)}(m) = h^{(l)}(m) - h^{(l)}(m-1)$$

is for large m the Hilbert function $h_{X \cap \Omega}$ of the intersection $X \cap \Omega$ of X with a general $(n-k)$-plane $\Omega \subset \mathbb{P}^n$. (Recall that this is true for any plane Ω such that the sum $(I(\Omega), I(X))$ of the ideals of X and Ω generates the ideal $I(X \cap \Omega)$ of the intersection locally.) The fact that the Hilbert polynomial of $X \cap \Omega$ is a constant led us to the conclusion that the Hilbert polynomial p_X of X was a polynomial of degree k. Similarly, the fact that the constant $P_{X \cap \Omega}$ in question is equal to the number d of points of $X \cap \Omega$ tells us now that the leading term of $p_X(m)$ is $d \cdot m^k/k!$; i.e., the degree d of X is $k!$ times the leading coefficient of the Hilbert polynomial p_X.

Note that this description of the degree allows us to deduce the statement of Exercise 18.2 in arbitrary characteristic, inasmuch as by Exercise 17.19 the general $(n-k)$-plane Ω has the property that the saturation of $(I(\Omega), I(X))$ is $I(X \cap \Omega)$.

We should mention a couple of other interpretations of the degree of a variety $X \subset \mathbb{P}^n$ that are special to the case of complex varieties. To begin with, suppose that X is smooth, that is, that it is a complex submanifold of $\mathbb{P}^n$. Then, being a compact, orientable real $2k$-manifold, X carries a fundamental class $[X]$, which we may view as being a class in the homology $H_{2k}(\mathbb{P}^n, \mathbb{Z})$. Now, the homology of complex projective space in even dimensions is just $\mathbb{Z}$, with generator $[\Phi]$ the class of a linear subspace of $\mathbb{P}^n$; so we can write

$$[X] = d \cdot [\Phi]$$

where Φ is a k-plane in $\mathbb{P}^n$. The coefficient d is then the degree of X.

The same definition in fact makes sense when X is not a complex submanifold of $\mathbb{P}^n$, though it's less easy to see this. What will be the case in general is that X admits a triangulation in which the nonmanifold points of X form a subcomplex ([Hi1]). By Exercise 14.3 the nonmanifold points of X form a proper subvariety of X, so that they will contain only simplices of dimension $2k-2$ and less. It is thus still true that X carries a fundamental class $[X] \in H_{2k}(X, \mathbb{Z}) \to H_{2k}(\mathbb{P}^n, \mathbb{Z})$ (and, though it is not a consequence of this, that $[X]$ is independent of the triangulation). We may accordingly define the degree of X the same way as before.

That a singular projective variety has a fundamental class also follows immediately from the resolution of singularities (Theorem 17.23), but to invoke this would truly be overkill. As a simpler alternative to actually triangulating X, to define $[X] \in H_{2k}(X, \mathbb{C})$ it is enough to show that X behaves like a manifold as far as integration is concerned, i.e., that the integral over X of any exact $2k$-form on $\mathbb{P}^n$ is zero; we then get a class η_X in the deRham cohomology group $H^{2k}(\mathbb{P}^n)^* = H_{2k}(\mathbb{P}^n, \mathbb{C})$. That the homological definition of degree agrees with the earlier ones follows from the fact that the maps

$$H_*(\mathbb{P}^n, \mathbb{Z}) \leftarrow H_*(\mathbb{P}^n - \{p\}, \mathbb{Z}) \to H_*(\mathbb{P}^{n-1}, \mathbb{Z})$$

induced by the inclusion $\mathbb{P}^n - \{p\} \subset \mathbb{P}^n$ and the projection $\pi_p \colon \mathbb{P}^n - \{p\} \to \mathbb{P}^{n-1}$ from a point are isomorphisms in degree $< 2n$.

Yet another interpretation of degree is truly peculiar. It is a special feature of a Kahler metric on a complex manifold that the area, or volume, of a complex submanifold X is given as the integral over X of the appropriate power of the associated (1, 1)-form of the metric. In the case of $\mathbb{P}^n$, the associated (1, 1)-form of the standard Fubini-Study metric (that is, the Hermitian metric on complex projective space invariant under the unitary group U_{n+1}) represents a generator of $H^2(\mathbb{P}^n, \mathbb{C})$ in deRham cohomology. Its powers correspondingly represent the generators of $H^{2k}(\mathbb{P}^n, \mathbb{C})$; so that if X is a k-dimensional subvariety of degree d in $\mathbb{P}^n$ we have

$$\int_X \omega^k = d \cdot \int_\Lambda \omega^k$$

for $\Lambda \cong \mathbb{P}^k \subset \mathbb{P}^n$ a k-plane. We can multiply the metric by a scalar to make the volume of a k-plane 1, so that the degree of a subvariety $X \subset \mathbb{P}^n$ is simply its area. In fact, complex subvarieties of $\mathbb{P}^n$ have minimal area among cycles in their homology class.

Finally, we define the degree of a reducible variety of dimension k to be, in general, the sum of the degrees of its k-dimensional irreducible components.

Bézout's Theorem

The basic fact about degree that we will want to know in order to compute examples is a simple one. First, to express Bézout's theorem in its simplest form, we need to introduce one bit of terminology. Suppose that X and $Y \subset \mathbb{P}^n$ are two subvarieties and that their intersection has irreducible components Z_i. We say that X and Y intersect *generically transversely* if, for each i, X and Y intersect transversely at a general point $p_i \in Z_i$, i.e., are smooth at p_i with tangent spaces spanning $\mathbb{T}_{p_i}(\mathbb{P}_n)$. Of course, if $\dim(X) + \dim(Y) = n$, saying that X and Y intersect generically transversely is the same as saying X and Y intersect transversely. We then have the following.

Theorem 18.3. *Let X and $Y \subset \mathbb{P}^n$ be subvarieties of pure dimensions k and l with $k + l \geq n$, and suppose they intersect generically transversely. Then*

$$\deg(X \cap Y) = \deg(X) \cdot \deg(Y).$$

In particular, if $k + l = n$, this says that $X \cap Y$ will consist of $\deg(X) \cdot \deg(Y)$ *points.*

We will give a proof of this theorem immediately following Example 18.17. We will be concerned in the meantime just with refinements and examples of these notions, so that the reader may skip directly to Example 18.17 and the proof.

We can strengthen the statement of Bézout substantially. Briefly, we say that a pair of pure-dimensional varieties X and $Y \subset \mathbb{P}^n$ *intersect properly* if their intersection has the expected dimension, i.e.,

$$\dim(X \cap Y) = \dim(X) + \dim(Y) - n.$$

Now, to any pair of varieties X, $Y \subset \mathbb{P}^n$ intersecting properly, and any irreducible variety $Z \subset \mathbb{P}^n$ of dimension $\dim(X) + \dim(Y) - n$, we can associate a nonnegative integer $m_Z(X, Y)$ called the *intersection multiplicity of* X and Y along Z, with the following properties:

(i) $m_Z(X, Y) \geq 1$ if $Z \subset X \cap Y$ (for formal reasons we set $m_Z(X, Y) = 0$ otherwise);

(ii) $m_Z(X, Y) = 1$ if and only if X and Y intersect transversely at a general point $p \in Z$;

(iii) $m_Z(X, Y)$ is additive, i.e., $m_Z(X \cup X', Y) = m_Z(X, Y) + m_Z(X', Y)$ for any X and X' as long as all three are defined and X and X' have no common components.

In terms of this multiplicity, we have the following.

Theorem 18.4. *Let X and $Y \subset \mathbb{P}^n$ be subvarieties of pure dimension, intersecting properly. Then*

$$\deg(X) \cdot \deg(Y) = \sum m_Z(X, Y) \cdot \deg(Z)$$

where the sum is over all irreducible subvarieties Z of the appropriate dimension (in effect, over all irreducible components Z of $X \cap Y$).

In case $\dim(X) + \dim(Y) = n$ and X and Y are local complete intersections intersecting properly, $m_Z(X, Y)$ has a straightforward description: for any isolated point p of intersection of X and Y it is the dimension (as vector space over K) of the quotient $\mathcal{O}_{\mathbb{P}^n, p}/(I_X + I_Y)$, where $\mathcal{O}_{\mathbb{P}^n, p}$ is the local ring of $\mathbb{P}^n$ at the point p and I_X and I_Y are the ideals of X and Y. Slightly more generally, if X and Y are local complete intersections of any dimensions intersecting properly, $m_Z(X, Y)$ may be defined by taking a normal slice; that is, choosing a general linear space Γ of codimension $\dim(X) + \dim(Y) - n$ in $\mathbb{P}^n$ intersecting Z transversely at a general point $p \in Z$, and taking the intersection multiplicity of X and Y along Z to be $m_p(X_0, Y_0)$, where $X_0 = X \cap \Gamma$ and $Y_0 = Y \cap \Gamma$. Equivalently, we can choose an affine open subset $U \subset \mathbb{P}^n$ not missing Z, setting $\tilde{X} = X \cap U$, $\tilde{Y} = Y \cap U$, and $\tilde{Z} = Z \cap U$, and defining $m_Z(X, Y)$ to be the rank of the $A(\tilde{Z})$-module $A(U)/(I_{\tilde{X}} + I_{\tilde{Y}})$.

To give a more explicit description of $m_Z(X, Y)$ in general would take us too far afield at present. Even so, from just the rudimentary properties listed here we may deduce a number of corollaries of the refined Bézout theorem. For example, we have the following.

Corollary 18.5. *Let X and Y be subvarieties of pure dimension in $\mathbb{P}^n$ intersecting properly. Then*

$$\deg(X \cap Y) \leq \deg(X) \cdot \deg(Y).$$

Corollary 18.6. *Let X and Y be subvarieties of pure dimension intersecting properly, and suppose that*

$$\deg(X \cap Y) = \deg(X) \cdot \deg(Y).$$

Then X and Y are both smooth at a general point of any component of $X \cap Y$. In particular, if X and Y have complementary dimension, they must be smooth at all points of $X \cap Y$.

Exercise 18.7. Let $X \subset \mathbb{P}^n$ be any subvariety of degree 1. Show that X is a linear subspace of $\mathbb{P}^n$.

As one immediate consequence of the Bézout theorem, we are able to prove a statement made much earlier: that every automorphism of projective space $\mathbb{P}^n$

is linear, that is, induced by an automorphism $A \in \mathrm{GL}_{n+1}K$ of K^{n+1}. The key ingredient supplied by Bézout is the observation that if H and $L \subset \mathbb{P}^n$ are subvarieties of dimensions $n-1$ and 1, respectively, meeting transversely at one point, then H must be a hyperplane and L a line. It follows that any automorphism of $\mathbb{P}^n$ carries hyperplanes into hyperplanes, and this is essentially all we need to know. We can write any such automorphism $\varphi: \mathbb{P}^n \to \mathbb{P}^n$ in terms of Euclidean coordinates $(z_1, \ldots, z_n)$ and $(w_1, \ldots, w_n)$ on source and target as

$$w_i = \frac{f_i(z_1, \ldots, z_n)}{g_i(z_1, \ldots, z_n)}$$

and then to say that a linear relation among the w_i implies a linear relation among the z_i amounts to saying that the degrees of all the f_i and g_i are at most 1, with all the g_i scalar multiples of one another. □

Note that in case $K = \mathbb{C}$ the fact that hyperplanes are carried into hyperplanes under any automorphism of $\mathbb{P}^n$ also follows directly from the topological characterization of hyperplanes as subvarieties whose fundamental classes are generators of $H^2(\mathbb{P}^n, \mathbb{Z})$.

Example 18.8. The Rational Normal Curve

Consider first the rational normal curve $C = v_d(\mathbb{P}^1) \subset \mathbb{P}^d$, that is, the image of the map

$$v_d: [X_0, X_1] \mapsto [X_0^d, X_0^{d-1}X_1, \ldots, X_1^d].$$

The intersection of C with a general hyperplane $(a_0Z_0 + \cdots + a_dZ_d = 0)$ is just the image of the zero locus of the general polynomial $a_0X_0^d + a_1X_0^{d-1}X_1 + \cdots + a_dX_1^d$, which of course is d points; thus the degree of C is d.

A number of the facts established directly in Part I become clear from this and Bézout's theorem. For example, the fact that the intersection of any two quadrics containing a twisted cubic curve $C \subset \mathbb{P}^3$ consists of the union of C and a line follows immediately; if two quadrics both contain a twisted cubic, they must intersect generically transversely, and so their intersection will have degree 4. We will point out other such applications as we go along.

Note that Corollaries 18.5 and 18.6 immediately imply a sort of converse to this degree computation; we have the following. Note: we say that a variety $X \subset \mathbb{P}^n$ is *nondegenerate* if it does not lie in any hyperplane.

Proposition 18.9. *Let $C \subset \mathbb{P}^d$ be any irreducible nondegenerate curve. Then* $\deg(C) \geq d$, *and if* $\deg(C) = d$ *then C is the rational normal curve.*

Proof. To see the first part, suppose just that C is irreducible of degree strictly less than d and choose any d points $p_1, \ldots, p_d \in C$. Any d points in $\mathbb{P}^d$ lie in a hyperplane $H \cong \mathbb{P}^{d-1}$; but then by Corollary 18.5 $\dim(H \cap C) = 1$ and hence $C \subset H$. As for the second, if C is irreducible of degree d and lies in no hyperplane, observe first that by the preceding any $d+1$ points of C must be linearly independent; and by

Corollary 18.6, C must be smooth. Now choose any points $p_1, \ldots, p_{d-1} \in C$; they span a plane $\Lambda \cong \mathbb{P}^{d-2} \subset \mathbb{P}^d$. For any hyperplane H containing Λ, then, the intersection $H \cap C$ will consist of $p_1, \ldots, p_{d-1}$ and one further point $p(H)$ (or just $p_1, \ldots, p_{d-1}$ if H is tangent to C at one of the p_i). In the other direction, for any point $p \in C$, let $H(p)$ be the hyperplane spanned by p and the points $p_1, \ldots, p_{d-1}$ (or the hyperplane spanned by $p_1, \ldots, p_{d-1}$ and the tangent line to C at p_i if $p = p_i$). This gives an isomorphism between C and the line $\Lambda^* = \{H: H \supset \Lambda\} \cong \mathbb{P}^1 \subset \mathbb{P}^{d*}$, and it follows from the description of rational curves in general given in Lecture 1 that C is a rational normal curve. □

We can extend the basic lower bound of Proposition 18.9 to varieties of higher dimension. The basic fact we need to do this is the following.

Proposition 18.10. *Let $X \subset \mathbb{P}^n$ be an irreducible nondegenerate variety of dimension $k \geq 1$, and let $Y = X \cap H \subset H \cong \mathbb{P}^{n-1}$ be a general hyperplane section of X. Then Y is nondegenerate in $\mathbb{P}^{n-1}$, and if $k \geq 2$ then Y is irreducible as well.*

PROOF. To begin, the nondegeneracy statement is relatively elementary; we will prove it for any H intersecting X generically transversely. Suppose that Y is contained in a subspace $\Lambda \cong \mathbb{P}^{n-2}$, and let $\{H_\lambda\}$ be the family of hyperplanes in $\mathbb{P}^n$ containing Λ. Since X is not contained in any finite union of hyperplanes, it will intersect a general member H_λ of this pencil in at least one point p not lying on Y; the point p will lie on an irreducible component Z_λ of $X \cap H_\lambda$ not contained in Y. By Bézout, then, we will have

$$\deg(X \cap H_\lambda) \geq \deg(Y) + \deg(Z_\lambda) > \deg(Y) = d,$$

a contradiction.

As for the irreducibility of $X \cap H$, this is more subtle; we will only sketch a proof here, and that over the complex numbers. For more details (and a much stronger theorem), see [FL].

To begin, just to make things easier to visualize, observe that it will suffice to prove the theorem in case X is a surface. Likewise, it is enough to prove it for a general hyperplane H containing a given $(n-3)$-plane Λ; so we might as well look at the projection map $f = \pi_\Lambda: X \to \mathbb{P}^2$ and prove that the inverse image $\pi^{-1}(L)$ of a general line $L \subset \mathbb{P}^2$ is irreducible.

Now consider the locus Z of points $p \in X$ that are singular or such that the differential df_p is singular—i.e., such that $\operatorname{Ker}(df_p) \neq 0$. This is a proper closed subset of X; let $B = f(Z)$ be its image in $\mathbb{P}^2$. B will a priori be the union of a plane curve of some degree b and a finite collection of points (in fact, there will be no 0-dimensional components of B, but we don't need to know this). Let $U = \mathbb{P}^2 - B$ be the complement of B and $V = f^{-1}(U) \subset X$ the inverse image of U in X.

Choose a general point $q \in U$. Let $\{L_\lambda\}_{\lambda \in \mathbb{P}^1}$ be the pencil of lines in $\mathbb{P}^2$ containing q and $\lambda_1, \ldots, \lambda_m$ the values of λ such that L_λ fails to meet B transversely in b points. Consider the incidence correspondence

$$\Sigma = \{(\lambda, p): f(p) \in L_\lambda\} \subset \mathbb{P}^1 \times V;$$

we may recognize this as simply the blow-up of V at the points $p_1, \ldots, p_d$ lying over q. As a final grooming step, throw away the points $\lambda_i \in \mathbb{P}^1$ and their inverse images in Σ; that is, set

$$W = \mathbb{P}^1 - \{\lambda_1, \ldots, \lambda_m\}$$
$$\tilde{\Sigma} = (\pi_1)^{-1}(W).$$

After all this, the map $\pi_1 \colon \tilde{\Sigma} \to W$ is a topological fiber bundle, with each fiber $(\pi_1)^{-1}(\lambda)$ a d-sheet covering space of the open set $L_\lambda - (L_\lambda \cap B) \subset L_\lambda$; the points p_i give sections of this bundle. Since X is irreducible, $\tilde{\Sigma}$ will be connected; since a fiber bundle with connected total space that admits a section must have connected fibers, we conclude that $(\pi_1)^{-1}(\lambda)$ is connected for all $\lambda \in W$. But $(\pi_1)^{-1}(\lambda)$ is an open dense subset of the curve $f^{-1}(L_\lambda) \subset X$, so this curve must be irreducible. □

Exercise 18.11. Give a proof of the nondegeneracy of the general hyperplane section of the variety X without invoking the notion of degree (or Bézout's theorem), as follows. First, show that if the general hyperplane section of X spans a k-plane, we have a rational map

$$\varphi \colon \mathbb{P}^{n*} \dashrightarrow \mathbb{G}(k, n)$$

defined by sending a general $H \in \mathbb{P}^{n*}$ to the span of $H \cap X$. Next, use the fact that the universal hyperplane section of X is irreducible to deduce that for any $H \in \mathbb{P}^{n*}$ and any point $\Lambda \in \varphi(H)$ (that is, any point in the image of the fiber of the graph Γ_φ over H), the hyperplane section $H \cap X$ lies on the k-plane Λ. It follows that if the general hyperplane section of X is degenerate, then all are—but any n independent points of X will span a hyperplane H with $X \cap H$ nondegenerate.

Corollary 18.12. *If $X \subset \mathbb{P}^d$ is any irreducible nondegenerate variety of dimension k, then the degree of X is at least $d - k + 1$.*

Proof. This is simply the amalgam of Propositions 18.9 and 18.10; by the latter, the intersection of X with a general plane of dimension $d - k + 1$ will be an irreducible nondegenerate curve in $\mathbb{P}^{d-k+1}$. □

We will give in Theorem 19.9 a complete list of varieties of this minimal degree.

More Examples of Degrees

Example 18.13. Veronese Varieties

Consider now Veronese varieties in general, that is, the image X of the Veronese map $v_d \colon \mathbb{P}^n \to \mathbb{P}^N$. Its degree is the number of points of intersection of the image with a general linear subspace of codimension n in $\mathbb{P}^N$, or equivalently with n general hyperplanes $H_i \subset \mathbb{P}^N$. But now the inverse image $v_d^{-1}(H_i)$ of a general hyperplane in $\mathbb{P}^N$ is a general hypersurface Y_i of degree d in $\mathbb{P}^n$; the intersection of n of these consists of d^n points by Bézout's theorem (by Bertini, the Y_i will intersect transversely). Thus the degree of the Veronese variety is d^n.

Note that we can avoid the use of Bézout's theorem in this computation if we use the fact that every hypersurface of degree d in $\mathbb{P}^n$ is the pullback $v_d^{-1}(H_i)$ of some hyperplane from $\mathbb{P}^N$. We let $Y_i \subset \mathbb{P}^n$ be the hypersurface consisting of the union of d general hyperplanes $\Lambda_{i,j} \subset \mathbb{P}^n$, so that the intersection $Y_1 \cap \cdots \cap Y_n$ will consist of the d^n points

$$p_j = \Lambda_{1,j(1)} \cap \cdots \cap \Lambda_{n,j(n)}$$

where j ranges over all functions $j: \{1, \ldots, n\} \to \{1, \ldots, d\}$. Of course, we have to verify in this case that the intersection of the Veronese variety X with the n hyperplanes H_i is transverse at each point $v_d(p_j)$.

Another way of arriving at this conclusion would be to use the characterization of degree in terms of topological cohomology. Explicitly, since the inverse image of a hyperplane in $\mathbb{P}^N$ is a hypersurface of degree d in $\mathbb{P}^n$, the pullback map

$$v_d^*: H^2(\mathbb{P}^N, \mathbb{Z}) \to H^2(\mathbb{P}^n, \mathbb{Z})$$

sends

$$v_d^*: \zeta_{\mathbb{P}^N} \mapsto d \cdot \zeta_{\mathbb{P}^n}$$

where $\zeta_{\mathbb{P}^n}$ is the generator of $H^2(\mathbb{P}^n, \mathbb{Z})$. It follows that

$$v_d^*: (\zeta_{\mathbb{P}^N})^n \mapsto d^n \cdot (\zeta_{\mathbb{P}^n})^n$$

and so we have

$$\begin{aligned} \deg(X) &= \langle (\zeta_{\mathbb{P}^N})^n, [X] \rangle \\ &= \langle (\zeta_{\mathbb{P}^N})^n, v_{d_*}[\mathbb{P}^n] \rangle \\ &= \langle v_d^*(\zeta_{\mathbb{P}^N})^n, [\mathbb{P}^n] \rangle \\ &= \langle d^n \cdot (\zeta_{\mathbb{P}^n})^n, [\mathbb{P}^n] \rangle \\ &= d^n. \end{aligned}$$

Yet another way to find the degree of the Veronese variety is via its Hilbert polynomial. Recall from Example 13.4 that the Hilbert polynomial of $X = v_d(\mathbb{P}^n)$ is

$$\begin{aligned} p_X(m) &= \binom{m \cdot d + n}{n} \\ &= \frac{(dm + n)(dm + n - 1) \cdot \ldots \cdot (dm + 1)}{n!} \\ &= \frac{d^n}{n!} m^n + \cdots \end{aligned}$$

so that the degree of X is d^n.

Exercise 18.14. Let $Y \subset \mathbb{P}^n$ be a variety of dimension k and degree e, and $Z = v_d(Y)$ its image in $\mathbb{P}^N$ under the Veronese map. Use each of the techniques in turn to compute the degree of Z.

Observe that the degree of the Veronese surface $S = \nu_2(\mathbb{P}^2) \subset \mathbb{P}^5$ is 4; in particular, it follows (using Propositions 18.9 and 18.10) that the general hyperplane section of S is a rational normal curve in $\mathbb{P}^4$ (this can also be seen directly). Note also, however, that this is the only case in which the general hyperplane section of a Veronese variety is again a Veronese variety.

Example 18.15. Segre Varieties

We consider next the degrees of the Segre varieties $\Sigma_{m,n} = \sigma(\mathbb{P}^m \times \mathbb{P}^n) \subset \mathbb{P}^N = \mathbb{P}^{mn+m+n}$. We can determine this by calculations analogous to three of the four given earlier (the one thing we cannot readily do is describe the intersection of $\Sigma_{m,n}$ with $m + n$ general hyperplanes in $\mathbb{P}^N$).

To begin with, we observe that among the hyperplane sections of $\Sigma_{m,n}$ is one consisting of the union of the pullback of a given hyperplane from each factor—if we think of the Segre variety as $\mathbb{P}V \times \mathbb{P}W \subset \mathbb{P}(V \otimes W)$, these correspond to the reducible elements $l \otimes m \in V^* \otimes W^*$. We can thus let $\Lambda_1, \ldots, \Lambda_{m+n} \subset \mathbb{P}^m$ and $\Gamma_1, \ldots, \Gamma_{m+n} \subset \mathbb{P}^n$ be general hyperplanes and $H_i \subset \mathbb{P}^N$ the hyperplane whose pullback

$$\sigma^{-1}(H_i) = \pi_1^{-1}(\Lambda_i) \cup \pi_2^{-1}(\Gamma_i).$$

Now, since the intersection of any m of the hyperplanes Λ_i will be a point and the intersection of any $m + 1$ of them empty, and similarly for the Γ_i, we see that the intersection

$$H_1 \cap \cdots \cap H_{m+n} \cap \Sigma_{m,n} = \{\sigma(p_I)\}$$

where $I = \{i_1, \ldots, i_m\}$ ranges over all subsets of order m in $\{1, \ldots, m + n\}$, $J = (j_1, \ldots, j_n\}$ is the complement of I, and

$$p_I = (\Lambda_{i_1} \cap \cdots \cap \Lambda_{i_m}) \times (\Gamma_{j_1} \cap \cdots \cap \Gamma_{j_n}).$$

As before, we have to check that the intersection of $\Sigma_{m,n}$ with the hyperplanes H_i is transverse at each point; assuming this, we deduce that the degree of the Segre variety $\Sigma_{m,n}$ is $\binom{m+n}{n}$.

A second approach is cohomological. If $\zeta_{\mathbb{P}^N}$ is the generator of $H^2(\mathbb{P}^N, \mathbb{Z})$ as earlier, and if we denote by α and β the pullbacks to $\mathbb{P}^m \times \mathbb{P}^n$ of the classes $\zeta_{\mathbb{P}^m}$ and $\zeta_{\mathbb{P}^n}$ respectively, then the pullback map

$$\sigma^*: \zeta_{\mathbb{P}^N} \mapsto \alpha + \beta.$$

It follows that

$$\begin{aligned}
\deg(\Sigma_{m,n}) &= \langle (\zeta_{\mathbb{P}^N})^{m+n}, [\Sigma_{m,n}] \rangle \\
&= \langle (\zeta_{\mathbb{P}^N})^{m+n}, \sigma_*[\mathbb{P}^m \times \mathbb{P}^n] \rangle \\
&= \langle \sigma^*(\zeta_{\mathbb{P}^N})^{m+n}, [\mathbb{P}^m \times \mathbb{P}^n] \rangle \\
&= \langle (\alpha + \beta)^{m+n}, [\mathbb{P}^m \times \mathbb{P}^n] \rangle.
\end{aligned}$$

Now, $\alpha^{m+1} = \pi_1^*((\zeta_{\mathbb{P}^m})^{m+1}) = 0$, and similarly for β^{n+1}; while by Kunneth we have $\langle \alpha^m \cdot \beta^n, [\mathbb{P}^m \times \mathbb{P}^n] \rangle = 1$. We may thus evaluate the last expression to give

$$\deg(\Sigma_{m,n}) = \binom{m+n}{n}.$$

As a final approach, we can write down the Hilbert polynomial of $\Sigma_{m,n}$ as in Exercise 13.6: since the homogeneous polynomials of degree l in $\mathbb{P}^N$ restrict to $\Sigma_{m,n}$ to give all bihomogeneous polynomials of bidegree (l, l) on $\mathbb{P}^m \times \mathbb{P}^n$, we have that the dimension of the lth graded piece of the ring $S(\Sigma_{m,n})$ is

$$\begin{aligned} p_{\Sigma_{m,n}}(l) &= \binom{m+l}{m}\binom{n+l}{n} \\ &= \frac{(l+m)(l+m-1)\cdot\ldots\cdot(l+1)(l+n)\cdot\ldots\cdot(l+1)}{m!n!} \\ &= \frac{1}{m!n!}l^{m+n} + \cdots. \end{aligned}$$

We conclude that the degree of $\Sigma_{m,n}$ is $(m+n)!$ times the leading coefficient of this polynomial, or in other words $\binom{m+n}{n}$.

Note that in case $n = 1$ the Segre variety $\Sigma_{m,1}$ is a variety of degree $m + 1$ in $\mathbb{P}^{2m+1}$, and in particular the degree of the Segre threefold $\Sigma_{2,1} \subset \mathbb{P}^5$ is 3. This tells us yet again that the general intersection of $\Sigma_{2,1}$ with a three-plane $\mathbb{P}^3 \subset \mathbb{P}^5$ is a twisted cubic curve. In general, observe that the varieties $\Sigma_{m,1}$ are varieties of minimal degree, in the sense of Corollary 18.12; we will see later that this is because they are special cases of rational normal scrolls.

Example 18.16. Degrees of Cones and Projections

We start with the simplest case. Let $X \subset \mathbb{P}^{n-1} \subset \mathbb{P}^n$ be a variety of dimension k and degree d, $p \in \mathbb{P}^n$ any point not lying on $\mathbb{P}^{n-1}$, and $Y = \overline{pX}$ the cone with vertex p over X. Then, inasmuch as the general hyperplane section of Y is projectively equivalent to X, we see that the degree of Y is equal to the degree d of X.

Next, suppose that $X \subset \mathbb{P}^n$, $p \notin X$, and let $\overline{X} = \pi_p(X) \subset \mathbb{P}^{n-1}$ be the projection of X from p. Assume $\operatorname{char}(K) = 0$. Now, since hyperplanes in the tangent space $\mathbb{P}^{n-1}$ correspond to hyperplanes in $\mathbb{P}^n$ through the point p, a general k-tuple of hyperplanes $\overline{H}_i \subset \mathbb{P}^{n-1}$ will come from a general k-tuple of hyperplanes H_i in $\mathbb{P}^n$ through p, which will by Bertini's theorem intersect X transversely; by Bézout they will intersect X in exactly d points p_α. The intersection of $\overline{X}$ with the hyperplanes $\overline{H}_i$ will thus consist of the images $\overline{p}_i = \pi_p(p_i) \in \mathbb{P}^{n-1}$. If we assume, then, that these are distinct—that is, that the general line joining p to a point of X meets X only once, or equivalently that the projection map is birational onto its image—it will follow that the degree of $\overline{X}$ will be equal to the degree d of X.

How is this argument affected if $p \in X$? If p is a smooth point of X, it's not hard to see how to amend it: it's immediate that a general k-tuple of hyperplanes H_i

through p will intersect X transversely at p. We can still apply Bertini to conclude that the hyperplanes H_i will intersect X transversely away from p. Still assuming that the general line $\overline{pq}$ joining p to another point $q \in X$ meets X only at p and q, we see that the points of intersection of $\overline{X}$ with the $\overline{H}_i$ correspond to the points of intersection of X with the H_i other than p, so that

$$\deg(\overline{X}) = \deg(X) - 1.$$

We will see in Lecture 20 how to deal with the case of $p \in X_{\text{sing}}$. Observe that in either case $p \notin X$ or $p \notin X_{\text{sm}}$, the calculation works in arbitrary characteristic as long as there exists an $(n - k)$-plane through p transverse to X (a non-trivial assumption in case $\operatorname{char}(K) > 0$, but one that can be verified in the following applications).

Note that, in view of the characterization of the rational normal scroll $X_{2,1} \subset \mathbb{P}^4$ as a projection of the Veronese surface $S \subset \mathbb{P}^5$ from a point of S, this tells us that the degree of $X_{2,1}$ is 3.

Example 18.17. Joins of Varieties

Recall the definition of the join of two disjoint varieties from Example 6.17: we have a pair of disjoint varieties $X, Y \subset \mathbb{P}^n$, and define the join of X and Y to be the union

$$J(X, Y) = \bigcup_{x \in X, y \in Y} \overline{x, y}$$

of the lines joining points of X to points of Y. As we saw in Example 11.36, the dimension of the join is 1 plus the sum of the dimensions $k = \dim(X)$ and $l = \dim(Y)$ of X and Y; we will now describe its degree.

We do this first in the apparently special case where X and Y live in complementary linear subspaces $\mathbb{P}^m$ and $\mathbb{P}^{n-m-1} \subset \mathbb{P}^n$; this is the important case, since as we have seen any join may be realized as the regular projection of such a join. We will give three ways of calculating the degree in this case.

Calculation I. Our first approach is to intersect with a somewhat special plane. We take $\Lambda \cong \mathbb{P}^{n-k-l}$ to be the subspace spanned by a general plane $\Lambda' \cong \mathbb{P}^{m-k} \subset \mathbb{P}^m$ and a general plane $\Lambda'' \cong \mathbb{P}^{n-m-1-l} \subset \mathbb{P}^{n-m-1}$. Λ' will then intersect X in exactly $\deg(X)$ points $p_1, \ldots, p_d$ and Λ'' will intersect Y in $\deg(Y)$ points $q_1, \ldots, q_e$; the intersection of Λ with the join $J(X, Y)$ will be the union of the $d \cdot e$ lines $L_{i,j}$ spanned by p_i and q_j. Moreover, if Λ' and Λ'' intersect X and Y transversely, then by the description of the tangent planes to $J(X, Y)$ given in Exercise 16.14, the intersection of Λ with $J(X, Y)$ is generically transverse; so the degree of $J(X, Y)$ is $d \cdot e = \deg(X) \cdot \deg(Y)$.

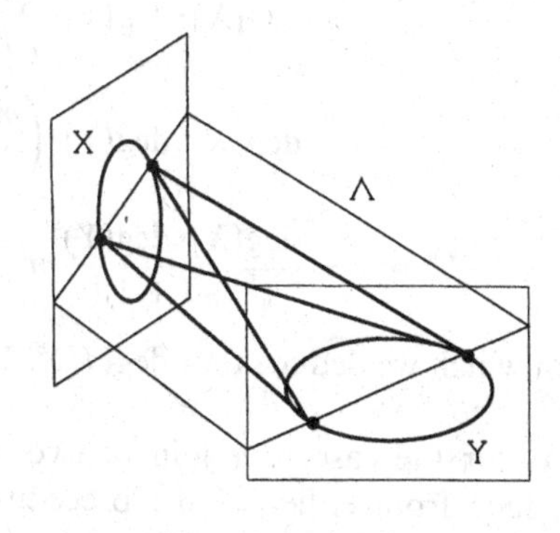

Calculation II. This is the least satisfactory proof; it works only in characteristic 0 and uses Bézout (since we are going to use the calculation of the degree of $J(X, Y)$ to prove Bézout later, this is not desirable). I include it mainly because I like Exercise 18.18.

Suppose that $\Gamma \subset \mathbb{P}^n$ is a general plane of complementary dimension $n - k - l - 1$ to the join $J(X, Y)$. We may describe the intersection $\Gamma \cap J(X, Y)$ by considering the projection π_Γ from $\mathbb{P}^n$ to $\mathbb{P}^{k+l}$. If the line $\overline{pq}$ joining X to Y intersects Γ (since Γ is disjoint from X and Y, it cannot lie in it), then the span $\overline{\Gamma, p} = \overline{\Gamma, q}$, i.e., $\pi_\Gamma(p) = \pi_\Gamma(q)$, and conversely. The intersection $\Gamma \cap J(X, Y)$ thus corresponds to the intersection $\pi_\Gamma(X) \cap \pi_\Gamma(Y)$; this intersection, if it is transverse, will consist of $\deg(X) \cdot \deg(Y)$ points. The result then follows from the following.

Exercise 18.18. (Characteristic 0 only). Let X, $Y \subset \mathbb{P}^n$ be disjoint subvarieties of dimensions k and l, and $\Gamma \subset \mathbb{P}^n$ a general plane of dimension $n - k - l - 1$. Show that the projections $\pi_\Gamma(X)$ and $\pi_\Gamma(Y)$ intersect transversely.

Calculation III. Finally, note that the calculation for the degree of the join $J(X, Y)$ of two varieties in disjoint linear subspaces follows from the characterization of degree in terms of Hilbert polynomials, without any intersection; very simply, if X, $Y \subset \mathbb{P}^n$ lies in disjoint linear spaces, then the homogeneous coordinate ring $S(J(X, Y))$ of the join $J(X, Y)$ will be the tensor product $S(X) \otimes S(Y)$ of the homogeneous coordinate rings of X and Y, as graded ring. We thus have

$$S(J(X, Y))_m = \bigoplus (S(X)_j \otimes S(Y)_{m-j})$$

so

$$\begin{aligned} h_{J(X,Y)}(m) &= \sum h_X(j) \cdot h_Y(m - j) \\ &= \sum_{j=0}^{m} \left(\deg(X) \cdot \binom{j+k}{k} + O(j^{k-1}) \right) \cdot \left(\deg(Y) \binom{m-j+l}{l} \right. \\ &\qquad \left. + O((m-j)^{l-1}) \right) \\ &= \deg(X) \cdot \deg(Y) \cdot \sum_{j=0}^{m} \binom{j+k}{k} \cdot \binom{m-j+l}{l} + O(m^{k+l}) \\ &= \deg(X) \cdot \deg(Y) \cdot \binom{m+k+l+1}{k+l+1} + O(m^{k+l}) \\ &= \frac{\deg(X) \cdot \deg(Y)}{(k+l+1)!} m^{k+l+1} + O(m^{k+l}) \end{aligned}$$

from which we deduce that $\deg(J(X, Y)) = \deg(X) \deg(Y)$.

As for the case of a join of two general varieties X, $Y \subset \mathbb{P}^n$, what we may conclude from either of the preceding arguments is simply that if the general point of a general line joining points of X and Y lies on exactly α lines joining X and Y, then the degree of the join $J(X, Y)$ is $\deg(X) \cdot \deg(Y)/\alpha$.

Proof of Bézout's Theorem. We can use the calculation of the degree of a join to prove the initial statement of Bézout's theorem (Theorem 18.3). First, observe that it is enough to prove it in case the varieties $X, Y \subset \mathbb{P}^n$ have complementary dimension, i.e., intersect transversely in points. The trick now is to view X and Y as subvarieties of two different ambient spaces $\mathbb{P}^n$, and embed these $\mathbb{P}^n$s into $\mathbb{P}^{2n+1}$ via the maps $i, j\colon \mathbb{P}^n \to \mathbb{P}^{2n+1}$ given by

$$i\colon [Z_0, \ldots, Z_n] \mapsto [Z_0, \ldots, Z_n, 0, \ldots, 0]$$

and

$$j\colon [Z_0, \ldots, Z_n] \mapsto [0, \ldots, 0, Z_0, \ldots, Z_n].$$

Let $\tilde{X}, \tilde{Y} \subset \mathbb{P}^{2n+1}$ be the images of X and Y under these embeddings and $J = J(\tilde{X}, \tilde{Y})$ their join.

Now let $L \cong \mathbb{P}^n \subset \mathbb{P}^{2n+1}$ be the linear subspace given by

$$L = (Z_0 - Z_{n+1}, Z_1 - Z_{n+2}, \ldots, Z_n - Z_{2n+1}).$$

Observe the intersection of L with J consists exactly of the points $[Z_0, \ldots, Z_n, Z_0, \ldots, Z_n]$ for $p = [Z_0, \ldots, Z_n] \in X \cap Y$, and that by the description of the tangent space to J given in Exercise 16.14 (and the hypothesis that X and Y intersect transversely) this intersection will be transverse. We thus have

$$\operatorname{card}(X \cap Y) = \deg(J) = \deg(X) \cdot \deg(Y). \qquad \square$$

In fact, we can use this trick to prove the stronger form of Bézout as well, once we have defined the intersection multiplicity; we just have to check that for any $p \in X \cap Y$ the intersection multiplicity of L and J at a point on the line $\overline{i(p), j(p)}$ is the same as the intersection multiplicity of X and Y at p.

Example 18.19. Unirationality of Cubic Hypersurfaces

To conclude this lecture, we will combine some of the ideas introduced in the last few lectures—specifically, tangent lines, order of contact and Bézout's theorem—to establish a statement made in Lecture 7.

Proposition 18.20. *A smooth cubic threefold $X \subset \mathbb{P}^4$ is unirational.*

Proof. We have already seen (Theorem 12.8) that X contains lines (in fact, the lines on X cover X); let L be a general one. Consider the subvariety $\Sigma \subset \mathcal{T}_1(X)$ of lines tangent to X at a point $p \in L$. We claim first of all that Σ is rational.

To see this, fix a general point $q \in \mathbb{P}^4$ and 2-plane $\Lambda \cong \mathbb{P}^2 \subset \mathbb{P}^4$, and consider the rational map

$$\varphi\colon L \times \Lambda \dashrightarrow \Sigma$$

given, for general $p \in L$ and $r \in \Lambda$, by

$$\varphi\colon (p, r) \mapsto \mathbb{T}_p(X) \cap \overline{pqr}.$$

To see that this is birational, note that for a general tangent line M to X at p, the plane $\overline{M, q}$ will intersect Λ in a point r; (p, r) will be the unique point of $\varphi^{-1}(M)$.

Now, as will be established in Exercise 18.21, a general tangent line to X will have order of contact 2 with X at its point of tangency, and therefore will intersect X in exactly one other point. We thus have a rational map

$$\pi: \Sigma \dashrightarrow X.$$

To see that this is dominant, note that for any point $q \in X$ not on L, the plane $\overline{L, q}$ will intersect X in a plane cubic curve; this curve will contain L, and hence will be singular somewhere along L (in fact, generically it will consist of the union of L and a smooth conic curve meeting L in two distinct points). The plane $\overline{L, q}$ will thus lie in the tangent plane $\mathbb{T}_p(X)$ for at least one $p \in L$ (again, generically two); the point q will correspondingly lie on at least one tangent line $\overline{qp}$ to X with $p \in L$. □

Exercise 18.21. Show that a general tangent line to X at a general point $p \in L$ has contact of order 2 with X at p.

Exercise 18.22. Show by the same argument that a smooth cubic hypersurface in $\mathbb{P}^n$ is unirational for all $n \geq 3$.

Exercise 18.23. Show that for any fixed d there exists an $n_0 = n_0(d)$ such that for all $n \geq n_0$ a smooth hypersurface of degree d in $\mathbb{P}^n$ is unirational.

LECTURE 19

Further Examples and Applications of Degree

Example 19.1. Multidegree of a Subvariety of a Product

Since finding out that a quadric surface $Q \subset \mathbb{P}^3$ is abstractly isomorphic to the product $\mathbb{P}^1 \times \mathbb{P}^1$ we have observed a number of times that, in describing a curve $C \subset Q$, it is much more useful to give its bidegree (a, b) in $\mathbb{P}^1 \times \mathbb{P}^1$ (that is, the bidegree of the bihomogeneous polynomial F defining it as a subvariety of $\mathbb{P}^1 \times \mathbb{P}^1$) than to give just its degree as a curve in $\mathbb{P}^3$ (the reader can check this is just $a + b$). We ask now what are the analogous numerical invariants of a k-dimensional subvariety $X \subset \mathbb{P}^m \times \mathbb{P}^n$ in general.

The answer to this is perhaps best suggested by the cohomological interpretation of the degree of a variety in $\mathbb{P}^n$. According to this definition, the degree of a variety $X \subset \mathbb{P}^n$ is simply the coefficient in the expression of the fundamental class $[X]$ of X as a multiple of a standard generator $[\mathbb{P}^k]$ of $H_{2k}(\mathbb{P}^n, \mathbb{Z})$. Now, by the Kunneth formula, the homology of the product $\mathbb{P}^m \times \mathbb{P}^n$ will be torsion-free, and generated by the fundamental classes of products $[\mathbb{P}^k \times \mathbb{P}^l] \in H_{2k+2l}(\mathbb{P}^m \times \mathbb{P}^n, \mathbb{Z})$. Thus, for any subvariety $X \subset \mathbb{P}^m \times \mathbb{P}^n$ of dimension k, we can write its fundamental class as

$$[X] = \sum d_i \cdot [\mathbb{P}^i \times \mathbb{P}^{k-i}].$$

This uniquely determines the integers d_i, which we will refer to as the *multidegree* of $X \subset \mathbb{P}^m \times \mathbb{P}^n$.

Equivalently, let η_X be the fundamental class of X in the de Rham cohomology group $H^{2k}(\mathbb{P}^m \times \mathbb{P}^n, \mathbb{Z})^* = H^{2n+2m-2k}(\mathbb{P}^m \times \mathbb{P}^n, \mathbb{Z})$. Then in terms of the classes α, $\beta \in H^2(\mathbb{P}^m \times \mathbb{P}^n, \mathbb{Z})$ introduced in Example 18.15, we can write

$$\eta_X = \sum d_i \cdot \alpha^{m-i} \beta^{n-k+i}.$$

A more geometric way to characterize the integers d_i is to say that d_i is the number of points of intersection of X with the product $\Gamma \times \Lambda \subset \mathbb{P}^m \times \mathbb{P}^n$, where $\Gamma \subset \mathbb{P}^m$ and $\Lambda \subset \mathbb{P}^n$ are a general $(m-i)$-plane and a general $(n-k+i)$-plane, respectively.

Exercise 19.2. Let $X \subset \mathbb{P}^m \times \mathbb{P}^n$ be any variety of dimension k. Show that the degree of X, as embedded in projective space via the Segre embedding of $\mathbb{P}^m \times \mathbb{P}^n$, is

$$\deg(X) = \sum \binom{k}{i} d_i.$$

Exercise 19.3. Use the preceding exercise to conclude that any linear space $\Lambda \subset \sigma(\mathbb{P}^m \times \mathbb{P}^n) \subset \mathbb{P}^{mn+m+n}$ lying on a Segre variety is contained in a fiber of a projection $\mathbb{P}^m \times \mathbb{P}^n \to \mathbb{P}^m$ or $\mathbb{P}^m \times \mathbb{P}^n \to \mathbb{P}^n$. Compare this with the proof of Theorem 9.22.

Example 19.4. Projective Degree of a Map

As a natural extension of the notion of the degree of a variety, we have the projective degree of a map $\varphi: X \to Y$ between two projective varieties $X \subset \mathbb{P}^n$ and $Y \subset \mathbb{P}^m$. This is easy enough to define: just as we define the degree of a variety to be the number of points of intersection with a plane of complementary dimension, the *projective degree* $d_0(\varphi)$ of a map $\varphi: X \to \mathbb{P}^m$ is the number of points in the inverse image of a plane $\mathbb{P}^{m-\dim(X)} \subset \mathbb{P}^m$. (It is called the projective degree to distinguish it from the degree of a map as defined in Lecture 7.) Equivalently, $d_0(\varphi)$ is zero if the dimension of the image $\varphi(X)$ is strictly less than the dimension of X (i.e., if the general fiber of the map has positive dimension) and otherwise is the degree of the image times the degree of the map (where the degree $\deg(\varphi)$ of the map is the number of points in the inverse image of a general point of $\varphi(X)$). Needless to say, watch out for the confusing terminology of degree versus projective degree. Note in particular that $d_0(\varphi)$ does not depend on the projective embedding $X \subset \mathbb{P}^n$.

In fact, the projective degree of a map is just the first (or 0th) of a series of degrees $d_i(\varphi)$, which we may call in general the *projective degrees* of the map φ. These are defined simply as the multidegree of the graph $\Gamma_\varphi \subset \mathbb{P}^n \times \mathbb{P}^m$; equivalently, if X has dimension k, $d_i(\varphi)$ may be defined simply as the number of points

$$d_i(\varphi) = \operatorname{card}(\varphi^{-1}(\mathbb{P}^{m-k+i}) \cap \mathbb{P}^{n-i}),$$

where $\mathbb{P}^{m-k+i} \subset \mathbb{P}^m$ and $\mathbb{P}^{n-i} \subset \mathbb{P}^n$ are general planes[6]. Note that this is the same thing as either the projective degree of the restriction of the map φ to the intersection of X with a plane $\mathbb{P}^{n-i} \subset \mathbb{P}^n$; or, equivalently, as the degree of the inverse image $\varphi^{-1}(\mathbb{P}^{m-k+i})$ times the degree of φ restricted to this inverse image. In particu-

[6] Note that the graph Γ_φ is here taken to lie in $\mathbb{P}^n \times \mathbb{P}^m$, i.e., the order of the factors is the reverse of the usual.

lar, of course, the projective degree of φ is $d_0(\varphi)$, while the degree of X itself is $\deg(X) = d_k(\varphi)$ (independently of φ).

Example 19.5. Joins of Corresponding Points

Recall the situation of Example 8.14: we let X and Y be subvarieties of $\mathbb{P}^n$, and suppose we are given a regular map $\varphi: X \to Y$ such that $\varphi(x) \neq x$ for all $x \in X$. We may then consider the union

$$K(\varphi) = \bigcup_{x \in X} \overline{x, \varphi(x)}$$

and ask for the degree of this variety.

We start with the case where X is a curve. As in the discussion of general joins in Example 18.17, we will assume for the moment that X and Y lie in complementary subspaces $\mathbb{P}^k, \mathbb{P}^l \subset \mathbb{P}^n$.

Now, to find the degree of $K(\varphi)$, we intersect with a particular hyperplane, namely, the plane Λ spanned by all of $\mathbb{P}^k$ and a general hyperplane Λ_0 in $\mathbb{P}^l$. It is easy enough to say what the intersection of Λ with $K(\varphi)$ is as a set, and in particular to see that it is proper: clearly, Λ contains $X \subset K(\varphi)$, and so $\Lambda \cap K(\varphi)$ will consist of X itself and the union of the $d_0(\varphi)$ lines $\overline{x, \varphi(x)}$ for all x such that $\varphi(x) \in \Lambda_0$. We can now use the analysis in Lecture 16 to see that this intersection is generically transverse (we leave this as an exercise for the reader). It follows that

$$\deg(K(\varphi)) = \deg(K(\varphi) \cap \Lambda) = \deg(X) + d_0(\varphi).$$

Observe finally that we can now answer the original question without the hypothesis that X and $\varphi(X)$ lie in disjoint subspaces, since any join $K(\varphi)$ may be obtained from one of this type by a regular projection: if a general point of a general line of the form $\overline{x, \varphi(x)}$ lies on exactly α lines of this form, then the degree of $K(\varphi)$ will be just $(\deg(X) + d_0(\varphi))/\alpha$.

Note, as a primary example of this, that the degree of a rational normal surface scroll $X_{a_1,a_2} \subset \mathbb{P}^n$ is $a_1 + a_2 = n - 1$.

Exercise 19.6. What happens in the preceding analysis if $x = \varphi(x)$ for some value of x?

Exercise 19.7. Let $C_1, \ldots, C_k$ be curves in $\mathbb{P}^n$ and $\varphi_i: C_1 \to C_i$ a map for $i = 2, \ldots, k$. Let

$$X = \bigcup_{p \in C_1} \overline{p, \varphi_2(p), \ldots, \varphi_k(p)}$$

be the union of the joins of corresponding points and assume that for each $p \in C_1$ the points $p, \varphi_2(p), \ldots, \varphi_k(p)$ are independent. Determine the degree of X and deduce in particular that the degree of a rational normal k-fold scroll $X \subset \mathbb{P}^n$ is $n - k + 1$. (*)

To return to our discussion of the degree of the variety $K(\varphi)$ in general, consider next the case of a map $\varphi: X \to Y$ where X is a surface. If we are to use a technique analogous to the one used earlier, that is, intersect $K(\varphi)$ with a linear subspace Λ such that X itself is a component of the intersection $\Lambda \cap K(\varphi)$, then the plane in question should be the hyperplane spanned by all of $\mathbb{P}^k$ and a general hyperplane Λ_0 in $\mathbb{P}^l$. The intersection of Λ with $K(\varphi)$ will then consist of X itself and the union K' of the lines $\overline{x, \varphi(x)}$ for all x such that $\varphi(x) \in \Lambda_0$. But K' itself is just the variety $K(\varphi')$, where φ' is the restriction of the map φ to the inverse image of the hyperplane $\Lambda_0 \subset \mathbb{P}^l$. The degree $d_0(\varphi')$ is of course just the projective degree of φ, and the degree of the inverse image $\deg(\varphi^{-1}(\Lambda_0))$ is the intermediate degree $d_1(\varphi)$. By the preceding analysis, we have

$$\begin{aligned}\deg(K(\varphi)) &= \deg(\varphi') + \deg(\varphi^{-1}(\Lambda_0)) + \deg(X)\\ &= d_0(\varphi) + d_1(\varphi) + d_2(\varphi).\end{aligned}$$

The general pattern follows this. For a map $\varphi: X \to Y$ on a variety X of any dimension k, assuming $\varphi(x) \neq x \; \forall x \in X$ and a general point of $K(\varphi)$ lies on a unique line $\overline{x, \varphi(x)}$ the degree of the join $K(\varphi)$ will be the sum $\sum d_\alpha(\varphi)$ of the degrees $d_\alpha(\varphi)$ of the map φ. As in the general case, if a general point of a general line of the form $\overline{x, \varphi(x)}$ lies on exactly α lines of this form, then the degree of $K(\varphi)$ will be just $(\sum d_i(\varphi))/\alpha$.

Example 19.8. Varieties of Minimal Degree

One consequence of Exercise 19.7 is that rational normal scrolls have minimal degree, as identified in Corollary 18.12. Indeed, we have now seen all varieties of minimal degree. We state this as the following.

Theorem 19.9. *Let $X \subset \mathbb{P}^n$ be any irreducible nondegenerate variety of dimension k having degree $n - k + 1$. Then X is*

(i) *a quadric hypersurface;*
(ii) *a cone over the Veronese surface $v_2(\mathbb{P}^2) \subset \mathbb{P}^5$; or*
(iii) *a rational normal scroll.*

Note that a quadric hypersurface is also a scroll if and only if it has rank 3 or 4, and that a Veronese surface may be distinguished from scrolls by the fact that it contains no lines.

We will not prove Theorem 19.9 here. We note, however, that many of the statements made earlier about scrolls (and some other varieties as well) follow from this theorem. For example,

(a) it follows that the general hyperplane section of a rational normal k-fold scroll in $\mathbb{P}^n$ is a rational normal $(k-1)$-fold scroll in $\mathbb{P}^{n-1}$.
(b) it follows that the projection of a rational normal scroll $X \subset \mathbb{P}^n$ from a point $p \in X$ is a rational normal scroll in $\mathbb{P}^{n-1}$.
(c) it follows that the projection of the Veronese surface $S \subset \mathbb{P}^5$ from a point $p \in S$ is the cubic scroll $X_{2,1} \subset \mathbb{P}^4$.

Example 19.10. Degrees of Determinantal Varieties

We will determine here the degrees of some more determinantal varieties, described initially in Lecture 9 and again in Examples 12.1 and 14.15. To recall the notation, we let M be the projective space $\mathbb{P}^{mn-1}$ associated to the vector space of $m \times n$ matrices and for each k let $M_k \subset M$ be the subset of matrices of rank k or less. As we observed at the time, the variety M_1 is just the Segre variety $\mathbb{P}^{n-1} \times \mathbb{P}^{m-1}$ whose degree we have already computed; we will now consider the other "extremal" case, where k is one less than the maximal rank $\min(m, n)$. To fix notation, we will say $m \geq n = k + 1$ for the following. Also, recall from Example 12.1 that the codimension of M_{n-1} is $m - n + 1$.

We will compute the degree of M_{n-1} by intersecting it with a linear subspace $\Lambda \cong \mathbb{P}^{m-1}$. Doing this is equivalent to specifying an $m \times n$ matrix of linear forms on $\mathbb{P}^{m-1}$. To specify such a matrix, let $(a_{i,j})$ be any $m \times n$ matrix of scalars all of whose $n \times n$ minors are nonzero, and take our matrix of linear forms to be the matrix $(a_{i,j} \cdot X_j)$, i.e.,

$$\begin{bmatrix} a_{1,1}X_1 & a_{1,2}X_2 & \cdots & a_{1,m}X_m \\ a_{2,1}X_1 & a_{2,2}X_2 & \cdots & a_{2,m}X_m \\ \vdots & & & \\ a_{n,1}X_1 & \cdots & \cdots & a_{n,m}X_m \end{bmatrix}.$$

The $n \times n$ minors of this matrix are exactly nonzero scalar multiples of the n-fold products of distinct coordinates $X_1, \dots, X_m$; explicitly, the minor involving the i_1st through i_nth columns is just $X_{i_1} \cdot \ldots \cdot X_{i_n}$ times the corresponding minor of $(a_{i,j})$. These will all vanish at a point $[X_1, \dots, X_m]$ if and only if $m - n + 1$ or more of the X_is are zero; so the intersection of Λ with the determinantal variety M_{n-1} is thus the union of the coordinate $(n-2)$-planes in $\mathbb{P}^{m-1}$. Moreover, at a point of $\Lambda \cap M_{n-1}$ at which exactly $n - 1$ of the coordinates X_i are nonzero—say $X_{i_1} \cdot \ldots \cdot X_{i_{n-1}} \neq 0$—we see that the minors $X_{i_1} \cdot \ldots \cdot X_{i_{n-1}} \cdot X_j$ for $j = n, \dots, m$ have independent differentials, so that the intersection $\Lambda \cap M_{n-1}$ is generically transverse. It follows that the degree of the determinantal variety M_{n-1} is given by

$$\begin{aligned} \deg(M_{n-1}) &= \deg(\Lambda \cap M_{n-1}) \\ &= \binom{m}{n-1}. \end{aligned}$$

Note that this agrees with the case $n = 2$ of the rational normal scroll discussed in Example 19.5 and Exercise 19.7, as well as with the "obvious" case $m = n$. Indeed, in the former case, combined with Theorem 19.9 it implies the statement (Example 9.10) that the rank 1 locus of a $2 \times m$ matrix of linear forms, if it is irreducible and has the expected codimension $m - 1$, is a rational normal scroll.

The degree of the determinantal variety M_k in the space of all $m \times n$ matrices may be worked out for all values of k, though I don't know of any way as simple as the one in the case $k = n - 1$. For reference (see [F1], example 14.4.14), the answer is

$$\deg(M_k) = \prod_{i=0}^{n-k-1} \frac{(m+i)! \cdot i!}{(k+i)! \cdot (m-k+i)!}$$

$$= \prod_{i=0}^{n-k-1} \frac{\binom{m+i}{k}}{\binom{k+i}{k}}.$$

Example 19.11. Degrees of Varieties Swept out by Linear Spaces

As another example of the computation of degree, we will consider a variety $X \subset \mathbb{P}^n$ given as the union of a one-parameter family of linear spaces; that is, we let $Z \subset \mathbb{G}(k, n)$ be a curve and consider

$$X = \bigcup_{\Lambda \in Z} \Lambda.$$

Assume for the moment that the planes Λ sweep out X once, that is, that a general point $p \in X$ lies on a unique plane $\Lambda \in Z$. It follows in particular that the dimension of X is $k + 1$, and hence that the degree of X is the cardinality of its intersection with a general plane $\Gamma \cong \mathbb{P}^{n-k-1} \subset \mathbb{P}^n$. Since each of the points of $X \cap \Gamma$ will lie on a unique plane $\Lambda \in Z$, this in turn will be equal to the number of planes $\Lambda \in Z$ that meet Γ.

Now, we have seen that the locus of k-planes $\Lambda \in \mathbb{G}(k, n)$ that meet a given $(n - k - 1)$-plane $\Gamma \subset \mathbb{P}^n$ is a hyperplane section $H_\Gamma \cap \mathbb{G}$ of the Grassmannian $\mathbb{G} = \mathbb{G}(k, n)$ under the Plücker embedding. We may therefore expect that the intersection of this locus with a curve $Z \subset \mathbb{G}$ is the degree of the curve Z under the Plücker embedding. We cannot deduce this immediately, however, since even if Γ is general, H_Γ is not a general hyperplane section of $\mathbb{G}$ (the space of hyperplanes in $\mathbb{P}^N = \mathbb{P}(\wedge^{k+1} K^{n+1})$ is the projective space $\mathbb{P}(\wedge^{k+1}(K^{n+1})^*)$, in which the set of hyperplanes of the form H_Γ sits as the subset of totally decomposable vectors $G(n - k, n + 1)$). We can, however, invoke the strong form of Bertini given in Theorem 17.22; since the group $\mathrm{PGL}_{n+1} K$ acts transitively on the space $\mathbb{G}$ and on the set of hyperplane sections $H_\Gamma \cap \mathbb{G}$, for general Γ it follows that H_Γ intersects a given curve Z transversely. We conclude, therefore, that the degree of a variety $X \subset \mathbb{P}^n$ swept out once by a one-parameter family of k-planes $Z \subset \mathbb{G}(k, n)$ is equal to the degree of Z under the Plücker embedding.

Exercise 19.12. Eliminate the need for any form of Bertini in the preceding argument by using the description of tangent spaces given in Example 16.12 to show directly that if $p \in X$ lies on a unique plane $\Lambda \in Z$, then a plane Γ will intersect the variety $X \subset \mathbb{P}^n$ transversely at the point p if and only if the hyperplane H_Γ intersects the curve Z transversely at Λ, and deducing that for general Γ the intersection $H_\Gamma \cap Z$ is transverse.

One application of this result would be another computation of the degree of a rational normal scroll: as we can see directly by writing down their equations, the $(k-1)$-planes of a rational normal k-fold scroll $X \subset \mathbb{P}^n$ correspond to the points of a rational normal curve $Z \subset \mathbb{G}(k-1, n)$ of degree $n-k+1$, and we may deduce from this that the degree of X is $n-k+1$.

Another example would be the tangential surface of a rational normal curve $C \subset \mathbb{P}^n$. Since C is given parametrically as

$$t \mapsto v(t) = [1, t, t^2, \ldots, t^n],$$

the Gauss map on C sends t to the vector $[v(t) \wedge v'(t)] \in \mathbb{P}(\wedge^2 K^{n+1})$, that is, the vector whose corrdinates are the 2×2 minors of the matrix

$$\begin{pmatrix} 1 & t & t^2 & \ldots & t^n \\ 0 & 1 & 2t & \ldots & nt^{n-1} \end{pmatrix}$$

These minors visibly span the space of polynomials of degree $2n-2$ in t, so that the image of $\mathscr{G}_C$ will be a rational normal curve of degree $2n-2$ in $\mathbb{G}(1, n) \subset \mathbb{P}(\wedge^2 K^{n+1})$; we conclude that the degree of the tangential surface of the rational normal curve is $2n-2$.

Exercise 19.13. Find the degree of the family $Z \subset \mathbb{G}(1, 5)$ of lines on the Segre threefold $X_{2,1} = \sigma(\mathbb{P}^2 \times \mathbb{P}^1) \subset \mathbb{P}^5$ not lying on a 2-plane of X. Use this to show that the degree of a variety $X \subset \mathbb{P}^n$ swept out once by a higher-dimensional family of k-planes $Z \subset \mathbb{G}(k, n)$ is not in general equal to the degree of Z.

Example 19.14. Degrees of Some Grassmannians

We will give an old-fashioned approach to the computation of the degrees of some Grassmannians $\mathbb{G}(k, n) \subset \mathbb{P}^N = \mathbb{P}(\wedge^{k+1} K^{n+1})$. To start, we make an observation analogous to that of the preceding example: among the hyperplane sections of the Grassmannian $\mathbb{G}(k, n) \subset \mathbb{P}^N$ are the special hyperplane sections H_Γ, defined for each $(n-k-1)$-plane $\Gamma \subset \mathbb{P}^n$ by

$$H_\Gamma = \{\Lambda \in \mathbb{G} : \Lambda \cap \Gamma \neq \emptyset\};$$

while these are not general hyperplane sections of $\mathbb{G}$, by the strong form of Bertini (Theorem 17.22), for a general collection $\{\Gamma_\alpha\}$ of $(n-k-1)$-planes the corresponding hyperplane sections H_Γ will intersect transversely. In particular, we have the interpretation: the degree of the Grassmannian $\mathbb{G}(k, n) \subset \mathbb{P}^N$ is the number of k-planes in $\mathbb{P}^n$ meeting each of $(k+1)(n-k)$ general $(n-k-1)$-planes.

Note that we can verify this directly in the first nontrivial case, that of the Grassmannian of lines $\mathbb{G}(1, 3)$ in $\mathbb{P}^3$. We have seen already that the family of lines meeting each of three mutually skew lines $L_1, L_2, L_3 \subset \mathbb{P}^3$ constitutes one ruling $\{M_\lambda\}$ of a quadric surface Q. A general fourth line L_4 will then meet Q at two distinct points p and q, through each of which will pass a single line of the ruling $\{M_\lambda\}$; these two lines are then the only lines meeting all four given lines L_i. This simply confirms what we knew already, since we remarked in the original discussion of Grassmannians in Lecture 6 that $\mathbb{G}(1, 3)$ was a quadric hypersurface in $\mathbb{P}^5$.

Another way of arriving at the statement that there are two lines meeting each of four given lines L_i in $\mathbb{P}^3$ would be to specialize the position of the lines L_i. For example, suppose we choose the lines so that L_1 and L_2 meet at a point p (and so span a plane $H \subset \mathbb{P}^3$), and likewise L_3 and L_4 meet at q and span a plane K. If $p \notin K$ and $q \notin H$, it is easy to see that any line meeting all four must be either the line $\overline{pq}$ or the line $H \cap K$.

This count does not in itself prove anything, since the observation depends on the assumption that the lines L_i are general. We can, however, verify directly in this case that the intersection of the hyperplane sections H_{L_i} is transverse. In Example 16.6 we showed that the tangent space to $H = H_{L_i}$ at a point L corresponding to a line meeting L_i in one point p consisted of the space of homomorphisms

$$T_L(H) = \{\varphi: L \to K^4/L: \varphi(p) \subset L + L_i\}.$$

Thus, for example, at the point corresponding to the line $L = \overline{pq}$, since $(L + L_1) \cap (L + L_2) = L$, the tangent spaces

$$T_L(H_{L_1}) \cap T_L(H_{L_2}) = \{\varphi: \varphi(p) = 0\}$$

and similarly

$$T_L(H_{L_3}) \cap T_L(H_{L_4}) = \{\varphi: \varphi(q) = 0\};$$

so the intersection of all four tangent spaces is zero, and the intersection is transverse. A similar argument applies at the point $L = H \cap K$.

Exercise 19.15. Let $L_1, \ldots, L_4$ be skew lines in $\mathbb{P}^3$, L a line meeting all four. Say $L \cap L_i = \{p_i\}$, and $l + L_i$ is the plane $H_i \subset \mathbb{P}^3$. Recall from Exercise 16.9 that the hyperplene sections H_{L_i} intersect transversely at L if and only if the cross-ratio of the four points $p_i \in L$ is not equal to the cross-ratio of the four planes $H_i \in \mathbb{P}(K^4/L)$. Verify this directly using the description of the lines meeting each of the three lines L_1, L_2, and L_3. Apply it to the preceding example (bear in mind that the cross-ratio of four points is still defined (if we allow ∞ as a value) in case no three coincide).

How do we calculate the degrees of other Grassmannians? In general, it is much less clear how to use the interpretation of $\deg(\mathbb{G})$, since we cannot determine directly the number of k-planes meeting each of $(k + 1)(n - k)$ general $(n - k - 1)$-planes. We can, however, use a specialization similar to the one used in the case of $\mathbb{G}(1, 3)$. Consider, for example, the case of the Grassmannian $\mathbb{G}(1, 4)$ of lines in $\mathbb{P}^4$. We want to know how many lines in $\mathbb{P}^4$ meet each of six two-planes $\Gamma_i \subset \mathbb{P}^4$; to answer this, suppose that we choose the Γ_i so that each pair $\Gamma_{2i-1}, \Gamma_{2i}$ meet in a line L_i and span a three-plane H_i. Every line meeting all six will then either meet the line L_i or lie in the hyperplane H_i, for $i = 1, 2$, and 3.

Now, the line of intersection of the hyperplanes H_i is the unique line lying in all three, and there is a unique line meeting L_i and L_j and lying in H_k (it is the line spanned by the points $L_i \cap H_k$ and $L_j \cap H_k$). In addition, there is a unique line meeting all three lines L_i (it may be characterized as the intersection of the three-planes $H_{i,j} = \overline{L_i L_j}$); but there will in general be no lines meeting L_i and lying in H_j and H_k. We conclude that there are exactly five lines meeting each of the Γ_i; upon checking the transversality of the intersection of the H_{Γ_i},

$$\deg(\mathbb{G}(1, 4)) = 5.$$

Exercise 19.16. Using the description of tangent spaces in Example 16.6, verify that the hyperplane sections H_{Γ_i} do in fact intersect transversely.

This is a fairly generally applicable technique. Indeed, in the 19th century a whole science (some might say art form) developed of such specializations. In particular, it is possible to use these techniques to find the degrees of Grassmannians in general. We will not do this here, but will give the general answer: the degree of the Grassmannian $\mathbb{G}(k, n)$ is

$$\deg(\mathbb{G}(k, n)) = ((k + 1)(n - k))! \cdot \prod_{i=0}^{k} \frac{i!}{(n - k + i)!}.$$

Example 19.17. Harnack's Theorem

As one application of Bézout's theorem in its simplest case—as it applies to the intersection of plane curves—we will prove the classical theorem of Harnack about the topology of real plane curves.

To begin, let us pose the problem. Suppose that $F(Z_0, Z_1, Z_2)$ is a homogeneous polynomial of degree d with real coefficients and such that the corresponding plane curve $X \subset \mathbb{P}^2$ is smooth. Consider now the locus $X(\mathbb{R})$ of real zeroes of F in $\mathbb{RP}^2$ (we will use the topological notation $\mathbb{RP}^2$ since we want to consider it not as a variety but as a real manifold). This is a one-dimensional real submanifold of $\mathbb{RP}^2$, and so consists of a disjoint union of a number δ of connected components, each homeomorphic to S^1 (these were called *circuits* or *ovals* in the classical language). A natural first question to ask, then, is how many circuits X may have.

Before answering this, we note one distinction between components of $X(\mathbb{R})$. Since the first homology $H_1(\mathbb{RP}^2, \mathbb{Z}) = \mathbb{Z}/2$, we may naturally distinguish between those ovals homologous to zero ("even circuits") and those not ("odd circuits"). One difference is that the complement in $\mathbb{PR}^2$ of an even oval has two connected components, with one component homeomorphic to a disc (this is called the *interior* of the oval) and the other homeomorphic to a Mobius strip; the complement of an odd oval, by contrast, is homeomorphic to a disc. Note in particular that any two odd ovals will necessarily intersect (a loop in the complement of an odd oval is necessarily homologous to zero in $\mathbb{RP}^2$), so that a smooth curve can possess at most one odd oval. It follows then that if the degree d of X is even, then $X(\mathbb{R})$ can possess only even ovals; while if d is odd then $X(\mathbb{R})$ will have exactly one odd oval.

With this said, the basic theorem concerning the number of ovals of a plane curve is the following.

Theorem 19.18 (Harnack's Theorem). *The maximum number of ovals a smooth plane curve of degree d can have is $(d-1)(d-2)/2+1$.*

PROOF. We will prove only one direction, that a smooth curve X of degree d cannot have more than $(d-1)(d-2)/2+1$ ovals. Even in that direction the proof will not be complete; we will use the fact that a plane curve Y defined by a polynomial with real coefficients and containing a point p of an even oval of X must intersect that oval at least twice, be tangent to it, or be singular at p. This is pretty clear topologically in case Y is smooth but less so if, for example, Y has singularities in the interior of the oval. It is also clear in case all of the oval can be included in an affine open $\mathbb{R}^2 \subset \mathbb{RP}^2$; in that case, we simply represent $Y(\mathbb{R})$ as the zeros of a real polynomial $f(x, y)$ on $\mathbb{R}^2$ and argue that f must have an even number of zeros on the oval, counting multiplicity. The latter argument can be extended to the general case, but we will not go through the details here.

In any event, to use this observation, suppose that $X(\mathbb{R})$ has $m = (d-1)(d-2)/2+2$ ovals $U_1, \ldots, U_m$, of which we can assume that $U_1, \ldots, U_{m-1}$ are even. Pick out points $p_1, \ldots, p_{m-1}$ with p_i on U_i, and choose points $q_1, \ldots, q_{d-4}$ on the oval U_m. Now recall that the homogeneous polynomials of degree $d-2$ on $\mathbb{P}^2$ form a vector space of dimension $d(d-1)/2$. Since

$$d(d-1)/2 > m + d - 4$$

this means that we can find a nonzero polynomial G of degree $d-2$ on $\mathbb{P}^2$ vanishing at all the points $p_1, \ldots, p_{m-1}$ and $q_1, \ldots, q_{d-4}$; or, equivalently, a plane curve Y of degree $d-2$ containing all these points. But now by our observation, Y must intersect each of the ovals $U_1, \ldots, U_{m-1}$ of $X(\mathbb{R})$ at least twice, and of course must meet the oval U_m at least $d-4$ times. The curve Y thus has at least

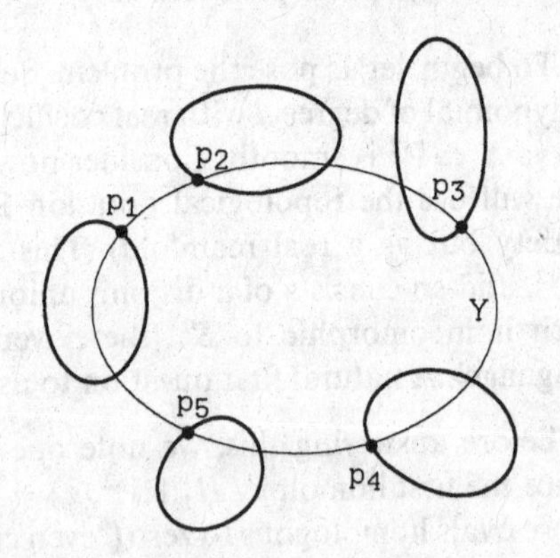

$$\begin{aligned} 2(m-1) + d - 4 \\ &= (d-1)(d-2) + 4 + d - 4 \\ &= d(d-2) + 2 \end{aligned}$$

points of intersection with X, contradicting Bézout's theorem. □

In fact, it is possible to exhibit plane curves of any degree d having the maximum number of ovals. This is usually done by starting with a reducible curve and deforming it slightly; for example, if $f(z_1, z_2)$ and $g(z_1, z_2)$ are quadratic polynomials whose loci are conics intersecting at four points, then the curves X given by $f \cdot g + \varepsilon$ for ε a suitable small constant is a quartic curve with the maximum number four of ovals. The construction of such maximal curves is in general fairly complicated, and we will not discuss it further here; see for example [C] for further discussion and references.

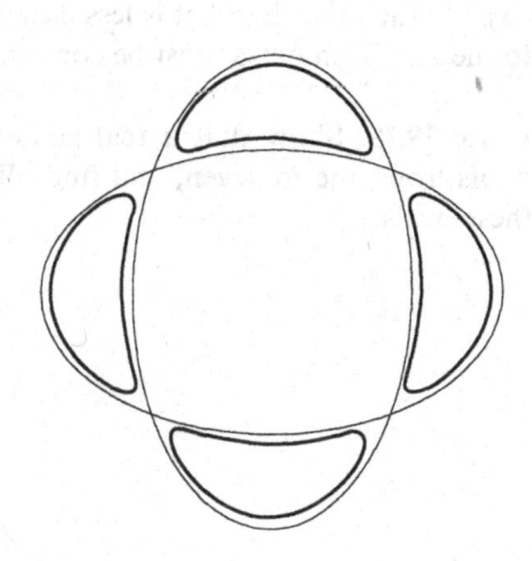

Of course, once we have settled the question of the number of connected components of the real locus $X(\mathbb{R})$ of a real plane curve X, many others follow. For example, the most natural follow-up is to ask about the isotopy class of the embedding $X(\mathbb{R}) \hookrightarrow \mathbb{RP}^2$. This amounts to the question of nesting: we say that two even ovals of a plane curve are *nested* if one lies in the interior of the other as defined earlier, and we may ask what configuration the ovals of a real plane curve may form with regard to nesting. For example, consider the first case in which there is any ambiguity at all, that of plane quartics. Note first that if a plane quartic has any nested pair of ovals, it cannot have no other ovals, since a line joining a point interior to the nested pair to any point of a third oval would have to meet the curve six times. Thus a quartic with three or four ovals has no nesting. On the other hand, a quartic with exactly two ovals may or may not be nested, as we can see by construction (just take two disjoint ellipses, either nested or not, and deform the equation as we did earlier). In general, however, the question of what configurations occur is far from completely solved; a summary of recent work may be found in [W].

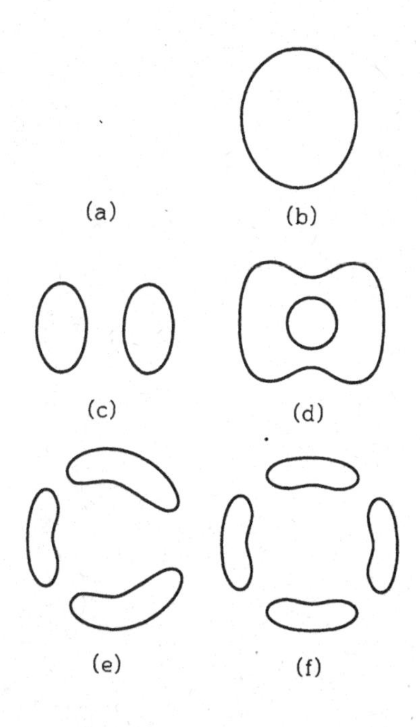

Beyond the question of nesting, there are questions about convexity of ovals, and so on; for example, we can see by construction that if a quartic has two nested ovals, both ovals may be convex, or the interior one may be convex and the exterior nonconvex (it is clear that the interior must be convex, since otherwise we could find a line meeting it in four points, and hence meeting the whole curve in six). On the other hand, it is less clear (but true) that if a quartic with two ovals is not nested, both ovals must be convex.

Exercise 19.19. Show that a real smooth quintic curve may have any number of ovals from one to seven, and find all configurations (with respect to nesting) of these ovals.

LECTURE 20

Singular Points and Tangent Cones

Tangent Cones

The Zariski tangent space to a variety $X \subset \mathbb{A}^n$ at a point p is described by taking the linear part of the expansion around p of all the functions on $\mathbb{A}^n$ vanishing on X. In case p is a singular point of X, however, this does not give us a very refined picture of the local geometry of X; for example, if $X \subset \mathbb{A}^2$ is a plane curve, the Zariski tangent space to X at any singular point p will be all of $T_p(\mathbb{A}^2) = K^2$. We will describe here the *tangent cone*, an object that, while it certainly does not give a complete description of the local structure of a variety at a singular point, is at least a partial refinement of the notion of tangent space.

The definition of the tangent cone is simple enough. In the definition of the tangent space of a variety $X \subset \mathbb{A}^n$ at a point p we took all $f \in I(X)$, expanded around p and took their linear parts; we defined $T_p(X)$ to be the zero locus of these homogeneous linear forms. The difference in the definition of the tangent cone $TC_p(X)$ to X at p is this: we take all $f \in I(X)$, expand around p, and look not at their linear terms but at their *leading terms* (that is, their terms of lowest degree), whatever the degree of those terms might be. We then define the tangent cone $TC_p(X)$ to be the subvariety of $\mathbb{A}^n$ defined by these leading terms. As we have defined it, the tangent cone is a subvariety of the ambient space $\mathbb{A}^n$; but since the linear forms defining $T_p(X)$ are among the leading terms of $F \in I(X)$, it will turn out to be a subvariety of the Zariski tangent space $T_p(X)$, via the inclusion of $T_p(X)$ in the tangent space $T_p(\mathbb{A}^n) = \mathbb{A}^n$.

In the simplest case, that of a hypersurface $X \subset \mathbb{A}^n$ given by one polynomial $f(x_1, \ldots, x_n)$ and $p = (0, \ldots, 0)$, we write

$$f(x) = f_m(x) + f_{m+1}(x) + \cdots$$

where $f_k(x)$ is homogeneous of degree k in $x_1, \ldots, x_n$; the tangent cone will be the cone of degree m given by the homogeneous polynomial f_m. Thus, for example, the tangent cone $TC_p(X)$ to a plane curve $X \subset \mathbb{A}^2$ with a node at p will be the union of the two lines tangent to its branches at p, while the tangent cone to a curve with a cusp will be a single line.

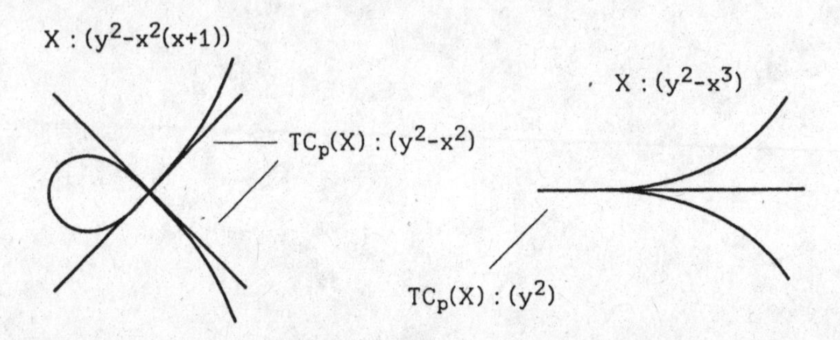

(Note that since the linear term of the sum of two power series is the sum of their linear terms, in defining the Zariski tangent space to a variety $X \subset \mathbb{A}^n$ we needed to look only at the linear parts of a set of generators of the ideal $I(X)$. By contrast, the leading term of a sum will not always be in the ideal generated by their leading terms; to describe the tangent cone in general we will have to take the leading terms of all $f \in I(X)$.)

After giving the initial definition of the Zariski tangent space to an embedded variety $X \subset \mathbb{A}^n$, we gave a more visibly intrinsic definition in terms of the ring of regular functions on X; we can do likewise for the tangent cone. Specifically, let $\mathcal{O} = \mathcal{O}_{X,p}$ be the ring of germs of regular functions on X at p and $\mathfrak{m} \subset \mathcal{O}$ the maximal ideal in this local ring. $\mathcal{O}$ has a filtration by powers of the ideal $\mathfrak{m}$:

$$\mathcal{O} \supset \mathfrak{m} \supset \mathfrak{m}^2 \supset \mathfrak{m}^3 \supset \cdots$$

and we define the ring B to be the associated graded ring, that is,

$$B = \bigoplus_{\alpha=0}^{\infty} \mathfrak{m}^\alpha/\mathfrak{m}^{\alpha+1}.$$

This is by definition generated by its first graded piece

$$B_1 = \mathfrak{m}/\mathfrak{m}^2$$

and so B is a quotient ring A/I of the symmetric algebra

$$A = \bigoplus_{\alpha=0}^{\infty} \mathrm{Sym}^\alpha(\mathfrak{m}/\mathfrak{m}^2)$$

$$= \bigoplus_{\alpha=0}^{\infty} \mathrm{Sym}^\alpha((T_p X)^*).$$

Now, A is naturally the ring of regular functions on the Zariski tangent space $T_p(X) = \mathfrak{m}/\mathfrak{m}^2$; we may define the tangent cone $TC_p(X)$ to be the subvariety of $T_p(X)$ defined by this quotient, that is, the common zero locus of the polynomials $g \in I \subset A$.

Exercise 20.1. Show that these two definitions of the tangent cone in fact agree.

Note that in either version, the polynomials that cut out the tangent cone to X at p—either the leading terms of $f \in I(X)$ at p, or the elements of the kernel of the ring homomorphism $A \to B$—need not generate the ideal of $TC_p(X) \subset T_p(X)$. In other words, the tangent cone $TC_p(X)$ comes to us not just as a subvariety of $T_p(X)$ but as a homogeneous ideal in the ring of polynomials on $T_p(X)$, which represents more data than its zero locus. While our definition of $TC_p(X)$ ignores this additional structure, as we will see later it is important, e.g., in the definition of the multiplicity of a singularity. (To put in one more plug for scheme theory, what we are saying here is that the tangent cone is naturally a subscheme of $T_p(X)$.)

The various constructions made earlier for the Zariski tangent space also apply to tangent cones. As in the case of tangent spaces, if $X \subset \mathbb{P}^n$ is a projective variety, we sometimes choose $\mathbb{A}^n \subset \mathbb{P}^n$ the complement of a hyperplane not containing p and take the closure in $\mathbb{P}^n$ of the affine tangent cone $T_p(X \cap \mathbb{A}^n)$ (viewed as a subvariety of $\mathbb{A}^n$). This projective variety will, by abuse of language, also be referred to as the *projective tangent cone* to X at p; it will be denoted, by analogy with the corresponding construction of the projective tangent space, $\mathbb{T}C_p(X)$. Also, since the tangent cone is defined by homogeneous polynomials on $T_p(X)$, there is associated to it a projective variety $\mathbb{P}TC_p(X) \subset \mathbb{P}T_p(X)$, called the *projectivized tangent cone.*

We can also realize the projective tangent cone $\mathbb{T}C_p(X) \subset \mathbb{P}^n$ as the union of the tangent lines to X at p, in the first sense of Example 15.17, and the projectivized tangent cone $\mathbb{P}TC_p(X) \subset \mathbb{P}T_p(X)$ as the set of these lines; this will be established in Exercise 20.4.

The most important fact about the tangent cone is that its dimension is always the local dimension of X at p. This follows from standard commutative algebra (cf. [AM], [E]), given that the coordinate ring of the tangent cone is the associated graded ring of the local ring of X at p with respect to the filtration given by the powers of the maximal ideal. It may seem less obvious from a naive geometric viewpoint (we will see a more geometric proof of it later). Certainly it does not jibe with the real picture of varieties; for example, it is hard to imagine from the picture alone what would be the tangent cone to the surface $z^2 + x^2 = y^4$ obtained by rotating the parabola $y^2 - x$ around its tangent line $x = 0$ at $(0, 0)$, though it's easy enough to look at the equation and see what it is.

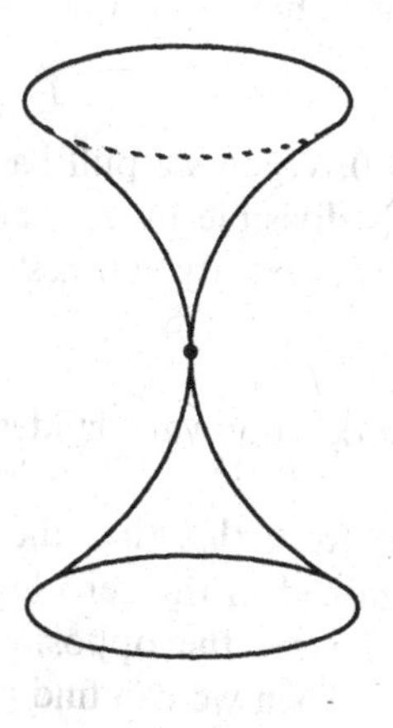

There are a number of geometric interpretations of the tangent cone, and especially of the projectivized tangent cone. Perhaps the central one involves the notion of the blow-up of a variety, introduced in Lecture 7; we will now discuss this.

To begin, recall the definition of the blow-up $\tilde{\mathbb{A}}^n$ of $\mathbb{A}^n$ at the origin; it is the graph of the projection map $\mathbb{A}^n \to \mathbb{P}^{n-1}$, which is to say the locus

$$\tilde{\mathbb{A}}^n = \{((z_1, \ldots, z_n), [W_1, \ldots, W_n]): z_i W_j = z_j W_i \ \forall i, j\}$$
$$\subset \mathbb{A}^n \times \mathbb{P}^{n-1}.$$

In this context, we think of $\tilde{\mathbb{A}}^n$ primarily in terms of the projection map $\pi: \tilde{\mathbb{A}}^n \to \mathbb{A}^n$, which is one to one except over the origin in $\mathbb{A}^n$, where the fiber is all of $\mathbb{P}^{n-1}$ (note that this fiber is naturally the projective space of lines through the origin in $\mathbb{A}^n$ itself, which may be identified with the projective space associated to the tangent space $T_0(\mathbb{A}^n)$). This fiber $E = \pi^{-1}(0)$ is called the *exceptional divisor* of the blow-up.

Now, for any subvariety $X \subset \mathbb{A}^n$, we define the proper transform of X in $\tilde{\mathbb{A}}^n$ to be the closure in $\tilde{\mathbb{A}}^n$ of the inverse image of $X \cap (\mathbb{A}^n - \{0\})$; equivalently, the closure of $\pi^{-1}(X) \cap U$, where U is the complement of the exceptional divisor in $\tilde{\mathbb{A}}^n$. In fact, this is isomorphic to the blow-up of X at the origin, as defined in Lecture 7; if we think of the blow-up as the graph of projection π from 0, it is just the graph of $\pi|_X$.

We can write down the equations of the proper transform in $\tilde{\mathbb{A}}^n$ of a variety $X \subset \mathbb{A}^n$, assuming we know them for X itself. To begin, let $U_i \subset \tilde{\mathbb{A}}^n$ be the open set given by $(W_i \neq 0)$; in terms of Euclidean coordinates $w_j = W_j/W_i$ we can write

$$U_i = \{((z_1, \ldots, z_n), (w_1, \ldots, \hat{w}_i, \ldots, w_n)): z_j = z_i w_j \ \forall j \neq i\}$$
$$\subset \mathbb{A}^n \times \mathbb{A}^{n-1}.$$

We see from this description that $U_i \cong \mathbb{A}^n$, with coordinates z_i and $w_1, \ldots, \hat{w}_i, \ldots, w_n$; the map $\pi: U_i \to \mathbb{A}^n$ is given by

$$(z_i, w_1, \ldots, \hat{w}_i, \ldots, w_n) \mapsto (z_i w_1, \ldots, z_i w_{i-1}, z_i, z_i w_{i+1}, \ldots, z_i w_n).$$

Note that the part $E_i = E \cap U_i$ of the exceptional divisor in U_i is given as the coordinate hyperplane $z_i = 0$; i.e., the open cover $\{U_i\}$ of $\tilde{\mathbb{A}}^n$ restricts to E to give the standard open cover of $\mathbb{P}^{n-1}$ by affine spaces $\mathbb{A}^{n-1}$, with $\{w_j\}_{j \neq i}$ the Euclidean coordinates on E_i.

Now suppose that $f \in I(X)$ is any polynomial vanishing on X. Write f as the sum of its homogeneous parts

$$f = f_m + f_{m+1} + f_{m+2} + \cdots$$

with $f_m \neq 0$. When we pull back f to $\tilde{\mathbb{A}}^n$ and restrict to U_i, we see that the term $\pi^*(f_k)|_{U_i}$ is divisible by z_i exactly k times; so that the pullback $\pi^*(f)|_{U_i}$ will be divisible by z_i exactly m times. We can thus write

$$\pi^*(f)|_{U_i} = z_i^m \cdot \tilde{f},$$

where $\tilde{f}$ does not vanish identically on E_i; $\tilde{f}$ will be called the proper transform of f.

We see from this that the part $\tilde{X} \cap U_i$ of the proper transform $\tilde{X}$ of X in U_i is contained in the zero locus of the proper transforms $\tilde{f}$ of the polynomials $f \in I(X)$. To see the opposite inclusion, suppose $p \in E_i$ is any point not lying on $\tilde{X} \cap U_i$. Then we can find a hypersurface in U_i containing $\tilde{X}$ but not p; taking the defining equation of this hypersurface and multiplying by a sufficiently high power of z_i yields a polynomial $f \in I(X)$ whose proper transform vanishes on $\tilde{X} \cap U_i$ but not at p.

Now restrict to E. Since $\pi^*(f_k)|_{U_i}$ is divisible by z_i $m+1$ times for $k > m$, we see that

$$\tilde{f}|_{E_i} = (\pi^*(f_m)/z_i^m)|_{E_i};$$

i.e., the zero locus of $\tilde{f}$ in E_i is just the intersection with E_i of the zero locus of the leading term f_m of f, viewed as a homogeneous polynomial on $E = \mathbb{P}^{n-1}$. In sum, then, we find that the projectivized tangent cone $\mathbb{P}TC_p(X)$ to X at p is simply the intersection $\tilde{X} \cap E$ of the proper transform of X with the exceptional divisor—equivalently, the exceptional divisor of the blow-up of X at the point p.

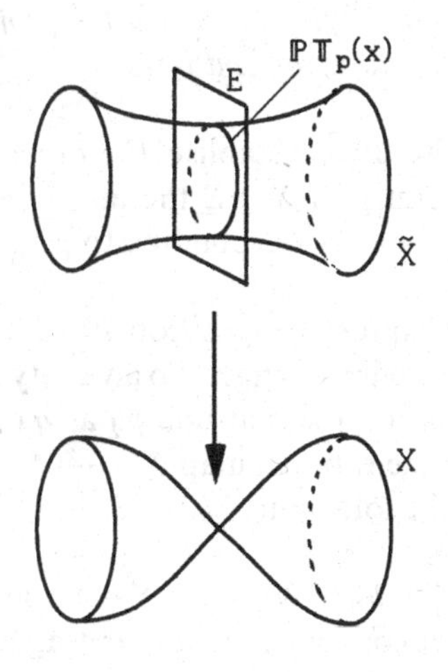

One immediate consequence of this description of the projectivized tangent cone is the observation that it does indeed have dimension one less than X itself: on one hand, since it is in the closure of $\tilde{X} - (\tilde{X} \cap E)$, it must have dimension strictly less than X; on the other hand, since E is locally defined by one equation, it can be at most one less. Likewise, it follows that the tangent cone has, as indicated, the same dimension as X. It also gives rise to other interpretations, which we will now discuss.

The most direct new interpretation of the tangent cone comes (for varieties over the complex numbers, at least) from the description, in Examples 17.1 and 17.5, of the tangent line to an arbitrary holomorphic arc. Recall specifically from Example 17.5 that if we have any arc $\gamma: \Delta \to X \subset \mathbb{P}^n$, the associated Gauss map $\mathscr{G}$ sending $t \in \Delta$ to the tangent line to the arc at t, which is defined a priori only for smooth points of the arc, in fact extends uniquely to all of Δ. Explicitly, if γ is written locally as

$$\gamma: t \mapsto [v(t)] = [v_0 + v_1 t + v_2 t^2 + \cdots]$$

with $v(0) \neq 0$, then $\mathscr{G}$ may be extended to $t = 0$ by taking $\mathscr{G}(0)$ to be the span of v_0 and the first coefficient vector v_k independent of v_0. This line will be called the tangent line to γ at $\gamma(0)$. Equivalently, if we choose local Euclidean coordinates $(z_1, \ldots, z_n)$ on an affine neighborhood $\mathbb{A}^n \subset \mathbb{P}^n$ such that $v(0) = v_0 = (0, 0, \ldots, 0)$ and write

$$\gamma: t \mapsto u_1 t + u_2 t^2 + \cdots$$

then the tangent line to γ at $t = 0$ is just the line through the origin $(0, 0, \ldots, 0) \in \mathbb{A}^n$ in the direction of the first nonzero coefficient vector u_k.

Now, suppose that the arc $\gamma: \Delta \to \mathbb{P}^n$ has image in the variety X, and that $f(z_1, \ldots, z_n)$ is any polynomial vanishing on X. Write f as

$$f = f_m + f_{m+1} + f_{m+2} + \cdots$$

and the arc γ as

$$\gamma: t \mapsto u_k t^k + u_{k+1} t^{k+1} + \cdots.$$

From the fact that $f(\gamma(t)) \equiv 0$ it follows that the leading term of f must vanish on the first nonzero coefficient vector u_k of u; in other words, the tangent line to the arc γ at $p = \gamma(0)$ lies in the tangent cone to X at p. We have thus established one half of the following.

Proposition 20.2. *Let $X \subset \mathbb{P}^n$ be any variety and $p \in X$ any point. The projectivized tangent cone $\mathbb{P}TC_p(X)$, viewed as a set of lines through p, is the set of tangent lines at p to arcs lying on X and the projective tangent cone $\mathbb{T}C_p(X)$ correspondingly the union of these tangent lines.*

Exercise 20.3. Establish the other half of Proposition 20.2 by showing that for any point p of $\tilde{X} \cap E$ the image in X of an arc $\gamma: \Delta \to \tilde{X}$ with $\gamma(0) = 0$ will have tangent line corresponding to p.

To express Proposition 20.2 another way, recall from Example 15.17 that one way to define a tangent to a variety X at a possibly singular point p is as the limiting position of a secant line $\overline{pq}$ as $q \in X$ approaches p—that is, as the image of $\{p\}$ under the rational map $X \dashrightarrow \mathbb{G}(1, n)$ sending q to $\overline{pq}$. From Proposition 20.2 we have the following.

Exercise 20.4. With $p \in X \subset \mathbb{P}^n$ as in the preceding, show that the projectivized tangent cone is the set of limiting positions of lines $\overline{pq}$ as $q \in X$ approaches p.

Example 20.5. Tangent Cones to Determinantal Varieties

We will describe here the tangent cones to the determinantal varieties M_k of $m \times n$ matrices of rank k, in a manner completely analogous with the discussion of their tangent spaces in Example 14.15. We start by choosing a point $A \in M_k$ corresponding to a matrix of rank exactly l, i.e., $A \in M_l - M_{l-1}$. We can choose bases for K^m and K^n so that A is represented by the matrix

$$\begin{bmatrix} 1 & 0 & 0 & \cdots & & \cdots & 0 \\ 0 & 1 & 0 & \cdots & & \cdots & 0 \\ \vdots & & & & & & \\ 0 & \cdots & & 1 & 0 & \cdots & 0 \\ 0 & \cdots & & 0 & 0 & \cdots & 0 \\ \vdots & & & & & & \\ 0 & \cdots & & \cdots & & \cdots & 0 \end{bmatrix}$$

and then, in the affine neighborhood U of A given by $(X_{11} \neq 0)$, we may take as Euclidean coordinates the functions $x_{i,j} = X_{i,j}/X_{1,1}$; in terms of these we may write a general element of U as

$$\begin{bmatrix} 1 & x_{1,2} & x_{1,3} & \cdots & & \cdots & \cdots & x_{1,m} \\ x_{2,1} & 1 + x_{2,2} & x_{2,3} & \cdots & & \cdots & \cdots & x_{2,m} \\ \vdots & & & & & & & \\ x_{l,1} & \cdots & \cdots & 1 + x_{l,l} & & x_{l,l+1} & \cdots & x_{l,m} \\ x_{l+1,1} & \cdots & \cdots & x_{l+1,l} & & x_{l+1,l+1} & \cdots & x_{l+1,m} \\ \vdots & & & & & & & \\ x_{n,1} & \cdots & \cdots & \cdots & & \cdots & \cdots & x_{n,m} \end{bmatrix}$$

with A of course corresponding to the origin in this coordinate system.

What are the leading terms of the $(k + 1) \times (k + 1)$ minors of this matrix? The answer is that they are the terms in the expansions of these minors involving the diagonal entries from the first $l \times l$ block of this matrix; they are thus the $(k + 1 - l) \times (k + 1 - l)$ minors of the lower right $(m - l) \times (n - l)$ block. The tangent cone to M_k at A is thus contained in the space of matrices φ whose lower right $(m - l) \times (n - l)$ block has rank at most $k - l$. On the other hand, this locus is irreducible of the right dimension, so in fact that tangent cone must be equal to it.

As before, we can give an intrinsic interpretation to this tangent cone. To do this, note that the bases $\{e_i\}$ and $\{f_j\}$ for K^m and K^n are chosen so that the kernel of A is exactly the span of $e_{l+1}, \ldots, e_m$ and the image of A exactly the span of $f_1, \ldots, f_l$. The lower right $(m - l) \times (n - l)$ block of φ thus represents the composition

$$\varphi' \colon \mathrm{Ker}(A) \hookrightarrow K^m \xrightarrow{\varphi} K^n \to K^n/\mathrm{Im}(A).$$

We thus have

$$\mathbb{T}_A(M_k) = \{\varphi \in \mathrm{Hom}(K^m, K^n) \colon \mathrm{rank}(\varphi' \colon \mathrm{Ker}(A) \to \mathrm{Coker}(A)) \leq k - l\};$$

i.e., we may say, without invoking coordinates, that the tangent cone to M_k at a point A corresponding to a map $A\colon K^m \to K^n$ of rank exactly l is the variety of maps $\varphi \in \mathrm{Hom}(K^m, K^n)$ whose image $\varphi' \in \mathrm{Hom}(\mathrm{Ker}(A), \mathrm{Coker}(A))$ has rank at most $k - l$.

Exercise 20.6. Let X and $Y \subset \mathbb{P}^n$ be any pair of varieties lying in disjoint subspaces of $\mathbb{P}^n$, and let $J = J(X, Y)$ be their join. Show that the tangent cone to J at a point $p \in X$ is the join of Y and the projective tangent cone $\mathbb{T}C_p(X)$ to X at p. What can happen if X and Y are disjoint but do not lie in disjoint linear spaces (so that, for example, $\mathbb{T}C_p(X)$ may intersect Y)? What happens if X actually meets Y?

Exercise 20.7. Let $C \subset \mathbb{P}^3$ be the twisted cubic curve and TC its tangential variety. What is the tangent cone to TC at a point $p \in C$? Is the same true if C is an arbitrary space curve, if we assume that the point p lies on no tangent line to C other than $\mathbb{T}_p C$?

Multiplicity

As we have remarked already, the tangent cone to a variety X at a point $p \in X$ comes to us, not just as a variety $TC_p(X) \subset T_p(X)$, but as a variety together with a homogeneous ideal $I \subset \mathrm{Sym}^*(T_p(X)^*)$ in the ring of polynomials on $T_p(X)$. By definition, the tangent cone $TC_p(X) = V(I)$ is the common zero locus of this ideal; but I need not be the ideal of all functions vanishing on $TC_p(X)$—in cases like the cusp $y^2 - x^3$ or the tacnode $y^2 - x^4$, the tangent cone is equal to the tangent line $y = 0$, but the ideal associated to it is the ideal (y^2). For the following, we will actually have to use this extra information.

We need to invoke the ideal in particular in order to define the *multiplicity* of a singular point in general. Multiplicity is easy to define in the case of a hypersurface $X \subset \mathbb{A}^n$, that is, the zero locus of a single polynomial $f(z_1, \ldots, z_n)$: the multiplicity of X at p is just the order of vanishing of f at p. Thus, if we write

$$f = f_m + f_{m+1} + \cdots$$

with f_α homogeneous of degree α and $f_m \neq 0$, we say that the multiplicity of X at the origin is m.

How do we generalize this to an arbitrary variety $X \subset \mathbb{A}^n$? The example of hypersurfaces suggests an answer: the multiplicity should be the degree of the tangent cone. The problem is that the multiplicity m of the hypersurface need not be the degree of its tangent cone if the polynomial f_m contains repeated factors, as it does in the cases of a cusp or a tacnode, for example.

The solution is to define the multiplicity in terms of the ideal $I \subset \mathrm{Sym}^*(T_p(X)^*)$ rather than in terms of the tangent cone. In other words, referring back to the definition of degree via the Hilbert polynomial, we define the multiplicity of a point p on a k-dimensional variety $X \subset \mathbb{A}^n$ to be $(k-1)!$ times the leading coefficient of the Hilbert polynomial of the ring $\mathrm{Sym}^*(T_p(X)^*)/I$, where I is the ideal generated by the leading terms of the expansions around p of polynomials vanishing on X.

(As we have suggested, this definition is an argument for the introduction of schemes; in that language, we simply define the tangent cone $\mathbb{T}_p(X)$ of a variety $X \subset \mathbb{A}^n$ at p to be the *subscheme* of $T_p(X)$ defined by the ideal I, and the multiplicity of X at p to be the degree of this scheme.)

Exercise 20.8. Let $M_k \subset M$ be the variety of matrices of rank at most k in the space M of all $m \times n$ matrices and let $A \in M_k$ be a point corresponding to a matrix of rank exactly l. What is the multiplicity of M_k at A?

Another way of describing the multiplicity of a point p on a variety X of dimension k over the complex numbers $\mathbb{C}$ is analogous to the definition of the degree of a variety via its fundamental class. Suppose that X is embedded in affine space $\mathbb{A}^n$ and let $\tilde{X}$ be the blow-up of X at the point p. This is naturally a subvariety of the blow-up $\tilde{\mathbb{A}}^n$ of $\mathbb{A}^n$ at the origin. Now $\tilde{\mathbb{A}}^n$ lives inside $\mathbb{A}^n \times \mathbb{P}^{n-1}$ as the graph of the projection map $\pi\colon \mathbb{A}^n \to \mathbb{P}^{n-1}$ and the projection map $\pi_2\colon \tilde{\mathbb{A}}^n \to \mathbb{P}^{n-1}$ is thereby a topological fiber bundle with fiber $\mathbb{A}^1 \cong \mathbb{C}$. It follows that the cohomology with compact support

$$H_c^m(\tilde{\mathbb{A}}^n, \mathbb{Z}) \cong H_c^{m+2}(\mathbb{P}^{n-1}, \mathbb{Z}).$$

Now $\tilde{X}$, being a closed subvariety of real dimension $2k$ of $\tilde{\mathbb{A}}^n$, defines a linear form η_X on $H_c^{2k}(\tilde{\mathbb{A}}^n, \mathbb{Z})$; we can write

$$\eta_X = m \cdot \varphi$$

where φ is a generator of the group $H_c^{2k}(\tilde{\mathbb{A}}^n, \mathbb{Z})^* \cong \mathbb{Z}$ and $m \geq 0$. The integer m is then the multiplicity of X at p.

(For those familiar with the language, η_X is just the fundamental class of $\tilde{X}$ in the Borel-Moore homology of $\tilde{\mathbb{A}}^n$. This is the homology theory dual to cohomology with compact supports; in it, we require chains to be locally finite, but not necessarily finite, linear combinations of simplices; see [BM].)

Suppose now that $X \subset \mathbb{P}^n$ is a projective variety of dimension k and degree d, $p \in X$ a point of multiplicity m, $\tilde{X}$ the blow-up of X at p, and $\tilde{\mathbb{P}}^n$ the blow-up of $\mathbb{P}^n$ at p. From the preceding, it follows that the fundamental class

$$[\tilde{X}] = d \cdot [\Lambda] - m \cdot [\Gamma]$$

where $\Lambda \cong \mathbb{P}^k$ is a k-plane in $\mathbb{P}^n$ not passing through p, and Γ is a k-plane contained in the exceptional divisor $E \cong \mathbb{P}^{n-1} \subset \tilde{\mathbb{P}}^n$ of the blow-up. From this follows another characterization of the multiplicity of the point $p \in X$: if we assume that the projection map $\pi_p: X \to \mathbb{P}^{n-1}$ is birational onto its image $\bar{X} \subset \mathbb{P}^{n-1}$, then the degree of $\bar{X}$ is the degree of X minus the multiplicity of X at p. More generally, if $\tilde{X}$ is the blow-up of X at p, then $\pi_p: \tilde{X} \to \mathbb{P}^{n-1}$ is regular and the multiplicity of X at p is the difference of the degree of X and the projective degree of π_p (see Example 19.4).

Exercise 20.9. Using the homological characterization of multiplicity, prove this.

Exercise 20.10. Let X and $Y \subset \mathbb{P}^n$ be varieties, $\bar{X}$ and $\bar{Y} \subset \mathbb{A}^{n+1}$ the cones over them. Show that $\bar{X}$ and $\bar{Y}$ are isomorphic if and only if X and Y are projectively equivalent.

Exercise 20.11. Let $X \subset \mathbb{P}^n$ be a hypersurface, $p \in X$ a smooth point of X, and

$$II_p: N_p(X)^* \to \mathrm{Sym}^2(T_p^*(X))$$

the second fundamental form, as described in Example 17.11. Show that the intersection $Y = X \cap \mathbb{T}_p(X)$ of X with its tangent hyperplane at p has a double point if and only if $II_p \neq 0$, and that in this case the tangent cone to Y at p is given by the image of II_p.

Exercise 20.12. Recall the circumstances of Exercise 20.6: we let X and $Y \subset \mathbb{P}^n$ be any pair of varieties lying in disjoint subspaces of $\mathbb{P}^n$, and let $J = J(X, Y)$ be their join. Show that the multiplicity of J at a point $p \in X$ is just the degree of Y times the multiplicity of p on X. As in Exercise 20.6, what can happen if X and

Y are disjoint but do not lie in disjoint linear spaces (so that, for example, $\mathbb{T}C_p(X)$ may intersect Y); and what happens if X actually meets Y?

Exercise 20.13. This time we will revisit Exercise 20.7. Let $C \subset \mathbb{P}^3$ be the twisted cubic curve and TC its tangential variety. What is the multiplicity of TC at a point $p \in C$? Is the same true if C is an arbitrary space curve, if we assume that the point p lies on no tangent line to C other than $\mathbb{T}_pC$?

Examples of Singularities

Ideally, one would like to be able to classify singular points of varieties, up to a reasonable equivalence relation. To this end, we can introduce invariants of the singularity; we have already introduced the Zariski tangent space, the tangent cone, and the multiplicity, and there are many others. We can likewise define classes of singularities with special properties and study these; for example, we can look at normal singularities, Gorenstein singularities, Cohen-Macaulay singularities, local complete intersection singularities, and so on. For that matter, which equivalence relation is appropriate to apply depends on context; people have studied singularities up to topological equivalence, corresponding (in case $K = \mathbb{C}$) to saying that two singularities $p \in X$, $p' \in X'$ are equivalent if there are balls B, B' in the analytic topology on $\mathbb{A}^n$ around p and p' such that the triples $(p, X \cap B, B)$ and $(p', X' \cap B', B')$ are homeomorphic, while others look at singularities up to analytic equivalence (i.e., saying that two singularities $p \in X$, $p' \in X'$ are equivalent if they have isomorphic neighborhoods in the analytic topology; equivalently, if the completions of the local rings $\mathcal{O}_{X,p}$ and $\mathcal{O}_{X',p'}$ are isomorphic).

With all this, the whole subject of singularities of algebraic varieties remains almost completely mysterious. This is not surprising: for example, since among singularities of varieties $X \subset \mathbb{A}^n$ are cones over varieties $Y \subset \mathbb{P}^{n-1}$, we could not ever achieve a classification of germs of singularities without having, as a very special and relatively simple subcase, a classification of all varieties in projective space. Rather than get enmeshed in the subject, we will here try to give a very rough idea of it by discussing some examples of the simplest case, $\dim(X) = 1$ and $\dim T_p(X) = 2$. Throughout, when we use the word "equivalent" without modification we will mean analytically equivalent, in the preceding sense.

To begin, the simplest of all curve singularities is the node. A *node* is a singularity analytically equivalent to the origin in the curves $xy = 0$ or $y^2 - x^2 = 0$ in $\mathbb{A}^2$, that is, it consists of the union of two smooth arcs meeting transversely. Such a singularity is characterized by saying that it is a double point (i.e., has multiplicity two) with tangent cone the union of two distinct lines.

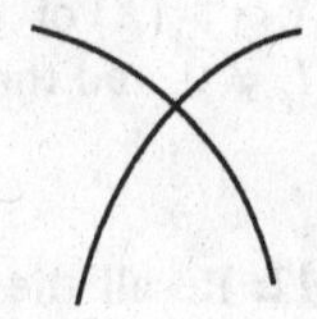

A *cusp* is a singularity analytically equivalent to the origin in the curve C given by $y^2 - x^3 = 0$ in $\mathbb{A}^2$. By contrast with the node, it is locally irreducible in the analytic topology (*unibranch*, in the terminology of singularities); that is, it is not the union of two or more distinct analytic arcs.

A cusp is easy to draw over the real numbers, but this may obscure some of the intricacy of even this relatively simple singularity. To get a better idea of the geometry of a cusp over $\mathbb{C}$, we may look at the link of the singularity, as follows. For any curve in $\mathbb{A}^2$ containing the origin, we may intersect the curve with the boundary of a ball of radius ε around the origin in $\mathbb{A}^2 = \mathbb{C}^2$. We have

$$\partial B_\varepsilon((0, 0)) = \{(x, y): |x|^2 + |y|^2 = \varepsilon\} \cong S^3,$$

and for small values of ε this should intersect the curve in a smooth real 1-manifold—that is, a disjoint union of copies of S^1 in S^3. For small values of ε the isotopy class of this submanifold of S^3 will be independent of ε and is called the *link* associated to the singularity of the curve at the origin.

To see what the link of the cuspidal curve $C \subset \mathbb{C}^2$ looks like, note that the absolute values of the coordinates x and y of any point $p \in C \cap \partial B_\varepsilon$ are c and $c^{3/2}$, respectively, where c is the unique real solution of the equation $c^2 + c^3 = \varepsilon$. We can thus write

$$\begin{aligned} X \cap \partial B_\varepsilon = \{(ce^{2i\vartheta}, c^{3/2}e^{3i\vartheta})\} &\subset \{(x, y): |x| = c, |y| = c^{3/2}\} \\ &\cong S^1 \times S^1 \subset S^3. \end{aligned}$$

In other words, the link of the cusp lies on a torus in the sphere $\partial B_\varepsilon \cong S^3$, winding twice around the torus in one direction and three times in the other; in particular, we see that the link of an ordinary cusp is a trefoil knot.

Note that the link of a smooth point is simply an unknotted $S^1 \subset S^3$; conversely, any point with such a link is smooth. In general, the connected components of the link of a singular point p of a curve correspond to the branches of the curve at p, with the intersection number of two such branches at p given by the linking number of the corresponding loops. In particular, note that the union of two unlinked simple closed curves cannot be the link of an algebraic curve singularity.

We can carry out the same analysis as earlier to deduce that the link of any singularity of the form $x^p + y^p = 0$, with p and q relatively prime, is a torus knot of type (p, q). More generally, the knot associated to any unibranch plane curve singularity is what is called an *iterated torus knot*; it is obtained by starting with a torus, drawing a torus knot on it, taking the boundary of a tubular neighborhood of this knot, which is again a torus, and repeating this procedure a finite number of times.

Exercise 20.14. Describe the link of the curve $x^p - y^q = 0$ at the origin in case p and q are not relatively prime. Verify (at least in case $q|p$, but preferably in general) that this agrees with the description of the linking number of the connected components of the link.

A *tacnode* is defined to be a singularity equivalent to the origin in the curve $y^2 - x^4 = 0$, that is, the union of two smooth arcs meeting with contact of order two. Similarly, an *oscnode* is a singularity equivalent to $y^2 - x^6 = 0$, i.e., consisting of two smooth branched with contact of order three, and so on.

Note that the branches of an oscnode do not have to have contact of order three with their affine tangent line, as they do in this example; for example, the automorphism of $\mathbb{A}^2$ given by $(x, y) \mapsto (x, y + x^2)$ destroys this property. In particular, a curve does not have to have degree six or greater to have an oscnode; an irreducible quartic curve, for example, can have one.

Exercise 20.15. Find the equation of an irreducible quartic plane curve $C \subset \mathbb{P}^2$ with an oscnode (*).

We should mention, however, that it is not known in general for what values of d and n an irreducible plane curve of degree d may have a singular point analytically equivalent to the origin in $y^2 - x^{2n}$.

A *ramphoid cusp* is a singularity equivalent to the origin in the curve $y^2 - x^5 = 0$. Like a cusp, it is unibranch. Unlike a cusp it is not resolved by blowing it up once; the blow-up of a curve with a ramphoid cusp has an ordinary cusp. As in the case of the oscnode, the standard form of the equation may give a false impression; a ramphoid cusp does not have to have contact or order five with its tangent line (the curve pictured is a quartic).

In fact, we have now listed all double points of curves. We will state the following theorem without proof; a reference is [BK].

Theorem 20.16. *Any singular point of multiplicity* 2 *on a curve* C *is equivalent to a singularity of the form* $y^2 - x^n = 0$ *for some* n.

After double points come triple points, and so on. As you might expect, things are more complicated here. One major difference is the possibility of continuously varying families of singularities, even in simple cases. For example, an ordinary m-fold point of a curve C is defined to be a singularity equivalent to a union of m concurrent distinct lines, that is, a point of multiplicity m with tangent cone consisting of m distinct lines. There is a unique ordinary triple point, but a continuous family of nonequivalent quadruple points; the singularities given by $xy(x - y)(x - \lambda y)$ and $xy(x - y) \times (x - \lambda' y)$ will be equivalent if and only if the projectivized tangent cones, viewed in each case as subsets of the projectivized tangent space $\mathbb{P}^1$, are projectively equivalent, that is, if and only if $j(\lambda) = j(\lambda')$, where j is the j-function introduced in Lecture 10. Thus, with singular points of higher multiplicity the topology of the singularity does not determine its equivalence class.

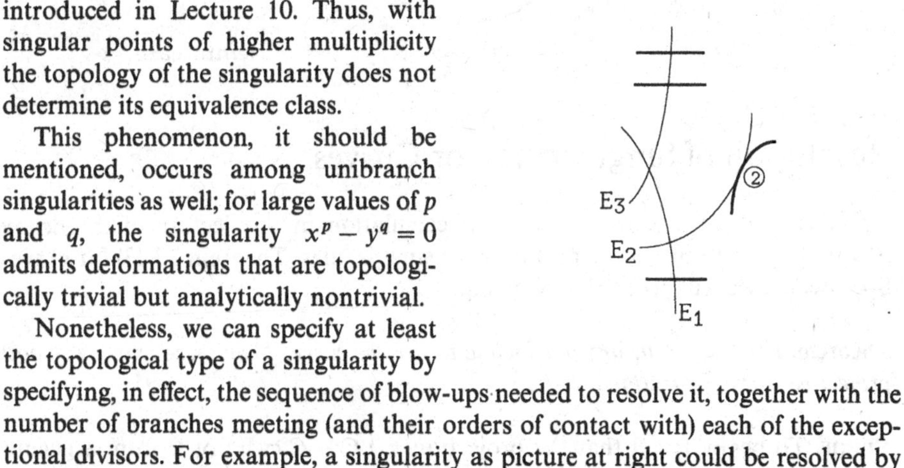

This phenomenon, it should be mentioned, occurs among unibranch singularities as well; for large values of p and q, the singularity $x^p - y^q = 0$ admits deformations that are topologically trivial but analytically nontrivial.

Nonetheless, we can specify at least the topological type of a singularity by specifying, in effect, the sequence of blow-ups needed to resolve it, together with the number of branches meeting (and their orders of contact with) each of the exceptional divisors. For example, a singularity as picture at right could be resolved by a sequence of three blow-ups, resulting in a union of disjoint smooth arcs as shown; specifying the latter diagram specifies the topological type of the singularity.

In classical terminology, a point of the exceptional divisor of the blow-up of a surface at a point p was called an "infinitely near point" to p, so that, for example, a tacnode could be described as a node with an infinitely near node. Similarly, the two singularities pictured to the right—consisting of a union of three smooth arcs with pairwise contact of order two (called a *triple tacnode*), and a union of three arcs of which two have contact of order two and the third is transverse—would be described in classical language as "a triple point with an infinitely near triple point," and "a triple point with an infinitely near node," respectively.

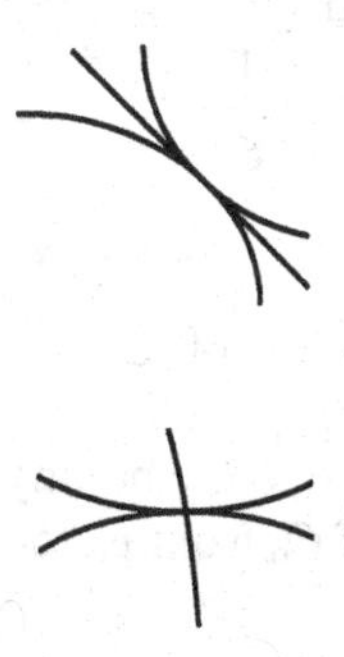

Exercise 20.17. Show that the following plane curves are algebraic, find their degrees and describe their singularities: $(*)$

(a) the cardioid, given in polar coordinates as $r = 1 + \cos(\vartheta)$.
(b) the limacon, given by the equation $r = 1 + 2\cos(\vartheta)$.

(c) the lemniscate, given by $r^2 = \cos(2\vartheta)$.
(d) the "four-leafed rose" $r = \cos(2\vartheta)$.

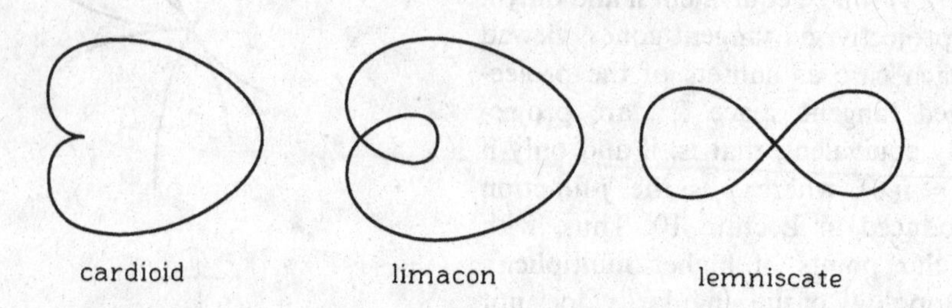

Resolution of Singularities for Curves

To finish this lecture, we will use the calculation in this lecture of the degree of a projection to prove the resolution of singularities (Theorem 17.23) for curves. Specifically, we will prove the following.

Theorem. *Let $C \subset \mathbb{P}^n$ be any irreducible projective curve. Then there exists a smooth projective curve $\tilde{C}$ birational to C.*

PROOF. To begin, recall that the projection $\pi_\Lambda: C \to \bar{C} \subset \mathbb{P}^2$ of C from a general $(n-3)$-plane $\Lambda \subset \mathbb{P}^n$ is birational, so we can start with $C \subset \mathbb{P}^2$. Say the degree of C is d. Choose n large, and consider the image of C under the nth Veronese map. The dimension of the vector space of homogeneous polynomials of degree n on $\mathbb{P}^2$ is $\binom{n+2}{2} = (n+2)(n+1)/2$, and the subspace of those vanishing on C of dimension $\binom{n-d+2}{2} = (n-d+2)(n-d+1)/2$, so that the dimension of the span of the image C_0 of C is

$$N = \dim(\overline{C_0}) = dn - d(d-3)/2$$

and the degree of C_0 is $D_0 = n \cdot d$. Assume n is chosen large enough that $N > D_0/2 + 1$.

Now, let $p_0 \in C_0$ be any singular point, and let $C_1 = \pi_{p_0}(C_0) \subset \mathbb{P}^{N-1}$ be the projection of C_0 from p_0 to a hyperplane. Let

$$D_1 \leq D_0 - 2 < 2N - 4$$

the projective degree of the map π_{p_0}. Note that C_1 is nondegenerate in $\mathbb{P}^{N-1}$ and D_1 is strictly less than $2(N-1)$, so that by Proposition 18.9 on the minimal degree of nondegenerate curves we see that π_{p_0} must be birational (and in particular we have $D_1 = \deg(C_1)$).

Now suppose C_1 has a singular point p_1. We can then repeat this process, projecting from p_1 to obtain a birational map from C_1 to a curve C_2 nondegenerate in a projective space of dimension $N - 2$ and having degree $D_2 < 2N - 6$. We continue doing this, generating a sequence of curves $C_i \subset \mathbb{P}^{N-i}$ of degrees $D_i < 2(N - i) - 2$ as long as C_i continues to have singular points. But this has to stop; we cannot arrive at a nondegenerate curve of degree 1 in $\mathbb{P}^2$. □

Exercise 20.18. Show that we can choose $n = d - 2$ in the preceding argument. Use this to show that a plane curve of degree d can have at most $(d - 1)(d - 2)/2$ singular points.

It is worth remarking that we do not know the maximum number of isolated singular points a surface of degree d in $\mathbb{P}^3$ may have.

LECTURE 21
Parameter Spaces and Moduli Spaces

Parameter Spaces

We can now give a slightly expanded introduction to the notion of parameter space, introduced in Lecture 4 and discussed occasionally since. This is a fairly delicate subject, and one that is clearly best understood from the point of view of scheme theory, so that in some sense this discussion violates our basic principle of dealing only with topics that can be reasonably well understood on an elementary level. Nevertheless, since it is one of the fundamental constructions of algebraic geometry, and since the constructions can at least be described in an elementary fashion, we will proceed. One unfortunate consequence of this sort of overreaching, however, is that the density of unproved assertions, high enough in the rest of the text, will reach truly appalling levels in this lecture.

With this understood, let us first say what we should mean by a parameter space. The basic situation is that we are given a collection of subvarieties X_α of a projective space $\mathbb{P}^n$—for example, the set of all varieties of a given dimension and degree or the subset of those with a given Hilbert polynomial. The problem is then to give a bijection between this set $\{X_\alpha\}$ and the points of an algebraic variety $\mathscr{H}$. Of course, not just any bijection will do; we want to choose a correspondence that is reasonably natural, in the sense that as the point X varies continuously in $\mathscr{H}$, the coefficients of the defining equations of the varieties $X \subset \mathbb{P}^n$ should likewise vary continuously, in whatever topology. Clearly, the first thing to do is to make precise this requirement.

One way to do this comes from the construction of the universal hypersurface: for example, we saw that if we associated to each hypersurface $X \subset \mathbb{P}^n$ of degree d a point $X \in \mathbb{P}^N$, then the subset of the product

$$\{(X, p) \colon p \in X\} \subset \mathbb{P}^N \times \mathbb{P}^n$$

is in fact a subvariety. In general, we can take the analogous statement as the definition of a parameter space: our first requirement in order that a bijection between a collection of subvarieties $\{X_\alpha \subset \mathbb{P}^n\}$ and the points of an algebraic variety $\mathscr{H}$ be an algebraic parametrization is that the subset

$$\mathscr{X} = \{(X, p) \colon p \in X\} \subset \mathscr{H} \times \mathbb{P}^n$$

is a subvariety. To put it another way, in terms of the notion of family of varieties introduced in Lecture 4 we require that the varieties parametrized form a family with base $\mathscr{H}$. Note that if this condition is satisfied, then in affine open sets $U \subset \mathscr{H}$ the defining equations of $\mathscr{X} \cap \pi^{-1}U \subset U \times \mathbb{P}^n$ will be homogeneous polynomials $F_i(Z_0, \ldots, Z_n)$ whose coefficients are regular functions on U.

This condition, however, is not enough to characterize $\mathscr{H}$ uniquely; for that, we need something more. To express this, note first that if we have a closed family $\mathscr{V} \subset B \times \mathbb{P}^n$ of subvarieties of $\mathbb{P}^n$ in the given set $\{X_\alpha\}$, we have a set-theoretic map

$$\varphi_{\mathscr{V}} \colon B \to \mathscr{H},$$

defined by sending each point $b \in B$ to the point of $\mathscr{H}$ corresponding to the fiber $V_b = \pi^{-1}(b)$ of the projection $\pi \colon \mathscr{V} \to B \subset \mathbb{P}^n$. We would like to require that, for any such family with base B, the associated map $\varphi_{\mathscr{V}}$ be a regular map of varieties.

Unfortunately, this is asking too much; we have to impose a condition on the family $\mathscr{V}$ before we can expect that it induces a regular map $B \to \mathscr{H}$. To see in a simple case why, let B be the cuspidal cubic curve $(y^2 - x^3) \subset \mathbb{A}^2$ and consider the graph $\Gamma \subset B \times \mathbb{A}^1 \subset B \times \mathbb{P}^1$ of the map $f \colon \mathbb{A}^1 \to B$ given by $t \mapsto (t^2, t^3)$. We can think of the projection $\Gamma \to B$ as a family of points in $\mathbb{P}^1$, but the map it induces from B to the parameter space of points in $\mathbb{P}^1$—that is, $\mathbb{P}^1$—is just f^{-1}, which is not regular.

To express a condition that will preclude this, given any family $\pi \colon \mathscr{V} \subset B \times \mathbb{P}^n \to B$ let $b \in B$ be any point and $p \in V_b \subset \mathscr{V}$ any point of the fiber over b. The map π then gives a pullback map $\mathscr{O}_{B,b} \to \mathscr{O}_{V,p}$. We then say that *the fiber of π is reduced at p* if the maximal ideal $\mathfrak{m}_b$ of $b \in B$ generates the ideal of V_b in $\mathscr{O}_{V,p}$—that is, if the pullbacks to V of regular functions on neighborhoods of b in B vanishing at b generate the ideal of V_b in some neighborhood of p in V. We will say that the family is *reduced* if this condition is satisfied for all $b \in B$ and $p \in V_b$; we will say that it is *generically reduced* if for all $b \in B$ the condition is satisfied at general points $p \in V_b$.

To relate this to the more standard terminology used in scheme theory, we say that a family of schemes $\pi \colon \mathscr{V} \to B$ is *flat* it for every point $p \in \mathscr{V}$ the local ring $\mathscr{O}_{\mathscr{V},p}$ is a flat $\mathscr{O}_{B,\pi(p)}$-module. If B is a connected variety, $\mathscr{V} \subset B \times \mathbb{P}^n$ a closed subvariety and $\pi \colon \mathscr{V} \to B$ the projection, this is equivalent to saying that the Hilbert polynomials of the scheme-theoretic fibers of π are all the same (see [EH], for example). Now, the condition that the family $\pi \colon \mathscr{V} \to B$ is reduced says that the scheme-theoretic fibers are the same as the fibers of π as varieties, so that for such a family flatness is equivalent to constancy of Hilbert polynomial of the set-theoretic fibers; and this is in turn implied by the constancy of the dimension and degree of the fibers. Thus a reduced family of varieties of the same dimension and degree is flat.

In general, the notion of flatness is absolutely central in the construction of the Hilbert scheme.)

Once we restrict ourselves to families satisfying one of these two conditions, it becomes reasonable to require that the map $B \to \mathscr{H}$ associated to a family be regular. In fact we can combine the two requirements neatly into one statement. Given that a universal family $\mathscr{X}$ exists over $\mathscr{H}$ and is reduced (resp., generically reduced), every map $\varphi: B \to \mathscr{H}$ will be the map $\varphi_{\mathscr{V}}$ associated to a reduced (resp., generically reduced) family $\mathscr{V} \subset B \times \mathbb{P}^n$: just take $\mathscr{V}$ the fiber product $\mathscr{V} = B \times_{\mathscr{H}} \mathscr{X}$. Conversely, if every map $\varphi: B \to \mathscr{H}$ comes from a family $\mathscr{V} \subset B \times \mathbb{P}^n$, the family $\mathscr{X} \subset \mathscr{H} \times \mathbb{P}^n$ associated to the identity map $i: \mathscr{H} \to \mathscr{H}$ will be the universal family. In sum, then, we can make the following definition.

Definition 21.1. We say that a variety $\mathscr{H}$, together with a bijection between the points of $\mathscr{H}$ and a collection of varieties $\{X_\alpha \subset \mathbb{P}^n\}$, is a *parameter space* for the collection $\{X_\alpha\}$ if, for any variety B, the association to each family $\mathscr{V} \subset B \times \mathbb{P}^n$ of varieties belonging to the collection $\{X_\alpha\}$ of the map $\varphi_{\mathscr{V}}: B \to \mathscr{H}$ induces a bijection

$$\left\{\begin{array}{c}\text{reduced closed families}\\ \text{with base } B\text{, whose}\\ \text{fibers are members of}\\ \text{the collection } \{X_\alpha\}\end{array}\right\} \to \{\text{regular maps } B \to \mathscr{H}\}.$$

We can similarly define a *cycle parameter space* to be a space satisfying this condition with respect to generically reduced families.

Most of the remainder of this lecture will be devoted to descriptions of two classical constructions, one showing the existence of a cycle parameter space for the collection of varieties $X \subset \mathbb{P}^n$ of given dimension and degree, and the other creating a parameter space for the set of varieties having given Hilbert polynomial. The varieties constructed are called the *open Chow variety* and the *open Hilbert variety*; in each case, we will sketch the construction, and say a word or two about what sort of objects are parametrized by their closures, the *Chow variety* and the *Hilbert variety*. We will also describe in more detail (although without proof) one elementary example of the two different approaches, the parametrization of curves of degree 2 in $\mathbb{P}^3$. Finally, at the end of the lecture we will give a very brief discussion of the related notion of *moduli space*.

Chow Varieties

The first construction of the parameter space for varieties of a given degree and dimension is called the Chow construction. The basic idea behind it is simply that the problem in parametrizing varieties $X \subset \mathbb{P}^n$ in general is that X is not generally given by a single polynomial, whose coefficients we can vary freely as in the example of hypersurfaces. The point of the construction is thus to associate to any such variety X a hypersurface Φ_X, albeit not one in a projective space.

There are two essentially equivalent ways of doing this. The one we will work with uses a product of projective spaces; as we will remark at the end of the section we can simply replace this product with a Grassmannian without altering the construction. Suppose first of all that X has pure dimension k, and consider the incidence correspondence consisting of points $p \in X$ together with $(k+1)$-tuples of hyperplanes containing p; that is,

$$\Gamma = \{(p, H_1, \ldots, H_{k+1}) : p \in H_i \ \forall i\}$$
$$\subset X \times \mathbb{P}^{n*} \times \cdots \times \mathbb{P}^{n*}.$$

The standard calculation yields the dimension of Γ readily enough: for each point $p \in X$, the set of hyperplanes containing p is a hyperplane $\mathbb{P}^{n-1} \subset \mathbb{P}^{n*}$, so that the set of $(k+1)$-tuples of hyperplanes is irreducible of dimension $(k+1)\cdot(n-1)$. We deduce that Γ is of pure dimension $k + (k+1)\cdot(n-1) = (k+1)\cdot n - 1$, with one irreducible component corresponding to each irreducible component of X. At the same time, for a general choice of point $p \in X$ and $H_1, \ldots, H_{k+1}$ containing p the intersection of the H_i with X will consist only of the point p, so that the projection map $\pi\colon \Gamma \to \mathbb{P}^{n*} \times \cdots \times \mathbb{P}^{n*}$ will be birational. It follows that the image of Γ under this projection is a hypersurface in $\mathbb{P}^{n*} \times \cdots \times \mathbb{P}^{n*}$ (this is where we need X to have pure dimension); we will call this hypersurface Φ_X.

Now, just as in the case of a product of two projective spaces, any hypersurface in $\mathbb{P}^{n*} \times \cdots \times \mathbb{P}^{n*}$ will be the zero locus of a single multihomogeneous polynomial F; F will be unique up to multiplication by scalars if we require as well that it have no repeated factors. We accordingly let F_X be the polynomial defining the hypersurface $\Phi_X \subset \mathbb{P}^{n*} \times \cdots \times \mathbb{P}^{n*}$. It is relatively easy to see what the multidegree of F_X is (of course, since Φ_X is symmetric with respect to permutation of the factors of $\mathbb{P}^{n*} \times \cdots \times \mathbb{P}^{n*}$, F_X must have the same degree with respect to each set of variables). If we fix k general hyperplanes $H_1, \ldots, H_k \subset \mathbb{P}^n$, the intersection of X with $H_1, \ldots, H_k$ will consist of exactly d points p_i, where d is the degree of X; and we will have $(H_1, \ldots, H_{k+1}) \in \Phi_X$ if and only if H_{k+1} contains one of the p_i. The intersection of Φ_X with the fiber $\{(H_1, \ldots, H_k)\} \times \mathbb{P}^{n*}$ will thus be the union of the d hyperplanes $p_i^* \subset \mathbb{P}^{n*}$, where p_i^* is the set of hyperplanes in $\mathbb{P}^n$ containing p_i. In particular, it has degree d, so that F_X must have degree $d = \deg(X)$ in each set of variables.

Now let V be the vector space of multihomogeneous polynomials of multidegree $(d, d, \ldots, d)$ in $k+1$ sets of $n+1$ variables. We have associated to a variety $X \subset \mathbb{P}^n$ of dimension k and degree d a well-defined element $[\Phi_X] \in \mathbb{P}V$ in the associated projective space; we thus have a set-theoretic map

$$\left\{\begin{matrix}\text{varieties of pure dimension} \\ k \text{ and degree } d \text{ in } \mathbb{P}^n\end{matrix}\right\} \xrightarrow{\xi} \mathbb{P}V.$$

The point $\xi(X) = [F_X]$ in $\mathbb{P}V$ corresponding to a given variety $X \subset \mathbb{P}^n$ is called the *Chow point* of X.

We claim next the map ξ is an injection, i.e., that the variety X is determined by its Chow point $[F_X] \in \mathbb{P}V$. To do this, we introduce another incidence correspondence Ψ in $\mathbb{P}^n \times (\mathbb{P}^{n*} \times \cdots \times \mathbb{P}^{n*})$, defined by

$$\Psi = \{(p, H_1, \ldots, H_{k+1}) : p \in H_i \ \forall i\}$$
$$\subset \mathbb{P}^n \times \mathbb{P}^{n*} \times \cdots \times \mathbb{P}^{n*};$$

we let $\pi: \Psi \to \mathbb{P}^n$ and $\eta: \Psi \to \mathbb{P}^{n*} \times \cdots \times \mathbb{P}^{n*}$ be the projection maps. Note that Ψ is just the fiber product of the standard incidence correspondence $\Gamma \subset \mathbb{P}^n \times \mathbb{P}^{n*}$ with itself $k+1$ times over $\mathbb{P}^n$; in particular, the fiber $\pi^{-1}(q)$ of Ψ over any point $q \in \mathbb{P}^n$ is just the product $q^* \times \cdots \times q^* \cong \mathbb{P}^{n-1} \times \cdots \times \mathbb{P}^{n-1}$ of the hyperplane $q^* \subset \mathbb{P}^{n*}$ with itself $k+1$ times (where again $q^* \subset \mathbb{P}^{n*}$ is the set of hyperplanes passing through q).

Now consider the inverse image $\eta^{-1}(\Phi_X)$. For any point $q \in X \subset \mathbb{P}^n$, this will contain the fiber $\pi^{-1}(q)$ of Ψ over q. By contrast, if $q \notin X$, $\eta^{-1}(\Phi_X)$ will intersect the fiber $\pi^{-1}(q) \cong \mathbb{P}^{n-1} \times \cdots \times \mathbb{P}^{n-1}$ of Ψ over q in a proper hypersurface; if we think of $q^* \cong (\mathbb{P}^{n-1})^*$ as the dual of the projective space of lines through q in $\mathbb{P}^n$, the intersection $\eta^{-1}(\Phi_X) \cap \pi^{-1}(q)$ will be the hypersurface $\Phi_{\bar{X}}$ in the product $(\mathbb{P}^{n-1})^* \times \cdots \times (\mathbb{P}^{n-1})^*$ associated to the image $\bar{X} = \pi_q(X)$ of X under projection from q. We may thus characterize

$$X = \{q \in \mathbb{P}^n : \pi^{-1}(q) \subset \eta^{-1}(\Phi_X)\};$$

equivalently, we can say that X is the set of points q such that the fiber of $\eta^{-1}(\Phi_X)$ over q has dimension $(k+1)(n-1)$.

The next claim in regard to the map ξ is that the image is a quasi-projective variety. This image is then called the *open Chow variety* of subvarieties of dimension k and degree d in $\mathbb{P}^n$ and is denoted $\tilde{\mathscr{C}}_{k,d} = \tilde{\mathscr{C}}_{k,d}(\mathbb{P}^n)$ (the closure $\mathscr{C}_{k,d}$ of $\tilde{\mathscr{C}}_{k,d}$ is called simply the *Chow variety*; we will discuss later to what the extra points in the closure correspond).

This claim is almost elementary. We will prove it first for the image under ξ of the subset of irreducible varieties $X \subset \mathbb{P}^n$ of dimension k and degree d; the general case is handled similarly. We first have to observe that we can use the preceding construction to associate to any hypersurface $\Phi \subset \mathbb{P}^{n*} \times \cdots \times \mathbb{P}^{n*}$ a subvariety $Z_\Phi \subset \mathbb{P}^n$: we set

$$Z_\Phi = \{q \in \mathbb{P}^n : \pi^{-1}(q) \subset \eta^{-1}(\Phi)\}$$
$$= \{q \in \mathbb{P}^n : \dim(\pi^{-1}(q) \cap \eta^{-1}(\Phi)) \geq (k+1)(n-1)\};$$

this is a subvariety of $\mathbb{P}^n$ since the fiber dimension of the map $\pi: \eta^{-1}(\Phi) \to \mathbb{P}^n$ is upper-semicontinuous. Indeed, the key point is that the association of Z_Φ to Φ defines a subvariety Ξ of the product space $\mathbb{P}V \times \mathbb{P}^n$, i.e., the set of pairs (Φ, q) such that $q \in Z_\Phi$ is a subvariety of $\mathbb{P}V \times \mathbb{P}^n$. To see this, we set up a diagram of incidence correspondences as indicated at left. To begin with, the subset

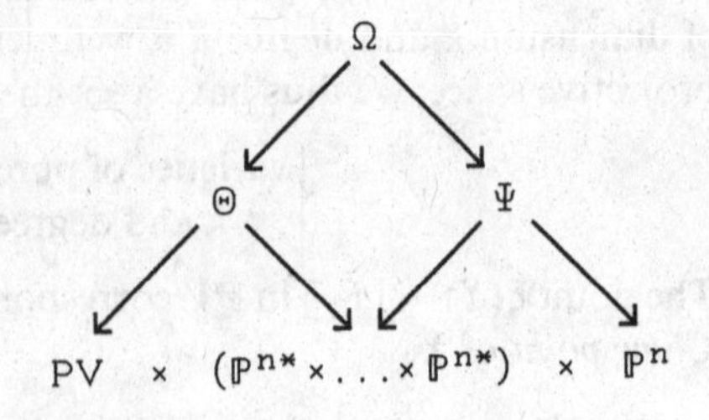

$$\Theta = \{(\Phi, H_1, \ldots, H_{k+1}) : (H_1, \ldots, H_{k+1}) \in \Phi\}$$
$$\subset \mathbb{P}V \times \mathbb{P}^{n*} \times \cdots \times \mathbb{P}^{n*}$$

is clearly a subvariety, and hence so is the subset

$$\Omega = \{(\Phi, H_1, \ldots, H_{k+1}, q) : (H_1, \ldots, H_{k+1}) \in \Phi \text{ and } q \in H_1 \cap \cdots \cap H_{k+1}\}$$
$$\subset \mathbb{P}V \times \mathbb{P}^{n*} \times \cdots \times \mathbb{P}^{n*} \times \mathbb{P}^n$$

(this is just the fiber product of $\Theta \subset \mathbb{P}V \times \mathbb{P}^{n*} \times \cdots \times \mathbb{P}^{n*}$ and $\Psi \subset \mathbb{P}^{n*} \times \cdots \times \mathbb{P}^{n*} \times \mathbb{P}^n$ over $\mathbb{P}^{n*} \times \cdots \times \mathbb{P}^{n*}$). We can then write

$$\Xi = \{(\Phi, q) : q \in Z_\Phi\}$$
$$= \{(\Phi, q) : \dim(\gamma^{-1}(\Phi, q)) \geq (k+1)(n-1)\}$$

where $\gamma: \Omega \to \mathbb{P}V \times \mathbb{P}^n$ is projection on the first and last factor; from the latter description it follows that Ξ is a subvariety of $\mathbb{P}V \times \mathbb{P}^n$.

Finally, we can check that the dimension of Z_Φ can never exceed k, and that among those irreducible Φ such that $\dim(Z_\Phi) = k$ the degree of Z_Φ will always equal d. Thus, if we restrict our attention to irreducible X, we are done; the image, under the Chow map, of the set of irreducible X will lie in the open subset $U \subset \mathbb{P}V$ of irreducible hypersurfaces of multidegree $(d, \ldots, d)$, and will be the locus of $\Phi \in U$ such that the fiber Ξ over Φ has dimension at least k.

If we want to include reducible varieties X as well (we do), we have to work a little harder, since for reducible Φ it is possible that $\dim(Z_\Phi) = k$ but $\deg(Z_\Phi) = d' < d$ (this will happen, for example, if $\Phi = \Phi_1 \cup \Phi_2$, with $\Phi_1 = \Phi_Y$ for some variety $Y \subset \mathbb{P}^n$ of dimension k and degree d', and Φ_2 a general hypersurface of multidegree $(d - d', \ldots, d - d')$. At this point, it would make sense to read one of the explicit descriptions of the equations defining the Chow variety, e.g., in [GM].

Note also that the variety $\Xi \subset \mathbb{P}V \times \mathbb{P}^n$ intersected with the inverse image $\tilde{\mathscr{C}}_{k,d} \times \mathbb{P}^n$ is the universal variety of dimension k and degree d over the open Chow variety. (This family is only generically reduced, as we will see in Exercise 21.6) What this all leads up to is the following theorem, which we will state without proof.

Theorem 21.2. *The open Chow variety is a cycle parameter space for the set of varieties of pure dimension k and degree d in* $\mathbb{P}^n$.

The existence of a universal family over the open Chow variety allows to us deduce a number of natural attributes of the Chow parametrization.

Exercise 21.3. Let $Z \subset \mathbb{P}^n$ be a fixed variety. Use the existence of a universal family over the open Chow variety $\tilde{\mathscr{C}}_{k,d}$ to deduce that the subset

$$\tilde{\mathscr{C}}_{k,d}(Z) = \{X \in \tilde{\mathscr{C}}_{k,d} : X \cap Z \neq \varnothing\}$$

is a subvariety of $\tilde{\mathscr{C}}_{k,d}$.

Exercise 21.4. As in the preceding exercise, let $Z \subset \mathbb{P}^n$ be a fixed variety. Use the existence of a universal family over the Chow variety $\tilde{\mathscr{C}}_{k,d}$ to deduce that the subset

$$\tilde{\mathscr{C}}_{k,d}(Z) = \{X \in \tilde{\mathscr{C}}_{k,d} : X \subset Z\}$$

is a subvariety of $\tilde{\mathscr{C}}_{k,d}$; this is called the Chow variety of cycles on Z. (Warning: you may want to do this only for the open subset of $\tilde{\mathscr{C}}_{k,d}$ parameterizing irreducible subvarieties $X \subset \mathbb{P}^n$.)

One question that comes up naturally, given that the open Chow variety $\tilde{\mathscr{C}}_{k,d}$ is a quasi-projective variety, is what its closure in the space $\mathbb{P}V$ is. The answer, as it turns out, is very straightforward, though we will have to state it here without proof. It is that if a polynomial $F \in V$ lies on the closure of $\tilde{\mathscr{C}}_{k,d}$ and we factor it into prime factors

$$F = \prod (F_i)^{a_i},$$

then each factor F_i is itself the Chow point of a subvariety $X_i \subset \mathbb{P}^n$. We must have, of course, that

$$\sum a_i \cdot \deg(X_i) = \sum a_i \cdot \deg(F_i) = \deg(F) = d.$$

We may thus say that the closure of $\tilde{\mathscr{C}}_{k,d}$ parameterizes *effective cycles* on $\mathbb{P}^n$, where an effective cycle of dimension k and degree d is defined to be a formal linear combination $\sum a_i \cdot X_i$ with X_i irreducible of dimension k, a_i positive integers, and $\sum a_i \cdot \deg(X_i) = d$. As we indicated, this closure is called simply the *Chow variety* and is denoted $\mathscr{C}_{k,d}$.

Exercise 21.5. Show that the Chow variety $\mathscr{C}_{k,d}$ is connected, since every component of $\mathscr{C}_{k,d}$ will contain the locus of points corresponding to cycles $d \cdot \Gamma$ where $\Gamma \cong \mathbb{P}^k \subset \mathbb{P}^n$ is a linear subspace. Find the number of irreducible components of the Chow varieties $\mathscr{C}_{1,2}(\mathbb{P}^3)$, $\mathscr{C}_{1,3}(\mathbb{P}^3)$, and $\mathscr{C}_{1,4}(\mathbb{P}^3)$ parameterizing curves of degrees 2, 3, and 4 in $\mathbb{P}^3$.

It is not known in general how many irreducible components the Chow variety has, even for curves in $\mathbb{P}^3$. As for the higher connectivity of $\mathscr{C}_{k,d}$, there has been some fascinating work by Lawson [L] on the homotopy groups of the "infinite" Chow variety $\mathscr{C}_{k,\infty} = \bigcup \mathscr{C}_{k,d}$ (where the inclusions are defined by adding a fixed k-plane Γ to each cycle $\sum a_i \cdot X_i \in \mathscr{C}_{k,d}$).

We mentioned at the outset that there are two equivalent approaches to the Chow construction. The second simply replaces the product $\mathbb{P}^{n*} \times \cdots \times \mathbb{P}^{n*}$ with the Grassmannian $\mathbb{G} = \mathbb{G}(n-k-1, n)$ of codimension $k+1$ planes in $\mathbb{P}^n$, and associates to a subvariety $X \subset \mathbb{P}^n$ of dimension k and degree d the corresponding hypersurface

$$\Phi_X = \{\Lambda : \Lambda \cap X \neq \emptyset\}.$$

This is of course simpler than the previous; the only additional step is that we have to remark that the hypersurface Φ_X is the intersection of the Grassmannian $\mathbb{G}$ with a hypersurface of degree d in $\mathbb{P}^N$, where $\mathbb{P}^N$ is the ambient space of the Plücker embedding. This hypersurface is given as the zeros of a homogeneous polynomial F_X of degree d on $\mathbb{P}^N$, defined modulo the dth graded piece $I(\mathbb{G})_d$ of the ideal of $\mathbb{G} \subset \mathbb{P}^N$ and modulo scalars. We may in this way associate to X the point $[F_X]$ in the projective space $\mathbb{P}(S(\mathbb{G})_d)$ of the dth graded piece of the homogeneous coordinate ring of $\mathbb{G}$; this is also called the Chow map and $[F_X]$ the Chow point of X.

Exercise 21.6. Consider the family of irreducible curves $C_t \subset \mathbb{P}^3$ where C_t is the image of the map

$$\varphi_t: \mathbb{P}^1 \to \mathbb{P}^3$$

given by

$$\varphi_t: [X_0, X_1] \mapsto [X_0^3, X_0^2X_1 + (1-t)X_0X_1^2, tX_0X_1^2, X_1^3].$$

C_t is just the image of the "standard" twisted cubic C_1 under the linear map $A_t: \mathbb{P}^3 \to \mathbb{P}^3$ given by

$$A_t = \begin{pmatrix} 1 & 0 & 0 & 0 \\ 0 & 1 & 0 & 0 \\ 0 & (1-t) & t & 0 \\ 0 & 0 & 0 & 1 \end{pmatrix},$$

so that C_t is a twisted cubic for $t \neq 0$ and a plane nodal cubic for $t = 0$. Show that the curves C_t form a closed family in $\mathbb{A}^1 \times \mathbb{P}^3$. Show that this family is generically reduced, but is not reduced. (In fact, no family whose general member is a twisted cubic specializing to the curve C_0 can be reduced.) Deduce that the universal family over the open Chow variety is only generically reduced and that the Chow variety is a cycle parameter space rather than a parameter space.

Hilbert Varieties

The second construction of a parameter space for subvarieties of $\mathbb{P}^n$, that of Hilbert, presents a nice contrast to the Chow construction. While the Chow variety can be defined in very elementary ways, its properties are harder to derive. By contrast, the Hilbert construction is in some sense much more naive, and the properties of the resulting varieties much easier to see, but in order to define it we will need to at least state a technical lemma whose proof is well beyond the scope of this book.

The Hilbert variety parametrizes varieties not just of fixed dimension and degree, but with a fixed Hilbert polynomial P (it also does not require the varieties in question to have pure dimension). We do need the following statement.

Lemma 21.7. *Given any polynomial p, there exists an integer m_0 such that for any variety X with Hilbert polynomial $p_X = p$,*

(i) *the Hilbert function $h_X(m) = p(m)$ for all $m \geq m_0$ and*
(ii) *the ideal $I(X)$ is generated in degree m_0.*

The second statement means that for any $m \geq m_0$ the elements of the mth graded piece of the ideal $I(X)$ generate the truncated ideal $\bigoplus_{l \geq m} I(X)_l$; in particular, this implies that $I(X)_m$ generates the ideal of X locally and a fortiori that the common zero locus of $I(X)_m$ is X.

Given this, the idea behind the construction is simplicity itself. The difficulty with parametrizing varieties $X \subset \mathbb{P}^n$ other than hypersurfaces is that their ideals I_X are generated not by single polynomials but by collections or vector spaces of them; the solution to associate to such a varety not a point in a projective space parametrizing single polynomials up to scalars, but a point in the Grassmannian parametrizing such vector spaces of polynomials. Thus, suppose $X \subset \mathbb{P}^n$ is a variety with Hilbert polynomial $p_X = p$. By Lemma 21.7 we know that for $m \geq m_0$ the ideal $I(X)_m$ has codimension exactly $p(m)$ in the vector space S_m of homogeneous polynomials of degree m on $\mathbb{P}^n$. Thus, setting $N(m) = \binom{m+n}{n}$ and $q(m) = N(m) - p(m)$, we have a set-theoretic map

$$\left\{\begin{matrix}\text{subvarieties } X \subset \mathbb{P}^n \text{ with} \\ \text{Hilbert polynomial } p_X = p\end{matrix}\right\} \to G = G(q(m), N(m))$$

called the Hilbert map; the point in the Grassmannian G associated to a variety $X \subset \mathbb{P}^n$ will be called its *Hilbert point*. The second part of Lemma 21.7 then says that this map is one to one; the image will be called the *open Hilbert variety* and denoted $\tilde{\mathscr{H}}_p$.

In fact, the open Hilbert variety is a quasi-projective subset of the Grassmannian G. To see this, observe that for any vector subspace $\Lambda \subset S_m$ and any positive integer k, we have a multiplication map

$$\Psi_k \colon \Lambda \otimes S_k \to S_{k+m};$$

if we have a subspace $\Lambda \subset S_m$ of codimension $p(m)$ it will be the mth graded piece of the ideal $I(X)$ of a variety $X \subset \mathbb{P}^n$ with Hilbert polynomial p if and only if for every k the rank of Ψ_k is what it should be, that is, $q(m + k)$. Now, in open sets $U \subset G$ we can represent $\Lambda \in G$ as spanned by polynomials $F_1, \ldots, F_{q(m)} \in S_m$, with the coefficients of F_i regular functions on G. The map Ψ_k may then be viewed as a linear map between the fixed vector spaces $(S_k)^{\oplus q(m)}$ and S_{k+m}, whose matrix entries are regular functions on U. The intersection $\tilde{\mathscr{H}}_p \cap U$ is correspondingly expressed in a neighborhood of a given point $I(x) \in \mathscr{H}_p$ as the locus in U where this map has rank exactly $q(k + m)$ for each k, showing that $\tilde{\mathscr{H}}_p \cap U$ is locally closed in U.

One apparent problem with this argument is that there are a priori infinitely many determinantal conditions associated to maps Ψ_k and an infinite intersection of locally closed subsets need not be locally closed. In fact, this is not serious, since only a finite number of Hilbert polynomials are possible in a small neighborhood of any point $\Lambda \in G$. More serious is the issue of whether the variety $\tilde{\mathscr{H}}_p$ constructed in this way is dependent on the choice of m. To show that in fact it is not requires techniques well beyond the scope of this book; a good reference for the construction of $\mathcal{H}$ in general, and this in particular, is [M1].

It is now immediate that a universal family $\mathcal{X} \subset \tilde{\mathcal{H}}_p \times \mathbb{P}^n$ exists: as we observed, in open sets $U \subset G$ we can represent $\Lambda \in G$ as spanned by polynomials $F_1, \ldots, F_{q(m)} \in S_m$, with the coefficients of F_i regular functions on G, and these F_i are just the defining equations of $\mathcal{X} \cap (U \times \mathbb{P}^n)$. Likewise, it's a good exercise to establish, on the basis of what we have said so far, the following.

Theorem 21.8. *The open Hilbert variety is a parameter space for the set of subvarieties $X \subset \mathbb{P}^n$ of given Hilbert polynomial.*

Finally we may ask, as we did in the case of the Chow variety, to what the points in the closure $\mathcal{H}_p$ of $\tilde{\mathcal{H}}_p$ correspond. The answer is that they correspond to subschemes of $\mathbb{P}^n$ with Hilbert polynomial p. In our (limited) language, we can take this to mean saturated ideals with Hilbert polynomial p, that is, ideals $I \subset S = K[Z_0, \ldots, Z_n]$ with quotient $A = S/I$ satisfying $\dim_K(A_m) = p(m)$ for m large. To see what this means in practice, it is probably best to refer to an example, which we will do now.

Curves of Degree 2

As an example of both Chow and Hilbert varieties, consider the case of curves of degree 2 in $\mathbb{P}^3$. Such a curve, if it is of pure dimension 1, will consist of either a plane conic or the union of two lines. There are thus two components $\mathcal{C}, \mathcal{C}'$ of the Chow variety $\mathcal{C}_{1,2}$, which intersect along the locus of pairs of incident lines. The latter component is isomorphic to the symmetric product of the Grassmannian $\mathbb{G}(1, 3)$ with itself (see Example 10.23), with the diagonal corresponding to cycles of the form $2 \cdot L$. Note in particular that it has dimension 8 and, as the reader can verify, is singular exactly along its diagonal. The first component, the closure of the locus of plane conics, is slightly trickier. Probably the best way to see what it looks like is to introduce the incidence correspondence

$$\Gamma = \{(C, H) : C \subset H\} \subset \mathcal{C}' \times \mathbb{P}^{3*};$$

Γ maps to the second factor $\mathbb{P}^{3*}$ with fibers isomorphic to the space $\mathbb{P}^5$ of conics in a plane. In fact, it's not hard to see that even in the Zariski topology it is a $\mathbb{P}^5$-bundle over $\mathbb{P}^3$; in particular, it is smooth of dimension 8. The map from Γ to $\mathcal{C}'$ is birational, since a general plane conic $C \in \mathcal{C}'$ lies on a unique plane H, but does collapse the locus where $C = 2 \cdot L$ is a line with multiplicity 2; specifically, for each $L \in \mathbb{G}(1, 3)$ the subvariety $\{(2L, H) : H \supset L\} \cong \mathbb{P}^1$ is mapped to a single point in $\mathcal{C}'$. We can see from this that $\mathcal{C}'$ is singular exactly along the locus of double lines (if $f: X \to Y$ is a regular birational map with X and Y both smooth, the locus in X where the differential df fails to have maximal rank will be either empty or of pure codimension 1). The picture, diagrammatically, is

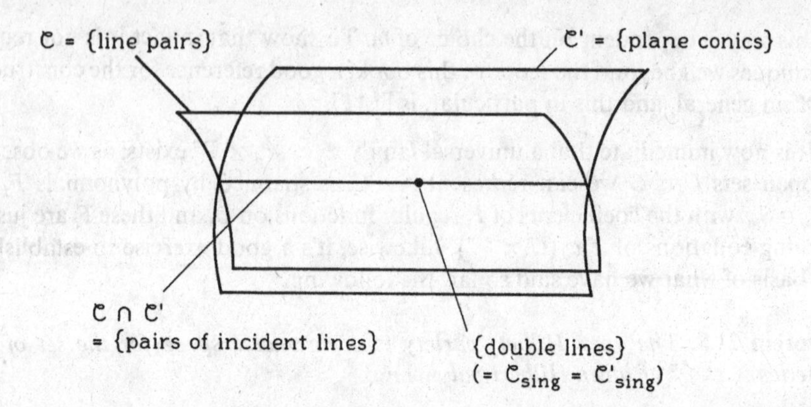

By way of contrast, consider the Hilbert varieties parametrizing curves of degree 2. To begin with, there are many varieties $\mathscr{H}_p$, corresponding to various polynomials $p(m) = 2m + c$. Of these, the smallest value of c for which $\mathscr{H}_p \neq \varnothing$ is $c = 1$; the corresponding component Hilbert variety $\mathscr{H}_{2m+1}$ parametrizes plane conics. Like the component $\mathscr{C}'$ of the Chow variety $\mathscr{C}_{1,2}$, $\mathscr{H}_{2m+1}$ has an open subset parametrizing smooth plane conics and pairs of distinct incident lines; indeed, this open subset is isomorphic to the corresponding one in $\mathscr{C}'$. The rest of $\mathscr{H}_{2m+1}$, however, is different; we can check that any saturated ideal $I \subset K[Z_0, \ldots, Z_3]$ with Hilbert polynomial $2m + 1$ is a complete intersection, that is, is generated by a linear and a quadratic polynomial with no common factor. It follows that the points of the Hilbert variety $\mathscr{H}_{2m+1}$ correspond to pairs consisting of a plane $H \subset \mathbb{P}^3$ and a conic curve in H—that is, a point of the incidence correspondence Γ. In fact, $\mathscr{H}_{2m+1}$ is isomorphic to Γ.

Next comes the Hilbert variety $\mathscr{H}_{2m+2}$. This contains a component $\mathscr{H}$ whose general member is a pair of skew lines, similar to the component $\mathscr{C}$ of the Chow variety $\mathscr{C}_{1,2}$. In fact, $\mathscr{H}_{2m+2}$ has an open subset isomorphic to the open subset of $\mathscr{C}$ parametrizing pairs of distinct lines, though in the case of a point of $\mathscr{H}_{2m+2}$ corresponding to two incident lines $L \cup M$ the ideal corresponding is not the ideal of the union; if $\{L_t\}$ and $\{M_t\}$ are families of lines, with L_t disjoint from M_t for $t \neq 0$ and $L_0 \cap M_0 = \{p\}$, then the limit of the ideals $I_t = I_{L_t \cup M_t}$ is not $I_{L_0 \cup M_0}$, but rather the ideal of polynomials vanishing on $L_0 \cup M_0$ and vanishing to order 2 at p.

Exercise 21.9. Verify this last statement, for example, in case L_t is given by $(Z_0 = Z_1 = 0)$ for all t and M_t is given by $(Z_0 - tZ_3 = Z_2 = 0)$. Check also that the ideal described is saturated and has Hilbert polynomial $p(m) = 2m + 2$.

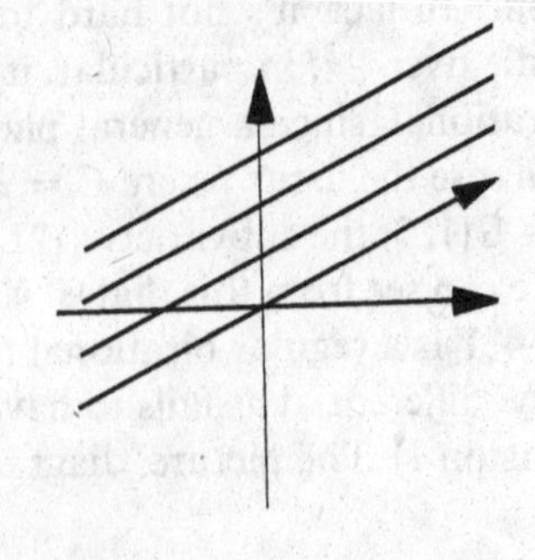

The really interesting part of $\mathscr{H}_{2m+2}$, however, is what happens to the variety $L_t \cup M_t$ when the two lines approach the same line $L_0 = M_0$. We arrive in this case at an ideal contained in the ideal I_{L_0} and containing the square $(I_{L_0})^2$, but not equal to either. Rather, it will consist, for some choice of linear map $\varphi: L_0 \to K^4/L_0$, of the ideal of polynomials F vanishing on L_0 and whose normal derivative at each point $p \in L_0$ in the direction of $\varphi(p)$ is equal to zero—for example, if L_0 is given as $Z_0 = Z_1 = 0$, it will be given as

$$I = (Z_0^2, Z_0Z_1, Z_1^2, \alpha(Z_0, Z_1)\cdot Z_2 + \beta(Z_0, Z_1)\cdot Z_3)$$

where α and β are homogeneous linear polynomials. (If φ fails to be an isomorphism, we require that F vanish to order 2 at the point $p = \text{Ker}(\varphi)$, so that the ideal will have codimension one in an ideal generated by a linear and a quadratic polynomial.) In particular, the limiting ideal I of the family $I_{L_t \cup M_t}$ will not depend only on the limiting position $L_0 = M_0$. One way to describe the limit is to think of L_t and M_t as points on the Grassmannian $\mathbb{G}(1, 3)$; as they approach their common limit they determine not only the point $L_0 \in \mathbb{G}(1, 3)$ but a tangent vector $\varphi: L_0 \to K^4/L_0$ at $\mathbb{G}(1, 3)$ at L_0 as well, and it is this tangent vector φ that determines the limit of the ideals.

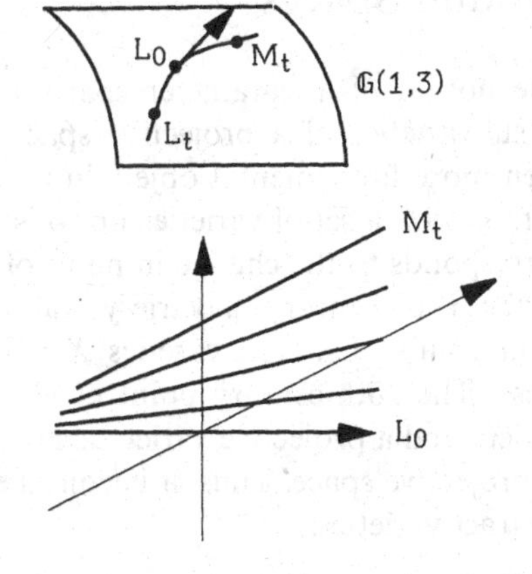

Exercise 21.10. Verify this last statement, for example, in case L_t is given by $(Z_0 = Z_1 = 0)$ for all t and M_t is given by $(Z_0 - tZ_3 = Z_1 - tZ_2 = 0)$. Check again that the ideal described is saturated and has Hilbert polynomial $p(m) = 2m + 2$.

The conclusion we may draw is that, at least set-theoretically, the component $\mathscr{H}$ of $\mathscr{H}_{2m+2}$ maps to $\mathscr{C}$ isomorphically away from the diagonal of $\mathscr{C}$, and with fiber over a point $2\cdot L \in \mathscr{C}$ the projectivized tangent space $\mathbb{P}T_L(\mathbb{G}(1, 3))$ to the Grassmannian at L. In fact, this map is regular, and $\mathscr{H}$ is just the blow-up of $\mathscr{C}$ along the diagonal. Another way of constructing $\mathscr{H}$ is to say that it is the quotient of the blow-up of $\mathbb{G}(1, 3) \times \mathbb{G}(1, 3)$ along the diagonal by the involution ι exchanging factors.

We are still not done with $\mathscr{H}_{2m+2}$, however: there is another irreducible component $\mathscr{H}'$ of $\mathscr{H}_{2m+2}$ to go. This is the component whose general member consists of the union of a plane conic C with a point $p \in \mathbb{P}^3$ not on C; as may be readily checked, this is also a subvariety of $\mathbb{P}^3$ with Hilbert polynomial $p(m) = 2m + 2$. We will not say much about this component except to remark on its existence and to observe that it intersects the component $\mathscr{H}$ along the union of the loci of pairs on incident planes, and of "double lines" $2\cdot L$ whose associated linear map $\varphi: L \to K^4/L$ has rank one.

Exercise 21.11. Describe the possible limits in $\mathscr{H}'$ of a union $C \cup p_t$ where p_t approaches a point $p_0 \in C$.

See [EH] for a discussion of limits of projective varieties, including a more detailed analysis of these examples.

Similarly, there is for every $c \geq 3$ a nonempty Hilbert variety $\mathcal{H}_{2m+c}$. It will have one irreducible component whose general member consists of a plane conic plus $c - 1$ points, and one component whose general member is the union of two skew lines and $c - 2$ points. (The Hilbert *scheme* will contain other components, whose general points correspond to saturated ideals $I \subset S$ whose quotients S/I have Hilbert polynomial $2m + c$ but that are not radical; these can get quite complicated.)

Moduli Spaces

The notion of a parameter space, a variety whose points parametrize the set of subvarieties of a projective space $\mathbb{P}^n$, is only the first half of the story. An even more fundamental object in many respects is a *moduli space*, whose points parametrize a set of varieties up to isomorphism. In a sense, the shift in emphasis corresponds to the change in point of view between the last century and this one: in the 19th century, a variety was a priori a subset of projective space, with isomorphism between vareties $X \subset \mathbb{P}^n$ and $Y \subset \mathbb{P}^m$ an equivalence relation on these. The 20th-century point of view is that the primary object is the abstract variety and a projective variety is an abstract variety with the extra data of a map to projective space. Thus, it becomes equally important to describe the families of abstract varieties.

The basic situation, then, is that we are given a collection of varieties $\{X_\alpha\}$ and ask whether the set of such varieties up to isomorphism may be given the structure of an algebraic variety in a natural way. Clearly, the first thing we have to do is say what we mean by "natural." In the case of parameter spaces, we had two answers to this. We first of all said that a bijection between a collection of subvarieties $X_\alpha \subset \mathbb{P}^n$ and the points of a variety $\mathcal{H}$ was an algebraic parametrization if the subset $\mathcal{X}$ of the product $\mathcal{H} \times \mathbb{P}^n$ defined by the relation $\mathcal{X} = \{(\alpha, p) : p \in X_\alpha\}$ is a subvariety. In our present circumstances, since the objects we are parametrizing are isomorphism classes of varieties, and not in any canonical way subvarieties of a fixed projective space, this does not make sense.

On the other hand, the requirement in basic Definition 21.1 of parameter spaces —that a family of varieties induces a map from the base of the family to the parameter space, and that this gives a bijection between such families and such maps—does have an analog in this setting. To begin with, we define a *family* of varieties of the collection $\{X_\alpha\}$ with base B to be a variety $\mathcal{V}$ and map $\pi: \mathcal{V} \to B$ such that for each point $b \in B$ the fiber $V_b = \pi^{-1}(b)$ is isomorphic to X_α for some α. Now, given a bijection between the set of isomorphism classes of X_α and the points of a variety $\mathcal{M}$, we may define a set-theoretic map

$$\varphi_\pi: B \to \mathcal{M}$$

by sending each point $b \in B$ to the point of $\mathcal{M}$ representing the isomorphism class $[X_b]$ of the fiber X_b over b. We will then say that the bijection is an algebraic parametrization of the set of isomorphism classes of X_α—or that $\mathcal{M}$ is a *coarse moduli space* for the varieties $\{X_\alpha\}$—if for any reduced family $\pi: \mathscr{V} \to B$ the induced map φ_π is a regular map of varieties and $\mathcal{M}$ is the maximal variety with these properties, i.e., if $\mathcal{M}' \to \mathcal{M}$ is any bijective morphism through which every φ_π factors, than $\mathcal{M}' \cong \mathcal{M}$.

We can ask for more. For example, by analogy with our definition of parameter space, we could ask that there exist a *tautological family* over $\mathcal{M}$—that is, a variety $\mathscr{X}$ and map $\pi: \mathscr{X} \to \mathcal{M}$ such that for any $p \in \mathcal{M}$ the fiber X_p is in the isomorphism class specified by p. We could require moreover that for any variety B the set map

$$\{\text{families with base } B\} \to \{\text{regular maps } B \to \mathcal{M}\}$$

sending a family $\pi: \mathscr{V} \to B$ to the map φ_π is in fact a bijection. If this additional condition is satisfied, the family $\pi: \mathscr{X} \to \mathcal{M}$ is called *universal.*[7] In this case, $\mathcal{M}$ is called a *fine moduli space* for the varities $\{X_\alpha\}$.

Fine moduli spaces are substantially rarer than coarse ones, as we will see in the following example.

Example 21.12. Plane Cubics

The fundamental example of a moduli space is one we encountered before in Example 10.16, that of plane cubics. Based on our previous discussion, we see that even a coarse moduli space does not exist for plane cubics. This is due to the various inclusions among closures of orbits of the action of $\mathrm{PGL}_3 K$ on the space $\mathbb{P}^9$ of plane cubics. For example, if $\mathcal{M}$ is the set of isomorphism classes of plane cubics, $\mathcal{M}$ will have one point p corresponding to irreducible plane cubics with a node, and another point q corresponding to cuspidal cubics. But by what we saw in Example 10.16, the point q would have to lie in the closure of the point p!

For another example, consider for any fixed value of λ the family $\mathscr{V} \to \mathbb{A}^1$ of plane cubics with base $\mathbb{A}^1$ given by the equation

$$y^2 = x \cdot (x - t) \cdot (x - \lambda t).$$

The fibers V_t of this family are all projectively equivalent for $t \neq 0$ but not isomorphic to the fiber V_0. If a moduli space $\mathcal{M}$ existed, then, the induced map $\mathbb{A}^1 \to \mathcal{M}$ would be constant on $\mathbb{A}^1 - \{0\}$, but not constant. Indeed, based on our description of the action of $\mathrm{PGL}_3 K$ on the space $\mathbb{P}^9$ of plane cubics in Example 10.16, we can see that in order to have a moduli space, we have to restrict to the open set consisting of smooth cubics and irreducible cubics with a node. Having

[7] In the case of parameter spaces $\mathscr{H}$ of subvarieties of $\mathbb{P}^n$ there was no distinction; the existence of a family $\mathscr{X} \subset \mathscr{H} \times \mathbb{P}^n$ of subvarieties of $\mathbb{P}^n$ whose fiber over $p \in \mathscr{H}$ was the subvariety corresponding to the point p implied the universal condition. Here, however, it does not, which is why we use the term "tautological" rather than "universal."

done this, a coarse moduli space does exist; it is the line $\mathbb{P}^1$ with Euclidean coordinate j.

Note that even after this restriction, a fine moduli space does not exist: for example, you can check that if $\pi: \mathscr{V} \to B$ is any family of curves isomorphic to smooth plane cubics, then the induced map $B \to \mathbb{P}^1$ will be ramified over the point $j = 0$, i.e., the j-function associated to the family cannot have simple zeros. For the same reason, no universal family exists over the j-line. If we exclude the two points $j = 0$ and $j = 1728$, a tautological family does exist over the complement, but even then it is not universal.

While this example is too simple to incorporate many of the subtleties of the theory in general, it does illustrate two aspects of one basic paradigm for the construction of a moduli space for a family of algebraic varieties $\{X_\alpha\}$. The first point is to realize the varieties in question as projective varieties, even if the embedding of the members of the family in $\mathbb{P}^n$ is only determined up to the action of $\mathrm{PGL}_{n+1}K$.

To express this, we may define a *polarization* of a variety X to be a projective equivalence class of embeddings of X in $\mathbb{P}^n$ (in general, the definition of polarization may be somewhat broader), and say that the first step in constructing a moduli space is to find a canonical choice of polarization for each object in our family. For example, in the case of the family of smooth curves of genus g we could take the *canonical embedding*, which is to say, for each curve C choose a basis $\omega_1, \ldots, \omega_g$ for the space of holomorphic 1-forms on C, write ω_α locally as $f_\alpha(z)\, dz$, and map

$$p \mapsto [f_1(p), \ldots, f_g(p)].$$

Equivalently, if we let V be the vector space of holomorphic differentials, we map C to the projective space $\mathbb{P}V^*$ by sending p to the hyperplane $\{\omega \in V : \omega(p) = 0\} \subset V$. This determines an embedding $C \hookrightarrow \mathbb{P}^{g-1}$ for any nonhyperelliptic curve of genus g; if we want to include hyperelliptic curves as well, we can use quadratic differentials instead of one-forms (or, in the case of genus 2, triple differentials).

If no canonical polarization of the objects in our family can be found, it may not be possible to construct a moduli space (at least in this manner). One possibility is to construct a moduli space for pairs (X_α, L) where X_α is a member of our family and L is a choice of polarization (with specified numerical invariants, such as degree) of X_α.

Having chosen an embedding $X_\alpha \hookrightarrow \mathbb{P}^n$ up to $\mathrm{PGL}_{n+1}K$ for each X_α, the problem is now converted from one of isomorphism to one of projective equivalence. Thus, the set of images of these embeddings will be a locally closed subset $\mathscr{H}$ of the Chow/Hilbert variety parametrizing such subvarieties of $\mathbb{P}^n$, and the moduli space we desire should then be the quotient of this quasi-projective variety $\mathscr{H}$ by the group $\mathrm{PGL}_{n+1}K$, if indeed this quotient exists. At this point, *geometric invariant theory* intervenes to give us in general the information we obtained by hand in the case of plane cubics, that is, whether such a quotient exists and, more generally, what is the largest open subset $U \subset \overline{\mathscr{H}}$ of the closure of $\mathscr{H}$ whose quotient

$U/\mathrm{PGL}_{n+1}K$ exists. Thus, for example, in the case of curves of genus g, if we choose the m-canonical embedding (that is, the polarization given by m-fold differentials) we find that for m large the quotient of the space $\mathscr{H}$ of m-canonically embedded curves does exist; indeed we can also include the locus of points in the closure of $\mathscr{H}$ corresponding to curves $C \subset \mathbb{P}^{m(2g-2)-g}$ whose only singularities are nodes and whose automorphism groups are finite (which amounts to saying that every rational component of C passes as least three times through the nodes of C). We arrive, ultimately, at what is a model of the success of this theory.

Definition. For $g \geq 2$, a *stable curve of genus g* is a connected curve of arithmetic genus g having only nodes as singularities and finite automorphism group.

(Given that a curve C is connected with only nodes as singularities, its arithmetic genus is given simply as the sum of the geometric genera of its components (that is, the genera of their desingularizations), plus the number of nodes, minus the number of components, plus one.) We then have the following.

Theorem 21.13. *There is a coarse moduli space $\overline{\mathscr{M}}_g$ for stable curves of genus $g \geq 2$; $\overline{\mathscr{M}}_g$ is an irreducible projective variety of dimension $3g - 3$.*

Of course, there is much more to the story. We can say that a universal family exists over the open subset $\overline{\mathscr{M}}_g^0 \subset \overline{\mathscr{M}}_g$ of curves with no automorphisms and that this open subset is smooth, what the singularities of $\overline{\mathscr{M}}_g$ look like, and so on. An excellent discussion of this construction is given in [MM], which includes references to other examples of the applications of geometric invariant theory as well.

LECTURE 22

Quadrics

In this final lecture, we will study in more detail what are perhaps the simplest and most fundamental of all varieties: quadric hypersurfaces. The idea is partly to become familiar with these basic objects and partly to see some of the ideas we have studied in the preceding lectures applied. In the course of this (somewhat lengthy) lecture, we will involve the notions of dimensions, degree, rational maps, smoothness and singularity, tangent spaces and tangent cones, Fano varieties, and families —all in the context of the analysis of one class of objects. This is a much less technically demanding lecture than the last; we are not pushing the boundaries of what we can do with available techniques here, but carrying out a classical and elementary investigation.

Needless to say, the geometry of individual quadrics does not provide that many surprises. Indeed, there are numerous overlaps with earlier material, as well as some topics that could have (and perhaps should have) been treated in earlier lectures. Things get more interesting when, toward the end of the lecture, we begin to investigate the geometry of families of quadrics; this is a much more subtle subject, and one that is still the object of study today.

One note: all our results and our techniques in this lecture are independent of characteristic, except for one thing: in case our ground field K has characteristic 2 the basic correspondence between quadrics and bilinear forms needs to be redefined. For the remainder of this lecture, accordingly, we will assume that the field K is of characteristic other than 2.

Generalities about Quadrics

To begin, we recall some of the notation and terminology initially introduced in Example 3.3. A quadric hypersurface $Q \subset \mathbb{P}V = \mathbb{P}^n$ is given as the zero locus of a homogeneous quadratic polynomial $Q: V \to K$, which may be thought of as the quadratic form associated to a bilinear form

$$Q_0: V \times V \to K,$$

or to the corresponding linear map

$$\tilde{Q}: V \to V^*.$$

In case $\tilde{Q}$ is an isomorphism, it induces an isomorphism from $\mathbb{P}V$ to $\mathbb{P}V^*$, which we also denote by $\tilde{Q}$. In general, the rank of the map $\tilde{Q}$ is called the *rank* of the quadric Q; we recall from Example 3.3 that a quadric $Q \subset \mathbb{P}^n$ of rank k may be described as the cone, with vertex $\Lambda \cong \mathbb{P}^{n-k}$, over a smooth quadric hypersurface in $\mathbb{P}^{k-1}$.

Tangent Spaces to Quadrics

Inasmuch as the quadratic form $Q(v)$ is defined by restricting the bilinear form $Q_0(v, v)$ to the diagonal, the differential dQ at any point v is just the linear form $2Q_0(v, \cdot)$, that is, twice $\tilde{Q}(v)$. Thus, for any point $[v] \in Q$

(i) if $\tilde{Q}(v) \neq 0 \in V^*$, then the quadric hypersurface $Q \subset \mathbb{P}V$ is smooth at $[v]$, with tangent plane $[\tilde{Q}(v)] \in \mathbb{P}V^*$; and
(ii) if $\tilde{Q}(v) = 0$, then Q is singular at $[v]$.[8]

In other words, the Gauss map $\mathscr{G}: Q \to \mathbb{P}V^*$ is the restriction to Q of the linear isomorphism $\tilde{Q}: \mathbb{P}V \to \mathbb{P}V^*$. In particular, the locus of points $P \in Q$ whose tangent planes contain a given point $R \in \mathbb{P}V$ is the intersection of Q with the hyperplane given by $\tilde{Q}(R)$; this is called the *polar* of R with respect to Q. Note also that if Q is singular with vertex Λ, the tangent planes to Q are constant along the lines of Q meeting Λ and all contain Λ.

One of the things we will be doing often in what follows is considering the intersection of a quadric Q with its projective tangent plane $\mathbb{T}_x Q$ at a smooth point $x \in Q$. Let

$$Q' = Q \cap \mathbb{T}_x Q \subset \mathbb{T}_x Q \cong \mathbb{P}^{n-1}.$$

We claim that the rank of Q' is the rank of Q minus 2. (Note that if $\operatorname{rank}(Q) = 1$, Q has no smooth points.) Indeed, the singular locus of Q' is the span of the vertex Λ of Q with the point x; the codimension of the singular locus in the quadric goes down by 2.

Exercise 22.1. Show that if $H \subset \mathbb{P}V$ is any hyperplane and $Q' = Q \cap H$, then

$$\operatorname{rank}(Q) - 2 \leq \operatorname{rank}(Q') \leq \operatorname{rank}(Q)$$

[8] If we are going to include "double planes" ($L^2 = 0$) among quadrics, then for the sake of consistency—for example, if we want it to be the case that a quadric Q is singular at a point p if all partials of its defining equations vanish at p—we have to adopt the convention that a quadric of rank 1 is singular everywhere

with equality holding if and only if H is tangent to Q. More generally, show that if $\Lambda \cong \mathbb{P}^{n-k}$ and $Q' = \Lambda \cap Q$, then

$$\operatorname{rank}(Q) - 2k \leq \operatorname{rank}(Q') \leq \operatorname{rank}(Q).$$

We can see this in the case of these plane sections of a smooth quadric in $\mathbb{P}^3$:

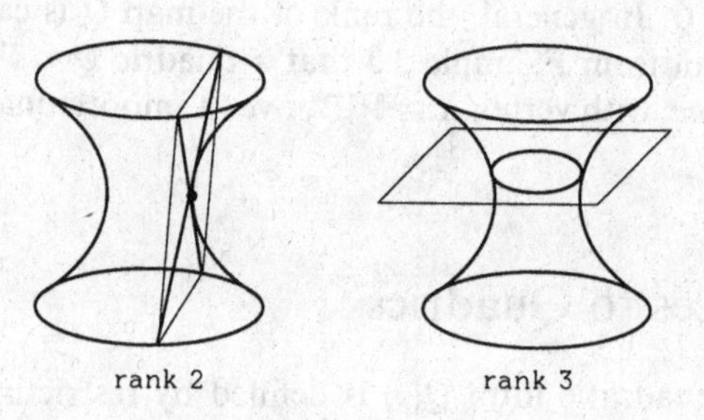

and these plane sections of a quadric cone:

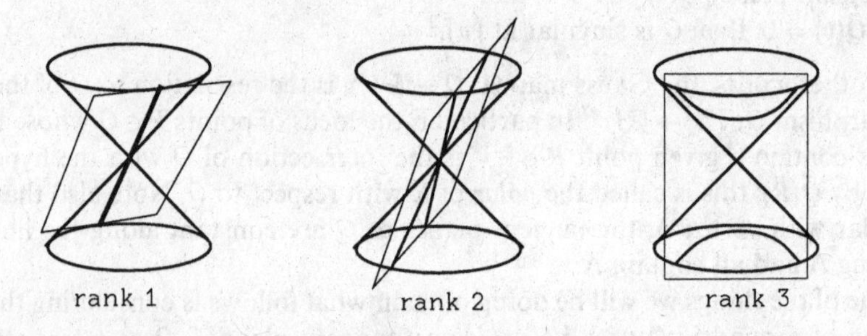

Plane Conics

We have already seen that the image of the quadratic Veronese map $v_2: \mathbb{P}^1 \to \mathbb{P}^2$ is a smooth conic in the plane; by what we have said any smooth conic is projectively equivalent to this one. More directly, we can see that a conic $Q \subset \mathbb{P}^2$ is isomorphic to $\mathbb{P}^1$ by projection from a point $x \in Q$; the map π_x, defined a priori only outside x, extends to an isomorphism by sending the point x to the intersection of the tangent line $\mathbb{T}_x Q$ with the line $\mathbb{P}^1$.

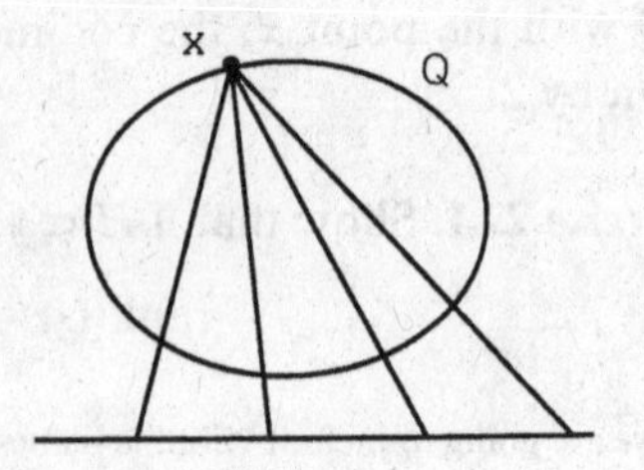

Quadric Surfaces

In similar fashion, we have already seen that the image of the Segre map $\sigma_{1,1}: \mathbb{P}^1 \times \mathbb{P}^1 \to \mathbb{P}^3$ is a smooth quadric Q in $\mathbb{P}^3$; it follows that every smooth quadric may be obtained in this way. We see, moreover, that the fibers of the two projection maps $\pi_i: Q \cong \mathbb{P}^1 \times \mathbb{P}^1 \to \mathbb{P}^1$ are lines in $\mathbb{P}^3$, so that the quadric has two rulings by lines with a unique line of each ruling passing through each point $x \in Q$.

It is fun, however to locate the rulings and the isomorphism $Q \cong \mathbb{P}^1 \times \mathbb{P}^1$ directly. The presence of the rulings can be deduced from the general remark that the intersection of a quadric with its tangent plane at any (smooth) point is a quadric of rank 2 less. In the present case, this means that for each $x \in Q$, the intersection $Q \cap \mathbb{T}_x Q$ is a union of two distinct lines. Since any line lying on Q and containing the point x must lie in this intersection, we see again that through each point of Q there pass exactly two lines.

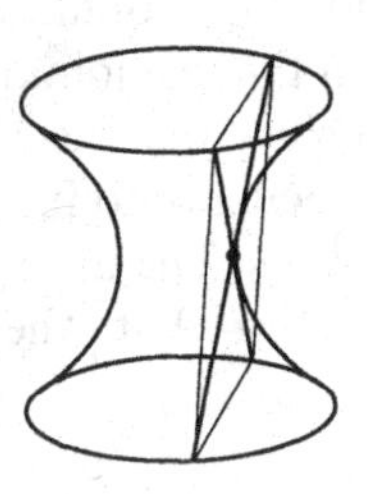

Now fix a point $x_0 \in Q$, and let L_0 and M_0 be the two lines of Q through x_0. Through every point $\lambda \in L_0$ there will pass one other line of Q, which we will call M_λ; likewise through every point $\mu \in M_0$ there will be one other line $L_\mu \subset Q$. Thus we see that Q has two rulings by lines, each ruling parametrized by $\mathbb{P}^1$. Note that (i) the lines of $\{L_\lambda\}$ are disjoint, since Q cannot contain three pairwise incident (and hence coplanar, since Q is smooth) lines; (ii) no line can be a line of both families for the same reason; and (iii) each line L_λ must meet each line M_μ, since the line $\mathbb{T}_\lambda Q \cap \mathbb{T}_\mu Q$ will intersect Q in two points, x_0 and a point of $L_\lambda \cap M_\mu$.

To finish, we claim that every point x of Q lies on exactly one line of each ruling. But this is clear: since the two lines L and M through x comprise the intersection of Q with a plane, one of them must meet L_0 and so be of the form M_μ, and one must meet M_0 and so be L_λ for some λ. Thus we arrive at an isomorphism of Q with $L_0 \times M_0 \cong \mathbb{P}^1 \times \mathbb{P}^1$.

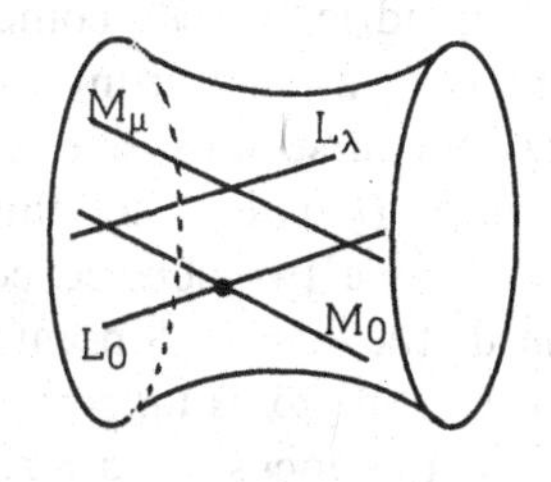

Recall also from Examples 7.11 and 7.22 our analysis of the projection map $\pi_x: Q \to H \cong \mathbb{P}^2$ to a plane H from a point $x \in Q$. Two things are different here from the case of a plane conic. First, the map π_x is only a rational map and cannot be extended to a regular map; as a point $y \in Q$ approaches x, the limiting position of the line $\overline{xy}$ may be any line in the tangent plane $\mathbb{T}_x Q$, and the limiting position of the image point $\pi_x(y)$ any point of the line $\mathbb{T}_x Q \cap H$. Also, whereas in the case of the conic every line $l \subset \mathbb{P}^2$ through the point x (with the exception of the tangent line) met Q in exactly one other point, there are as we have seen two lines L, M

through x entirely contained in Q. Thus, in terms of the graph Γ of π_x, the map π_1 from Γ to Q is an isomorphism except over the point x, where the fiber $E \cong \mathbb{P}^1$. The map $\pi_2: \Gamma \to \mathbb{P}^2$ is an isomorphism except over the points p, q of intersection of the lines L and M with the plane $\mathbb{P}^2$, and carries the curve E to the line $\mathbb{T}_x Q \cap \mathbb{P}^2 = \overline{pq}$. As we expressed in Example 7.22, the quadric Q is obtained from $\mathbb{P}^2$ by blowing up the two points p and q and blowing down the line joining them.

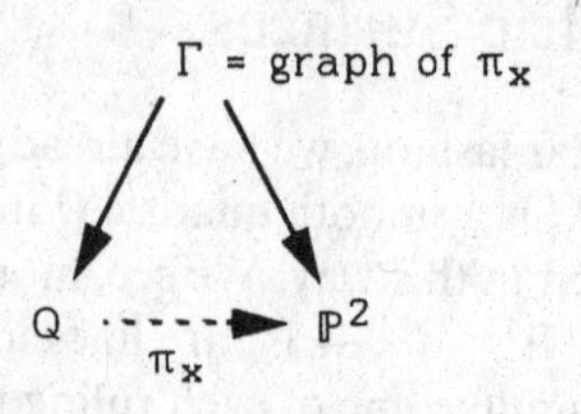

Note that since the line L is carried to the point $p \in \mathbb{P}^2$, the lines of the ruling $\{M_\mu\}$ are projected onto lines through p, and the lines L_λ likewise onto the pencil of lines through q. The inverse map from $\mathbb{P}^2$ to Q may thus be realized as the map sending a point $r \in \mathbb{P}^2$ to the pair (slope of $\overline{pr}$, slope of $\overline{qr}$).

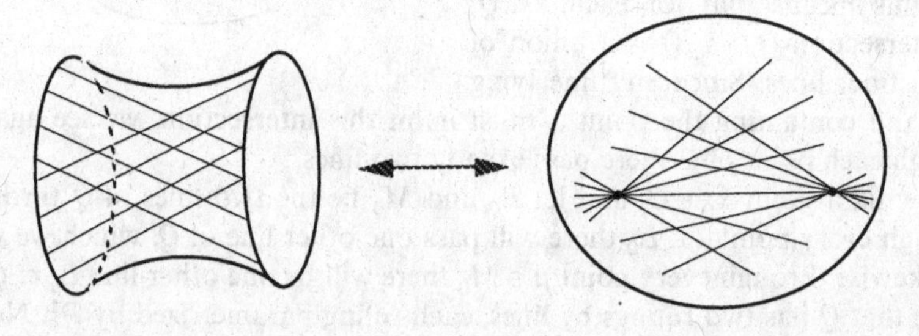

As the reader can also verify, the inverse to the map π_x may be given as the composition of the quadratic Veronese map $v_2: \mathbb{P}^2 \to \mathbb{P}^5$ with the projection of the image from the two points $v_2(p)$ and $v_2(q)$ (compare this with Exercise 22.3).

Finally, another way of representing a smooth quadric $Q \subset \mathbb{P}^3$ comes from projecting Q to a plane from a point x not on Q. Of course, most lines through x will meet Q twice, so this map expresses Q as a two-sheeted cover of $\mathbb{P}^2$. Indeed, the locus of points $p \in Q$ such that the line $\overline{px}$ is tangent to Q at p—that is, the locus of p such that $x \in \mathbb{T}_p Q$—is, as we have seen, a plane section $C = Q \cap H$ of Q; so π_x expresses Q as a double cover of the plane $\mathbb{P}^2$, branched along a conic curve $\overline{C} \subset \mathbb{P}^2$.

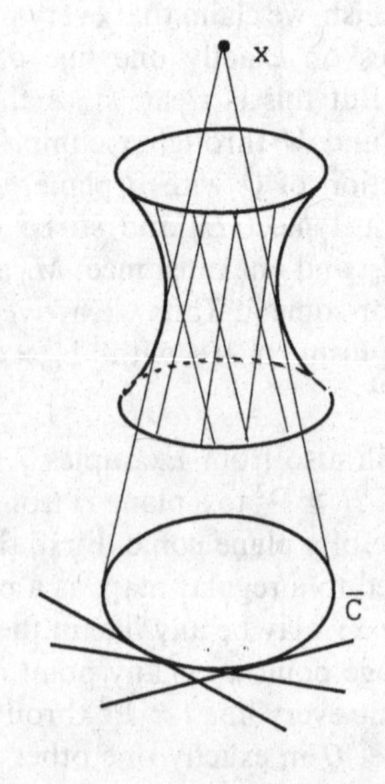

The rulings of Q are once more visible in this form: every line lying on Q will meet the plane section C of Q exactly once, so its image on the plane will be a line meeting $\overline{C}$ exactly once—that is, a tangent line to $\bar{C}$. Indeed, the inverse image of the tangent line to $\bar{C}$ at a point $\pi_x(p)$ will be the intersection $\mathbb{T}_pQ \cap Q$, which will consist of one line of each of the two rulings.

Actually, we have seen this map before in another guise: the map $\pi_x: Q \cong \mathbb{P}^1 \times \mathbb{P}^1 \to \mathbb{P}^2$ is the map from the product of $\mathbb{P}^1$ with itself to its symmetric product, which we have seen (Example 10.23) is isomorphic to $\mathbb{P}^2$, taking an ordered pair (p, q) of points of $\mathbb{P}^1$ to the unordered pair $p + q$.

Quadrics in $\mathbb{P}^n$

Thus far, the smooth quadrics we have dealt with have been varieties encountered in other contexts. In fact, the isomorphisms of a plane conic with $\mathbb{P}^1$ and of a quadric surface in $\mathbb{P}^3$ with $\mathbb{P}^1 \times \mathbb{P}^1$ are reflections of the coincidences of complex Lie groups $SO_3K \cong PSL_2K$ and $SO_4K/\{\pm I\} \cong PSL_2K \times PSL_2K$, respectively (of course, the latter isomorphism is not valid for real Lie groups, corresponding to the fact that a real quadric need not have real rulings). The further coincidence $SO_5K \cong Sp_4K$ and $SO_6K/\{\pm I\} \cong PSL_4K$ will, as we will see in due course, provide alternate descriptions of quadrics in $\mathbb{P}^4$ and $\mathbb{P}^5$. From that point on, however, the coincidences stop, and so do the isomorphisms of quadrics with other varieties naturally encountered (there is in a sense one more, corresponding to triality on SO_8K; see Exercise 22.20). The two representations of a smooth quadric Q via projection, however, still have analogs; we will describe them here.

We start with the simpler of the two representations, that given by projection from a point x not in Q. Again, any line through such a point x will meet Q in either one or two points, so that π_x expresses Q as a two-sheeted branched cover of a hyperplane $H \cong \mathbb{P}^{n-1}$. As in the case of a quadric surface, the fibers of π_x consisting of only one point correspond to the locus of points $p \in Q$ such that the line $\overline{xp}$ is tangent to Q at p. This is the same thing as saying that $x \in \mathbb{T}_pQ$, or equivalently, that p lies in the hyperplane Λ given by $\tilde{Q}(x) \in \mathbb{P}V^*$. Since $x \notin Q$, the hyperplane $\tilde{Q}(x)$ is not tangent to Q, and so intersects Q in a smooth quadric $C \subset \Lambda \cong \mathbb{P}^{n-1}$, which in turn projects isomophically to a smooth quadric $\bar{C} \subset H$. We may say, then, that a smooth quadric $Q \subset \mathbb{P}^n$ is a double cover of $\mathbb{P}^{n-1}$ branched over a smooth quadric $\bar{C} \subset \mathbb{P}^{n-1}$.

All this can of course be seen directly from the equation of Q: if the point $x = [0, \ldots, 0, 1] \notin Q$, we can write

$$Q(Z_0, \ldots, Z_n) = Z_n^2 + a(Z_0, \ldots, Z_{n-1}) \cdot Z_n + b(Z_0, \ldots, Z_{n-1})$$

where a and b are homogeneous of degrees 1 and 2; Q is then the double cover of $\mathbb{P}^{n-1}$ branched over the hypersurface given by $a^2 - 4b = 0$.

Note that, as in the case of the quadric surface, lines on Q will map under the projection π_x to lines in H either tangent to, or contained in, the quadric $\bar{C}$. For a line $l \subset H$ tangent to $\bar{C}$, $\pi_x^{-1}(l)$ will consist of exactly two lines of Q, while if $l \subset \bar{C}$ there will of course be one.

Exercise 22.2. Use this representation of a smooth quadric $Q \subset \mathbb{P}^n$ to find the dimension of the Fano variety $F_1(Q)$ of lines on Q and to show that this family is irreducible if $n \geq 4$.

The projection of a smooth quadric $Q \subset \mathbb{P}^n$ to a hyperplane $H \cong \mathbb{P}^{n-1}$ from a point $x \in Q$ gives a very different picture. As before, this is a rational map, not a regular one; the limiting position of $\pi_x(p)$ as p approaches x on Q could be any point of $H \cap \mathbb{T}_x Q$. The graph Γ of the map is thus the blow-up of the quadric Q at the point x, with the projection carrying the exceptional divisor E of this blow-up to the hyperplane $\mathbb{T}_x Q \cap H$.

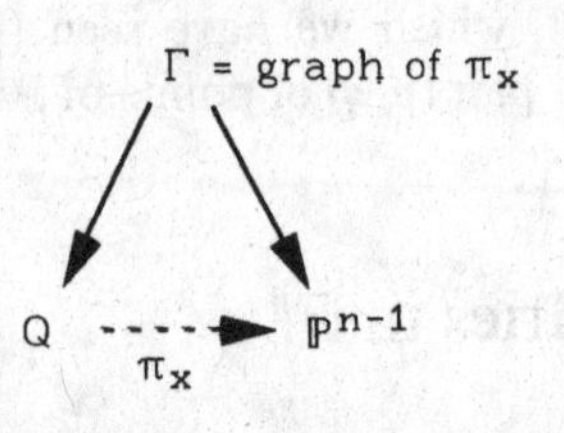

At the same time, π_2 will collapse all lines on Q through the point x. Now, any such line will lie in the tangent plane $\mathbb{T}_x Q$, and so in the intersection $\mathbb{T}_x Q \cap Q$. This intersection, by what we have said before, is of rank $n - 1$; that is, it is a cone over a smooth quadric hypersurface C inside the plane $\mathbb{T}_x Q \cap H \cong \mathbb{P}^{n-2} \subset H \cong \mathbb{P}^{n-1}$. In the graph, then, all these lines correspond to disjoint curves, which are then collapsed to the points of the quadric C; in fact Γ is the blow-up of the hyperplane $\mathbb{P}^{n-1}$ along the quadric C. In sum, then, a smooth quadric $Q \subset \mathbb{P}^n$ may be obtained by blowing up $\mathbb{P}^{n-1}$ along a smooth quadric C in a hyperplane $\mathbb{P}^{n-2} \subset \mathbb{P}^{n-1}$ and then blowing down to a point the proper transform of the plane $\mathbb{P}^{n-2}$ containing C.

Exercise 22.3. Verify the last statement by showing that Q is the image of $\mathbb{P}^{n-1}$ under the rational map

$$\varphi: \mathbb{P}^{n-1} \to \mathbb{P}^n$$
$$: [Z_0, \ldots, Z_{n-1}] \mapsto [F_0(Z), \ldots, F_n(Z)]$$

where $\{F_0, \ldots, F_n\}$ is a basis for the vector space of quadrics in $\mathbb{P}^{n-1}$ vanishing on a quadric hypersurface $C \subset \mathbb{P}^{n-2} \subset \mathbb{P}^{n-1}$ (equivalently, Q is the image of the Veronese variety $v_2(\mathbb{P}^{n-1}) \subset \mathbb{P}^N$ under projection from the plane $\Lambda \subset \mathbb{P}^N$ spanned by C).

The following diagram is an attempt to represent the map π_x in the case of a quadric $Q \subset \mathbb{P}^4$.

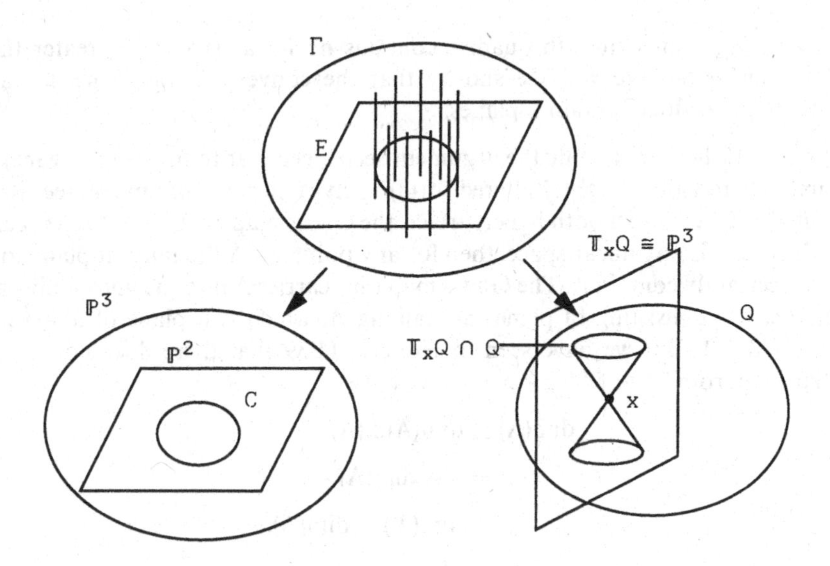

Exercise 22.4 Show that the lines of Q not passing through the point X correspond, via this projection, to the lines of $\mathbb{P}^{n-1}$ meeting the quadric C. Use this to derive once more the dimension of the family of lines on Q and to show that it is irreducible in case $n \geq 4$.

Linear Spaces on Quadrics

We next want to undertake in general a description of the linear spaces contained in a smooth quadric, that is, to describe the Fano variety

$$F_k(Q) = \{\Lambda: \Lambda \subset Q\} \subset \mathbb{G}(k, n)$$

of k-planes contained in Q (since all smooth quadrics in $\mathbb{P}^n$ are isomorphic, we will often omit the Q and write $F_{k,m}$ for the variety of k-planes in a smooth m-dimensional quadric Q). We then ask when $F_{k,m}$ is nonempty, what its dimension is and whether it is irreducible, etc.

The question of existence can be answered easily. To say that a plane $\Lambda = \mathbb{P}W$ lies in a quadric $Q \subset \mathbb{P}V$ is to say that the corresponding linear subspace $W \subset V$ is *isotropic* for the form Q, that is, that $Q|_W \equiv 0$, or equivalently that $\tilde{Q}(W) \subset \operatorname{Ann}(W)$. If Q is smooth, however, $\tilde{Q}$ is an isomorphism, so we must have

$$\dim(W) \leq \dim(\operatorname{Ann}(W)) = \dim(V) - \dim(W);$$

i.e.,

$$2 \cdot \dim(W) \leq \dim(V).$$

Since the dimension of the quadric Q is the dimension of the vector space V minus 2 and the dimension of Λ one less than the dimension of W, we can express this by saying that a smooth quadric contains no linear spaces of greater than half its dimension. We will see shortly that the converse is also true: for any $k \leq \dim(Q)/2$, Q does contain k-planes.

We remark here that while the argument seems peculiar to quadrics, it generalizes immediately (in a slightly altered form) to hypersurfaces of any degree. Basically, if $X \subset \mathbb{P}^n$ is any smooth hypersurface, the Gauss map $\mathscr{G}: X \to \mathbb{P}^{n*}$ is a regular map. If $\Lambda \subset X$ is any linear space, then for any point $x \in \Lambda$ the tangent plane to X at x will certainly contain Λ. The Guass map thus carries Λ into the linear subspace $\operatorname{Ann}(\Lambda) \subset \mathbb{P}^{n*}$ consisting of planes containing Λ, which is a plane of dimension $n - \dim(\Lambda) - 1$. But we have seen in Exercise 11.39 that there does not exist a regular map from $\mathbb{P}^k$ to $\mathbb{P}^l$ for $k > l$; so we have

$$\begin{aligned} \dim(\Lambda) &\leq \dim(\operatorname{Ann}(\Lambda)) \\ &= n - \dim(\Lambda) - 1 \\ &= \dim(X) - \dim(\Lambda). \end{aligned}$$

Thus, X contains no linear subspace of greater than half its dimension.

Example 22.5. Lines on Quadrics

We have already done one example; as we have seen a number of times, the variety $F_{1,2}$ of lines on a quadric surface is isomorphic to two disjoint copies of $\mathbb{P}^1$. We next consider the case of the variety $\Gamma = F_{1,3}$ of lines on a smooth quadric threefold Q in $\mathbb{P}^4$. Of course, the industrious reader will already have computed the dimension of Γ and determined its irreducibility in two ways in Exercises 22.2 and 22.4; we will do it a different way here.

Our approach will be to consider the incidence correspondence Ψ of pairs of points $x \in Q$ and lines l through them; that is, we set

$$\Psi = \{(x, l): x \in l \subset Q\} \subset Q \times \Gamma.$$

What is the fiber of Ψ over a point x? Since every line on Q through x must lie in the tangent hyperplane $\mathbb{T}_x Q$, and since the intersection $Q \cap \mathbb{T}_x Q$ is a quadric cone in $\mathbb{T}_x Q \cong \mathbb{P}^3$, we see that the fibers of Ψ over Q are all isomorphic to $\mathbb{P}^1$. It follows then that Ψ is irreducible of dimension $\dim(Q) + 1 = 4$. Finally, since the fiber of Ψ over a point $l \in \Gamma$ is just a copy of the line l, it follows that Γ is irreducible of dimension $\dim(\Psi) - 1 = 3$.

In fact, Γ turns out to be a very familiar variety; it is nothing other than projective three-space. The following exercise outlines a proof of this fact.

Exercise 22.6. Let V be a four-dimensional vector space over K and $\Omega: V \times V \to K$ a nondegenerate skew-symmetric bilinear pairing. Show that the set Q of two-dimensional isotropic linear subspaces of V is a subvariety of the Grassmannian $G(2, 4) = \mathbb{G}(1, 3)$, and that it is isomorphic to a smooth quadric hypersurface in $\mathbb{P}^4$. Show that for every point $p \in \mathbb{P}V \cong \mathbb{P}^3$ the variety of isotropic subspaces containing p is a line in Q, and that conversely every line in Q is of this form for a unique point p. Conclude that $F_{1,3} \cong \mathbb{P}^3$.

Note that we can use an incidence correspondence like Ψ to describe the family of lines in a quadric Q in $\mathbb{P}^n$ for any n; we let $F = F_1(Q)$ be the Fano variety of lines on Q, and as before set

$$\Psi = \{(x, l): x \in l \subset Q\} \subset Q \times F.$$

Then the fibers of Ψ over Q are, by the same analysis, isomorphic to smooth quadrics Q' in $\mathbb{P}^{n-2}$, so that Ψ must be irreducible of dimension $2n - 4$ and F irreducible of dimension $2n - 5$.

Example 22.7. Planes on Four-Dimensional Quadrics

Next, we look at a smooth quadric hypersurface Q in $\mathbb{P}^5$ and ask about the variety $F = F_2(Q)$ of 2-planes on Q. Once again, we use the incidence correspondence Ψ between points and planes containing them; that is, we set

$$\Psi = \{(x, \Lambda): x \in \Lambda \subset Q\} \subset Q \times F.$$

To describe the fiber of Ψ over a point $x \in Q$, observe that any two-plane in Q passing through x will lie in the intersection $\mathbb{T}_xQ \cap Q$, which is a cone over a smooth quadric surface Q'. The 2-planes in Q through x yhus correspond to lines on the quadric Q', which we have seen are parametrized by $\Gamma_{1,2} \cong \mathbb{P}^1 \amalg \mathbb{P}^1$. It follows that Ψ has dimension $\dim(Q) + 1 = 5$, and hence that F has dimension 3.

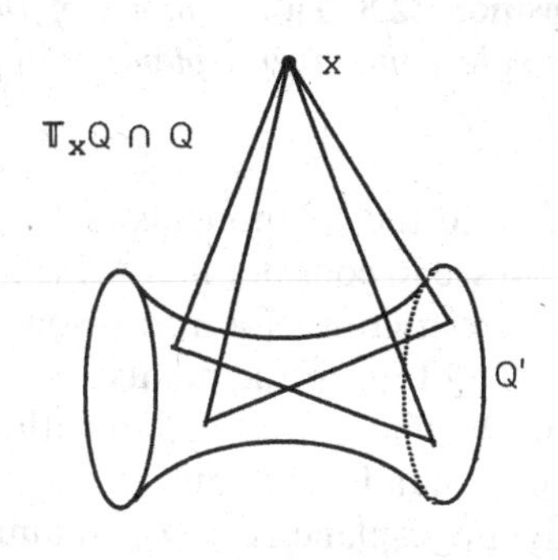

What about the irreducibility of F? Here the situation is very different; since every fiber of Ψ over Q has exactly two connected components, the components of the fibers of Ψ—that is, the set of rulings of tangent hyperplane sections of Q— form a two-sheeted covering space of Q. There are thus two possibilities: F could be irreducible or it could have exactly two irreducible connected components, with the two rulings of $\mathbb{T}_xQ \cap Q$ lying in those two components. In fact, the latter turns out to be the case; there are two distinct rulings of Q by 2-planes. We will proceed on that assumption, and then at the end of this discussion indicate three ways of proving this.

In the case of the lines on a quadric surface in $\mathbb{P}^3$, we saw that two lines of the same ruling were either disjoint or coincident, while lines of opposite rulings always intersected in a point. We ask the analogous question for $Q \subset \mathbb{P}^5$: how may the various 2-planes on Q intersect each other? The answer, in fact, comes from the answer for the quadric surface, via the correspondence given earlier.

Specifically, suppose that two planes $\Lambda, \Lambda' \subset Q$ have at least one point x in common. Then they both lie in the intersection $\mathbb{T}_x Q \cap Q$ and so are cones over lines l, l' of the quadric surface Q'; Λ and Λ' will belong to the same ruling of Q if and only if l and l' belong to the same ruling of Q'. Since the intersection $\Lambda \cap \Lambda'$ has dimension 1 greater than $l \cap l'$, we see that if Λ and Λ' belong to the same family, $\Lambda \cap \Lambda'$ will be either a point or a two-plane; while if they belong to opposite families (always assuming they are incident) it will be a line.

What if Λ and Λ' are disjoint? In this case, let Φ be any 3-plane containing Λ. The intersection $\Phi \cap Q$, since it contains a two-plane, can have rank at most 2; so by Exercise 22.1 it must have rank exactly 2, i.e., it must consist of Λ plus another 2-plane Λ''. Of course, Λ'' meets Λ in a line, so by what we have said, they must belong to opposite families. On the other hand, Φ must meet the plane Λ', and since that intersection is contained in $\Lambda \cup \Lambda''$ and disjoint from Λ, it can only be a single point of Λ''. It follows that Λ' and Λ'' lie in the same ruling, and so Λ and Λ' belong to opposite families. In sum, then we have the following.

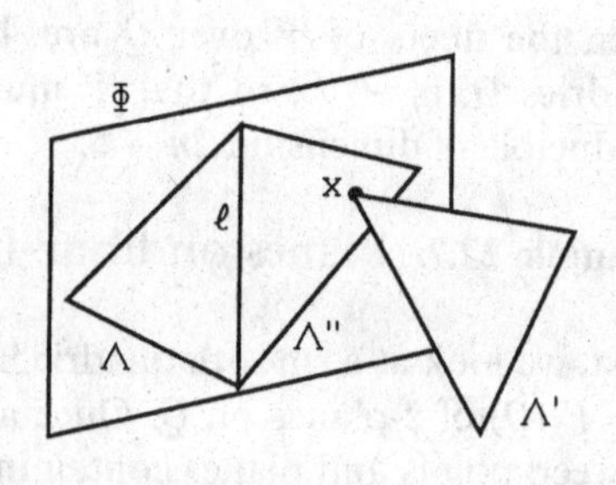

Proposition 22.8. *Two 2-planes of the same ruling in Q either coincide or intersect in a single point; two 2-planes of opposite ruling either intersect in a line or are disjoint.*

What do the components of F actually look like? The answer turns out to be easy. To see it, consider first a line $l \subset Q$. As we have observed, the Gauss map on Q is the restriction of a linear isomorphism $\mathbb{P}^5 \to \mathbb{P}^{5*}$, so that the tangent hyperplanes $\mathbb{T}_x Q$ to Q at the points $x \in l$ form a line in $\mathbb{P}^{5*}$. Their intersection is thus a 3-plane Φ_l whose intersection with Q is singular at every point of l. $\Phi_l \cap Q$ must thus be a quadric of rank 2, i.e., it must consist of two distinct 2-planes; conversely any 2-plane $\Lambda \subset Q$ containing l will lie in $\mathbb{T}_x Q$ for each $x \in l$ and hence in $\Phi_l \cap Q$. We have thus shown that every line $l \subset Q$ lies on exactly two 2-planes of Q. Of course, since these two 2-planes intersect in a line, they must belong to opposite rulings.

Now, let H be a general hyperplane of Q, or indeed any hyperplane whose intersection $Q' = H \cap Q$ with Q is smooth. Since Q' will contain no 2-planes, every 2-plane Λ of Q will intersect H exactly in a line l_Λ. Conversely, by the preceding we see that there is a unique 2-plane of each ruling containing a given line l; so that the association $\Lambda \mapsto l_\Lambda$ gives an isomorphism between each connected component of F and the variety $F_1(Q') = F_{1,3}$ of lines on Q'. Given the fact established in Exercise 22.6 that this variety is isomorphic to $\mathbb{P}^3$, we conclude that the variety F of 2-planes on a smooth quadric $Q \subset \mathbb{P}^5$ is isomorphic to two disjoint copies of $\mathbb{P}^3$.

The promised three proofs of the fact that Q has two rulings by planes are suggested in the following three exercises.

Exercise 22.9. Show that the variety F of 2-planes on Q has two components by showing that the subfamily of planes meeting a given 2-plane Λ_0 in a single point or equal to Λ_0 is an irreducible component of F.

Exercise 22.10. Show that the group $O(6) = O(V, Q)$ acts transitively on the set of maximal isotropic 2-planes for Q and that the stabilizer of a given isotropic plane lies in SO(6). Deduce that $F_{2,4}$ has two components.

Exercise 22.11. Show that $F_{2,4}$ has two components by using the isomorphism of Q with the Grassmannian $\mathbb{G}(1, 3)$ and applying Exercise 6.5.

Example 22.12. Fano Varieties of Quadrics in General

We have spent a fair amount of time on the case of quadrics in $\mathbb{P}^5$, but in fact the general case will now come pretty readily. To understand the variety $F_k(Q) = F_{k,n-1}$ of k-planes lying on a quadric in $\mathbb{P}^n$ (from our present point of view, at any rate), the key is the incidence correspondence Ψ between points and k-planes on Q introduced earlier, that is,

$$\Psi = \{(x, \Lambda): x \in \Lambda \subset Q\} \subset Q \times F_{k,n-1}.$$

As before, the fiber of Ψ over $x \in Q$ is simply the variety of k-planes $\Lambda \subset Q$ containing x, which is the same as the variety of k-planes in the intersection $\mathbb{T}_x Q \cap Q$ through x. But $\mathbb{T}_x Q \cap Q$ is a cone, with vertex x, over a smooth quadric $Q' \subset \mathbb{P}^{n-2}$, and so the k-planes in $\mathbb{T}_x Q \cap Q$ through x correspond exactly to $(k-1)$-planes on Q'. In other words, the fibers of Ψ over Q are isomorphic to the variety $F_{k-1}(Q') = F_{k-1,n-3}$ of $(k-1)$-planes on a smooth quadric of dimension 2 less. We thus have

$$\dim(\Psi) = \dim(F_{k-1,n-3}) + n - 1$$

and hence

$$\dim(F_{k,n-1}) = \dim(F_{k-1,n-3}) + n - k - 1.$$

Moreover, whenever the variety $F_{k-1,n-3}$ is irreducible, we may deduce that Ψ, and hence $F_{k,n-1}$, is too. This covers all cases except the case of maximal-dimensional linear spaces on quadrics of even dimension, that is, $F_{k,2k}$. Summing up, we have the following.

Theorem 22.13. *The variety $F_{k,m}$ of k-planes on a smooth m-dimensional quadric hypersurface is smooth of dimension*

$$\dim(\Gamma_{k,m}) = (k+1)\left(m - \frac{3k}{2}\right)$$

when $k \leq m/2$ and empty otherwise; when $k < m/2$ it is irreducible.

The situation in the special case $m = 2k$ is—again as might be expected from the examples—richer. Here we have the following.

Theorem 22.14. *Let $Q \subset \mathbb{P}^{2k+1}$ be a smooth quadric. Then*

(i) *the variety $F_k(Q) = F_{k,2k}$ has two connected components;*
(ii) *for any two k-planes $\Lambda, \Lambda' \subset Q$ we have*

$$\dim(\Lambda \cap \Lambda') \equiv k \pmod 2$$

if and only if Λ and Λ' belong to the same connected component of $F_k(Q)$; and
(iii) *for every $(k-1)$-plane contained in Q there are exactly two k-planes in Q containing it, and these belong to opposite families; so that*
(iv) *each connected component of $F_{k,2k}$ is isomorphic to the variety $F_{k-1,2k-1}$.*

Exercise 22.15. Prove Theorem 22.14.

Exercise 22.16. Let $Q \subset \mathbb{P}^n$ be a smooth quadric and consider the variety $F_{k,l}(Q) \subset \mathbb{G}(k, n)$ defined by

$$F_{k,l}(Q) = \{\Lambda \subset \mathbb{P}^n: \operatorname{rank}(\Lambda \cap Q) \leq l\}.$$

Find the dimension of $F_{k,l}(Q)$ and its irreducible components. Is it smooth?

Exercise 22.17. Establish Theorem 22.13 by representing the quadric Q as a double cover of $\mathbb{P}^{n-1}$ branched along a quadric. Can you prove Theorem 22.14 the same way?

Exercise 22.18. Same as Exercise 22.17, but use the representation of Q obtained by projection from a point $x \in Q$.

Exercise 22.19. Denote by $\mathcal{Q}$ the projective space of all quadrics in $\mathbb{P}^n$. Establish the dimension counts of Theorems 22.13 and 22.14 by considering the incidence correspondence $\Xi \subset \mathcal{Q} \times \mathbb{G}(k, n)$ defined by

$$\Xi = \{(Q, \Lambda): \Lambda \subset Q\},$$

estimating the dimension of the fibers of Ξ over $\mathbb{G}(k, n)$ to arrive at the dimension of Ξ and hence at the dimension of the general fiber $F_{k,n-1}$ of Ξ over $\mathcal{Q}$. What goes wrong when $k > (n-1)/2$?

Exercise 22.20. Here is one more coincidence. Show that either component of the variety $F_{3,6}$ of 3-planes on a smooth quadric $Q \subset \mathbb{P}^7$ is isomorphic to the quadric Q itself. (This is a reflection of *triality*, the automorphism of order three of the Dynkin diagram of SO_8K; cf. [FH]).

Families of Quadrics

As we indicated at the beginning of this lecture, things get even more interesting when we consider not individual quadrics but families of them. This will be the focus of the remainder of the lecture; we will start by considering the family of all quadrics in $\mathbb{P}^n$, starting with $n = 1$ and 2.

Example 22.21. The Variety of Quadrics in $\mathbb{P}^1$

We consider the variety parametrizing all quadrics in $\mathbb{P}^1$. As we have seen, writing the equation of an arbitrary quadric $C \subset \mathbb{P}^1$ as

$$C = (aX^2 + bXY + cY^2)$$

we see that the space Ψ of all quadrics is just the projective plane $\mathbb{P}^2$, with homogeneous coordinates a, b, and c. Note that this is a special case of the isomorphism mentioned in Example 10.23 between the nth symmetric product of $\mathbb{P}^1$ and $\mathbb{P}^n$.

As suggested in Lecture 4, a natural question to ask is to describe the locus of points in this plane corresponding to each of the two types of quadrics. The answer in this case is immediate; the locus $\Sigma \subset \Psi$ of singular quadrics is given by the discriminant

$$\Sigma = (b^2 - 4ac),$$

which we may observe is a smooth plane conic in $\Psi \cong \mathbb{P}^2$. Alternatively, observe that the map

$$v\colon \mathbb{P}^1 \to \Psi \cong \mathbb{P}^2$$

given by sending a point $P \in \mathbb{P}^1$ to the quadric $2P \in \Psi$ is algebraic; it takes the point $[\alpha, \beta]$, which is the locus of the linear polynomial $\beta X - \alpha Y$, to the quadric $(\beta X - \alpha Y)^2 = \beta^2 X^2 - 2\alpha\beta XY + \alpha^2 Y^2$, so that in coordinates we have

$$v\colon [\alpha, \beta] \mapsto [\beta^2, 2\alpha\beta, \alpha^2].$$

Next, we may ask, for example, to describe the locus L_P of quadrics $C \subset \mathbb{P}^1$ containing a given point $P \in \mathbb{P}^1$—that is, the locus of quadrics of the form $P + Q$, with P fixed. It's not hard to see that this must be a line in $\Psi \cong \mathbb{P}^2$; for one thing, for a given $P = [\alpha, \beta]$ the condition

$$a\alpha^2 + b\alpha\beta + c\beta^2$$

that the quadric C vanish at P is clearly a linear condition on the coefficients a, b, and c of C, so the locus of such C will be a line. Alternatively, we can write down the locus of such C parametrically. We have

$$L_P = \{(\beta X - \alpha Y)\cdot(\delta X - \gamma Y)\}_{[\gamma, \delta] \in \mathbb{P}^1},$$

so parametrically L_P is given as the line

$$L_P = \{[\delta\beta, -\delta\alpha - \gamma\beta, \gamma\alpha]\}_{[\gamma, \delta] \in \mathbb{P}^1}.$$

What relation does this line bear to the conic curve $\Sigma \subset \Psi$? The answer is clear: since the line L_P can meet the conic Σ at only one point—the quadric $2P$—the line L_P must be the tangent line to Σ at $2P$. We can arrive at the same conclusion directly via analytic, rather than synthetic, means: the tangent line to the conic

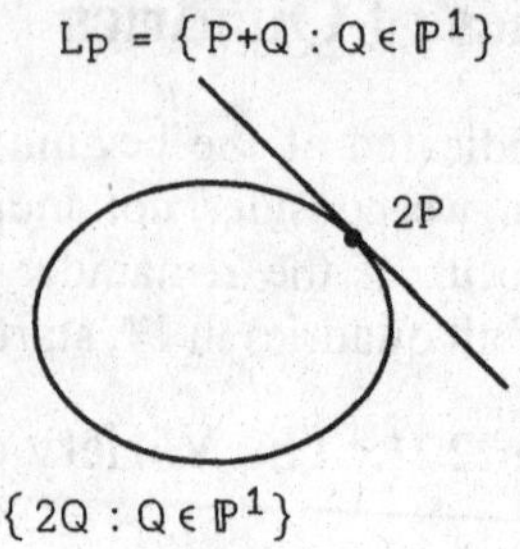

$L_P = \{P+Q : Q \in \mathbb{P}^1\}$

2P

$\{2Q : Q \in \mathbb{P}^1\}$

$$\Sigma = (b^2 - 4ac)$$

at the point $[U, V, W]$ is

$$-4W \cdot a + 2V \cdot b - 4U \cdot c = 0;$$

so at the point

$$2P = [\beta^2, -2\alpha\beta, \alpha^2]$$

it is the line

$$\mathbb{T}_{2P}(\Sigma) = (-4\alpha^2 \cdot a - 4\alpha\beta \cdot b - 4\beta^2 \cdot c),$$

which coincides with the line L_P as given earlier.

Note finally that the two tangent lines to Σ at the points $2P$ and $2Q$ necessarily meet at the point of Ψ corresponding to the quadric $P + Q$; conversely, any point $P + Q$ of Ψ not on Σ will lie on exactly two tangent lines to Σ, corresponding to the fact that the dual of Σ is again a quadric hypersurface.

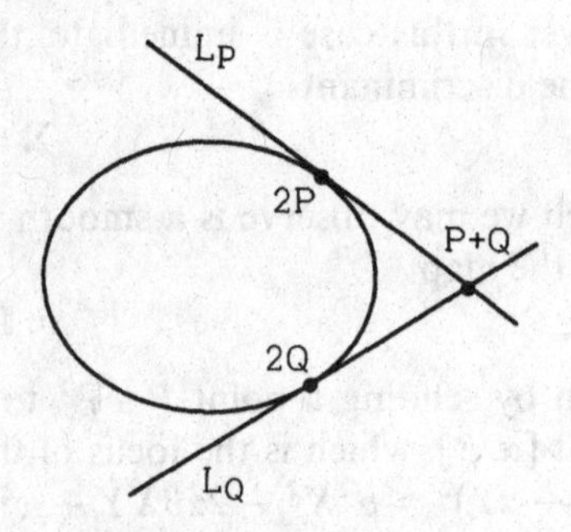

Example 22.22. The Variety of Quadrics in $\mathbb{P}^2$

The geometry of the space $\mathbb{P}^5$ of plane conics is a fascinating one. We have already dealt with it on several occasions; here we will summarize what we have said and give some further statements as exercises.

To begin, there are two distinguished subvarieties of the $\mathbb{P}^5$ of quadrics: the locus Φ of double lines (quadrics of rank 1) and the locus Σ of singular conics. The first of these is just the Veronese surface—it is the image of the map $\mathbb{P}^{2*} \to \mathbb{P}^5$ sending a line $L \in \mathbb{P}^{2*}$ to the corresponding double line, which is the Veronese map v_2. The second may be realized as either the secant variety to Φ or the tangent variety to Φ; it is a cubic hypersurface. If we realize $\mathbb{P}^5$ as the space of nonzero 3×3 symmetric matrices up to scalars, Φ and Σ are the loci of matrices of ranks 1 and ≤ 2, respectively.

Exercise 22.23. Show that Σ is singular exactly along the surface Φ.

Exercise 22.24. Let L and M be distinct lines in $\mathbb{P}^2$. Show that the projective tangent space to Φ at the point $2L$ is the space of conics containing L. Show that the projective tangent space to Σ at $L + M$ is the space of conics containing the point $p = L \cap M$. Deduce from this that the dual variety of Φ is isomorphic to Σ and the dual variety to Σ is isomorphic to Φ.

Exercise 22.25. Let $L \subset \mathbb{P}^2$ be any line and let $\Gamma_L \subset \mathbb{P}^5$ be the space of conics tangent to or containing L. Show that Γ_L is a quadric hypersurface of rank 3, with vertex the linear space Λ_L of conics containing L.

Exercise 22.26. Let $L \subset \mathbb{P}^2$ be a line as earlier. Show that the projective tangent cone $\mathbb{T}C_{2L}(\Sigma)$ to Σ at the point $2L$ is the variety of conics tangent to or containing L and that the multiplicity of Σ at $2L$ is 2.

Example 22.27. Complete Conics

We can see some very interesting phenomena associated to the dual of a variety in the simplest possible case, that of a plane conic. As we have remarked in the case of a quadric hypersurface $X \subset \mathbb{P}^n$ in general, the Gauss map $\mathscr{G}_X: X \to \mathbb{P}^{n*}$ is the restriction of a linear map from $\mathbb{P}^n$ to $\mathbb{P}^{n*}$, so that the dual of a smooth conic curve $C \subset \mathbb{P}^2$ will again be a smooth conic C^* in $\mathbb{P}^{2*}$. We now consider what happens when the conic C varies, and in particular when C becomes singular.

We can actually see a lot of what goes on just from pictures. For example, consider first what happens when the smooth conic C_λ approaches a conic C_0 of rank 1, that is, the union of two distinct lines $L, M \subset \mathbb{P}^2$:

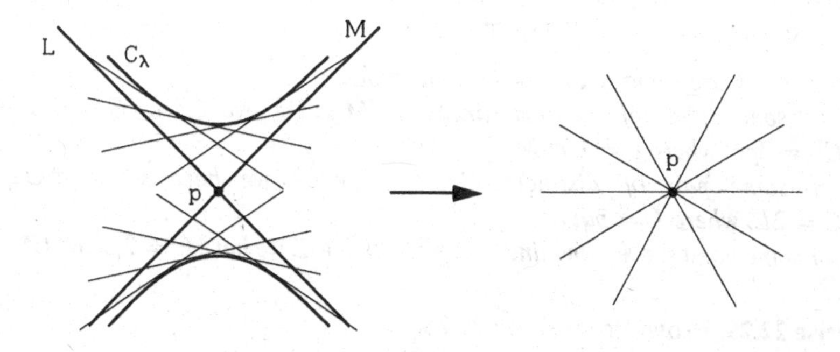

What this picture suggests is that all limiting positions of tangent lines to C_λ as $\lambda \to 0$ pass through the point $p = L \cap M$. If this is the case, the limit as $\lambda \to 0$ of the dual conic must be simply the double line $2p^*$ in $\mathbb{P}^{2*}$.

Next, suppose we have a suitably general family of conics C_λ approaching a double line $C_0 = 2L$. ("General" here should mean that the conics C_λ intersect C_0 transversely in two points p_λ, q_λ that approach distinct points p, $q \in L$ as $\lambda \to 0$.) Again, the picture is suggestive: the limiting positions as $\lambda \to 0$ of the tangent lines to C_λ should be the locus of lines through p or q, and the limit of the dual of C_λ accordingly the sum of the two lines $p^*, q^* \subset \mathbb{P}^{2*}$.

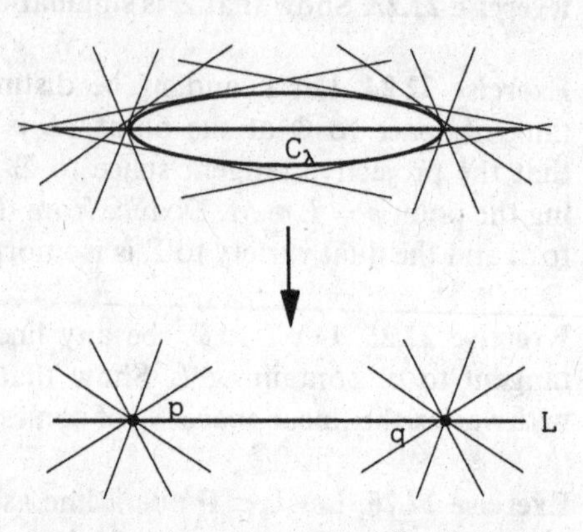

In fact, we can make this precise. To set it up, we identify as before the set of all conics in $\mathbb{P}V \cong \mathbb{P}^2$ with $\mathbb{P}(\mathrm{Sym}^2 V^*) \cong \mathbb{P}^5$; we will simultaneously identify the space of conics in the dual plane $\mathbb{P}V^* = \mathbb{P}^{2*}$ with the dual projective space $\mathbb{P}(\mathrm{Sym}^2 V) = \mathbb{P}^{5*}$. Let $U \subset \mathbb{P}^5$ and $U^* \subset \mathbb{P}^{5*}$ be the open subsets corresponding to smooth conics. We consider the bijection

$$\tilde{\varphi}: U \to U^*$$

given by sending a conic C to its dual C^*. This extends to a rational map

$$\varphi: \mathbb{P}^5 \dashrightarrow \mathbb{P}^{5*};$$

we let $\Gamma \subset \mathbb{P}^5 \times \mathbb{P}^{5*}$ be the graph of this map. Γ is called the variety of *complete conics*, and the questions suggested earlier about the behavior of duals of conics in families can all be expressed in terms of Γ and its projection maps. The answers, bearing out the pictures, are expressed as follows.

Proposition 22.28. *A pair $(C, C^*) \in \mathbb{P}^5 \times \mathbb{P}^{5*}$ of conics lies in Γ if and only if it is one of the following four types:*

(i) *C and C^* are smooth and dual to one another;*
(ii) *for some pair of distinct lines L, $M \subset \mathbb{P}^2$ we have $C = L \cup M$ and $C^* = 2p^*$, where $p = L \cap M$;*
(iii) *for some pair of distinct points p, $q \in \mathbb{P}^2$ we have $C^* = p^* \cup q^*$ and $C = 2L$, where $L = \overline{pq}$; or*
(iv) *for some point $p \in \mathbb{P}^2$ and line $L \subset \mathbb{P}^2$ with $p \in L$ we have $C = 2L$ and $C^* = 2p^*$.*

Exercise 22.29. Prove Proposition 22.28.

Exercise 22.30. Note that by Proposition 22.28, the projection $\Gamma \to \mathbb{P}^5$ is one to one except over the subvariety Φ of double lines, where the fibers are isomorphic to $\mathbb{P}^2$. Show that in fact Γ is the blow-up of $\mathbb{P}^5$ along the Veronese surface Φ.

Example 22.31. Quadrics in $\mathbb{P}^n$

In general, the space of quadric hypersurfaces in $\mathbb{P}^n$ is a projective space of dimension $N = n(n+1)/2 - 1$; we may think of this as the projective space associated to the vector space of symmetric $(n+1) \times (n+1)$ matrices M. Inside $\mathbb{P}^N$ we will let Φ_k denote the variety of all quadrics of rank at most k, that is, the subvariety given by the vanishing of the $(k+1) \times (k+1)$ minors of M. Equivalently, Φ_k is the set of quadrics Q expressible as cones, with vertex $\Lambda \cong \mathbb{P}^{n-k}$, over a quadric $\bar{Q}$ in $\mathbb{P}^{k-1}$. Note that the points of $\Phi_k - \Phi_{k-1}$ correspond to such cones where the quadric $\bar{Q}$ is smooth, or equivalently, where the vertex Λ is unique.

This characterization gives us a way of determining the dimension of the Φ_k via an incidence correspondence; as usual, we let $\mathbb{G} = \mathbb{G}(n-k, n)$ be the Grassmannian of k-planes in $\mathbb{P}^n$ and we let $\Psi_k \subset \mathbb{G} \times \mathbb{P}^N$ be the locus

$$\Psi_k = \{(\Lambda, Q) \colon \Lambda \subset Q_{\text{sing}}\}.$$

For any given $\Lambda = \mathbb{P}W \subset \mathbb{P}^n$, the fiber of Ψ_k over Λ is just the space of quadrics in the projective space associated to the quotient $\mathbb{C}^{n+1}/W$, which has dimension $k(k+1)/2 - 1$; thus

$$\begin{aligned}\dim \Psi_k &= \dim \mathbb{G} + k(k+1)/2 - 1\\ &= kn - (k-1)(k-2)/2.\end{aligned}$$

Since Ψ_k maps one to one onto Φ_k except over Φ_{k-1}, this is also the dimension of Φ_k; the easiest way to remember this is by observing that

$$\operatorname{codim}(\Phi_{n+1-l} \subset \mathbb{P}^N) = l(l+1)/2,$$

or in other words the codimension of the locus of quadrics whose rank is l less than the maximum is the binomial coefficient $l(l+1)/2$. The most visible examples of this are

(i) The locus of singular quadrics is of course the hypersurface given by the vanishing of the determinant of M. Note that since the determinant of M is an irreducible polynomial of degree $n+1$ in the entries of M, this hypersurface has degree $n+1$.

(ii) At the other extreme, the locus of rank 1 quadrics is just the Veronese variety; in intrinsic language, if we view $\mathbb{P}^n$ as the projectivization $\mathbb{P}V$ and $\mathbb{P}^N$ as the projective space $\mathbb{P}(\mathrm{Sym}^2 V^*)$, then Φ_1 is just the image of the map

$$v_2 \colon \mathbb{P}V^* \to \mathbb{P}(\mathrm{Sym}^2 V^*)$$

given by sending a linear form to its square. As we have seen, this is just the Veronese map on $\mathbb{P}V^*$. Note in particular that the degree of Φ_1 is thus 2^n.

(iii) One more case where we can describe Φ_k directly is the case $k = 2$. Here Φ_2 is the locus of pairs of hyperplanes in $\mathbb{P}^n$, that is, the image of the map

$$s \colon \mathbb{P}V^* \times \mathbb{P}V^* \to \mathbb{P}(\mathrm{Sym}^2 V^*)$$

given by sending a pair of linear forms to their product. Observing that this is just the Segre map $\mathbb{P}V^* \times \mathbb{P}V^* \to \mathbb{P}(V^* \otimes V^*)$ followed by the projection $V^* \otimes V^* \to \mathrm{Sym}^2 V^*$; since the latter map is generically two to one on the image of the former, the degree of Φ_2 is half the degree of the Segre variety; i.e.,

$$\deg(\Phi_k) = \frac{1}{2}\binom{2n}{n}.$$

The first example of a variety Φ_k not described in one of these ways is the variety Φ_3 of quadrics of rank at most 3 in $\mathbb{P}^4$. Its degree, as it turns out, is 10; I don't know of any simpler way of determining this than deriving the following general proposition.

Proposition 22.32. *The degree of the variety Φ_k of quadrics of rank at most k in the space $\mathbb{P}^N$ of all quadrics in $\mathbb{P}^n$ is*

$$\deg(\Phi_k) = \prod_{\alpha=0}^{n-k} \frac{\binom{n+\alpha+1}{n-k-\alpha}}{\binom{2\alpha+1}{\alpha}}.$$

We won't describe the derivation of this formula; see [F1] example 14.4.11 or [JLP].

The next question to ask about the varieties Φ_k, after dimension and degree, is about their smooth and singular loci and about their tangent spaces and tangent cones. The answers are exactly analogous to those for the determinantal subvarieties of the space of general $m \times n$ matrices.

Theorem 22.33. *Let $Q_0 \in \Phi_m - \Phi_{m-1}$, i.e., let Q_0 be a cone with vertex $\Lambda \cong \mathbb{P}^{n-m}$ over a smooth quadric $\overline{Q} \subset \mathbb{P}^{m-1}$. Then Φ_m is smooth at Q_0, with tangent space*

$$T_{Q_0}(\Phi_m) = \{Q \colon Q \supset \Lambda\}.$$

More generally, if $Q_0 \in \Phi_{m-k} - \Phi_{m-k-1}$—i.e., Q_0 is a cone with vertex $\Lambda \cong \mathbb{P}^{n-m+k}$ over a smooth quadric—then the tangent cone to Φ_m at Q_0 is given by

$$TC_{Q_0}(\Phi_m) = \{Q \colon \mathrm{rank}(Q|_\Lambda) \le k\}$$

and the multiplicity of Φ_m at Q_0 is correspondingly the degree of the variety Φ_k in the space of quadrics in $\mathbb{P}^{n-m+k}$, as given by Proposition 22.32.

As we indicated, the proof is identical in form to that given in Examples 14.15 and 20.5 in the case of general determinantal varieties, and we will not go through it again. We should remark that, as in the previous case, in order to deduce the multiplicity from the description of the tangent cone given in the statement of the theorem and the formula for the degrees of symmetric determinantal varieties we need to know that the ideals of the varieties Φ_k really are generated locally by $(k + 1) \times (k + 1)$ minors.

Pencils of Quadrics

In a sense, the study of the geometry of the space $\mathbb{P}^N$ of all quadrics in $\mathbb{P}^n$ is just the prelude to the study of more general families of quadrics, which correspond to subvarieties B of $\mathbb{P}^N$ (or, more accurately, varieties B with maps $B \to \mathbb{P}^N$). We will illustrate this with a discussion of the simplest such families, namely, those corresponding to a line L in $\mathbb{P}^N$. These are called *pencils* of quadrics. The set of all pencils is parametrized by the Grassmannian $\mathbb{G}(1, N)$, and for the time being we will focus on the behavior of a general pencil.

Of course, there is a unique line through two points in a projective space, and if Q_0 and $Q_1 \in \mathbb{P}^N$ are two quadrics, the corresponding pencil is the family of quadrics $L = \{Q_\lambda : \lambda \in \mathbb{P}^1\}$, where

$$Q_\lambda = \lambda_0 \cdot Q_0 + \lambda_1 \cdot Q_1$$

for each $\lambda = [\lambda_0, \lambda_1] \in \mathbb{P}^1$. We introduce their intersection, called the *base locus* of the pencil

$$X = Q_0 \cap Q_1 = \bigcap_{\lambda \in \mathbb{P}^1} Q_\lambda.$$

Since the pencil is assumed general, Q_0 and Q_1 will intersect transversely in X, that is, X will be empty if $n < 2$ and a smooth variety of degree 4 and dimension $n - 2$ if $n \geq 2$. In case $n \geq 2$, we will have, by Proposition 17.18,

$$L = \{Q : Q \supset X\}.$$

The first thing to ask about a family of quadrics is about its singular members: what they look like, and how many there are. In the present case, since $L \subset \mathbb{P}^N$ has been taken to be a general line, it will miss the subvarieties Φ_k for $k \leq n - 1$, since they all have codimension of at least three, and will intersect the hypersurface Φ_n of singular quadrics transversely in $n + 1$ points.

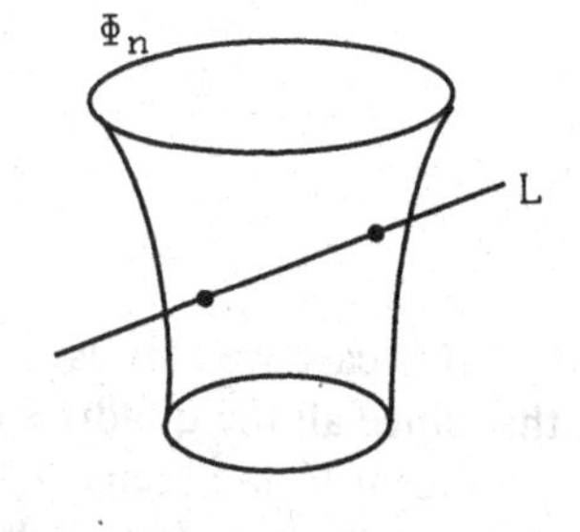

In the case $n = 1$, this says that a general pencil of quadrics in $\mathbb{P}^1$ will have two members consisting of a double point. Note that the only way a pencil L could fail to have exactly two singular members is to be tangent to the conic of singular quadrics—that is, to have a nonempty base locus.

In case $n = 2$, the singular elements of a general pencil of conics are even more visible. The base locus X of the pencil will be four points A, B, C, and D, no three of which may be collinear; if X is to be contained in a union of two lines, each of the lines will have to contain exactly two of these points. The three singular conics in the pencil L are thus the line-pairs $\overline{AB} + \overline{CD}$, $\overline{AC} + \overline{BD}$, and $\overline{AD} + \overline{BC}$.

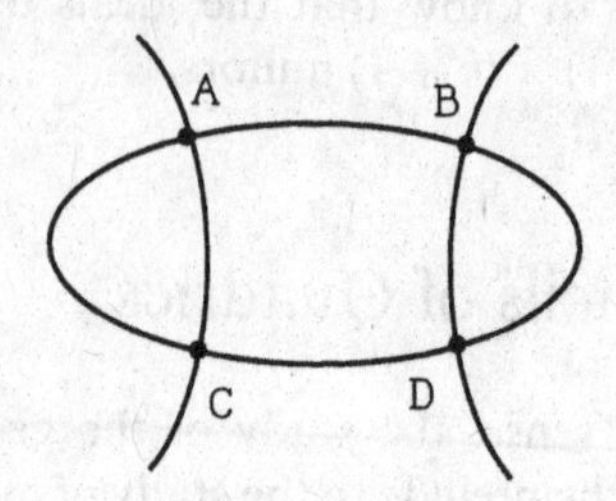

Note in particular that whenever the base locus X consists of four distinct points, the pencil will have exactly three singular elements. There are many ways to see the converse; for example, if the quadrics Q_0 and Q_1 are tangent at a point C in addition to meeting transversely at points A and B, the pencil L will consist of all quadrics containing A, B, and C, and either singular at C or tangent to Q_0 and Q_1 at C. The singular quadrics in L are then just the sum Q of the lines $\overline{AB}$ and $\overline{AC}$, and the sum Q' of the line $\overline{AB}$ with the line through C tangent to Q_0 and Q_1.

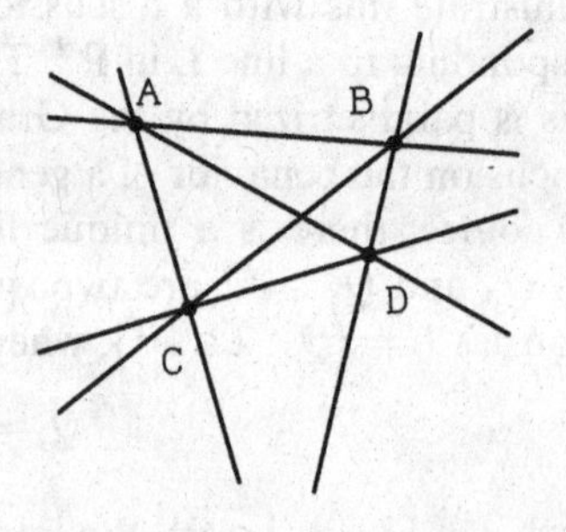

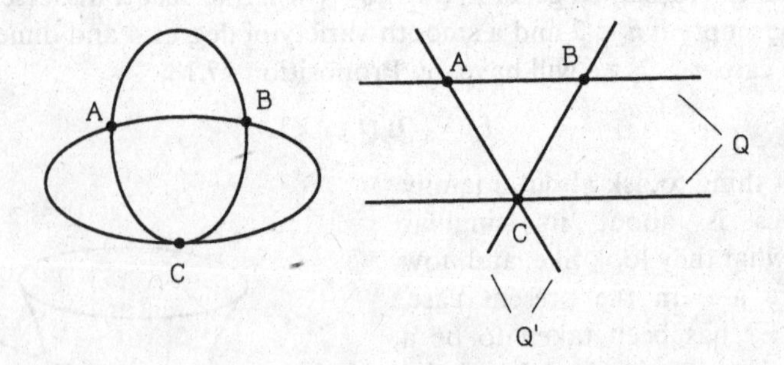

Indeed, in this case we can use Exercise 22.24 (or more generally Theorem 22.33) to see that since all the quadrics of the pencil contain the singular point of Q', the line L is tangent to the locus Φ_2 of singular quadrics at Q', and so Q' is a double point of intersection of L with Φ_2 (that is, the polynomial $\det(M)$ on the space of quadrics restricted to L, vanishes to order at least 2 at Q').

The preceding are examples of a general statement, namely, the following.

Proposition 22.34. *The pencil L spanned by Q_0 and Q_1 contains exactly $n + 1$ singular elements if and only if Q_0 and Q_1 intersect transversely.*

Proof. We observe first that the condition that Q_0 and Q_1 intersect transversely is independent of the choice of generators Q_0 and Q_1; it is equivalent in any case to the condition that the differentials dQ_λ of the polynomials $Q_\lambda \in L$ are independent at each $P \in X$.

This said, suppose first that L contains other than $n + 1$ singular elements, i.e., L fails to intersect Φ_n transversely. This can happen in one of two ways; either L can contain a point Q of the singular locus of Φ_n, which we know by Theorem 22.33 to be Φ_{n-1}, or L may be tangent to Φ_n at a smooth point Q. In the former case, since the singular locus of Q is positive dimensional, any other $Q' \in L$ will meet it, and so will fail to meet Q transversely. In the latter case, by Theorem 22.33 all quadrics $Q' \in L$ will contain the singular point of Q, and so again will fail to meet Q transversely.

Conversely, if two generators of the pencil fail to meet transversely at a point p, then some element Q of the pencil will be singular at p, i.e., for some pair $Q, Q' \in L$, Q' meets the singular locus of Q; then either $Q' \in \Phi_{n-1} = (\Phi_n)_{\text{sing}}$ or L is tangent to Φ_n at Q. □

Exercise 22.35. Find a pencil of conics in $\mathbb{P}^2$ that contains only one singular element but does not include a double line. In general, does there exist a pencil of quadrics in $\mathbb{P}^n$ having only one singular element, but not meeting Φ_{n-1}? (*)

In view of Proposition 22.34 we will say that a pencil of quadrics is *simple* if the equivalent conditions of the proposition are satisfied, and we will restrict our attention for the time being to such pencils. The main problem we will set ourselves is the classification of simple pencils—that is, we ask whether all simple pencils are projectively equivalent to one another, and if not what distinguishes them?

Consider first the case $n = 2$, that is, pencils of conics in the plane. Here we have seen that a simple pencil consists of all conics passing through four points in the plane, no three of which are collinear; since any such set of points can be carried into any other, it follows that all simple pencils in the plane can likewise be carried into each other by an automorphism of $\mathbb{P}^2$.

The situation in $\mathbb{P}^3$ is different. For one thing, we observe that not all base loci X are isomorphic; as we will see in the next two Exercises, if we project X from a point $P \in X$ into a plane, X is mapped isomorphically onto a smooth plane cubic curve $\overline{X}$, which has a j-invariant, and any plane cubic may arise in this way.

Exercise 22.36. Let X be the smooth intersection of two quadrics in $\mathbb{P}^3$ and $P \in X$ any point. Show that the projection π_P from P to a plane gives an isomorphism of X with a plane cubic.

Exercise 22.37. Now let $E \subset \mathbb{P}^2$ be a plane cubic, with equation

$$y^2 = x^3 + ax + b.$$

Show that the map $E \to \mathbb{P}^3$ given by sending (x, y) to $[1, x, x^2, y]$ maps E isomorphically to the intersection of the two quadrics

$$Z_0Z_2 - Z_1^2 = 0 \quad \text{and} \quad Z_3^2 = Z_1Z_2 + aZ_0Z_1 + bZ_0^2.$$

At the same time, a pencil of quadrics in $\mathbb{P}^3$ has a directly observable invariant. If $L = \{Q_\lambda\}$ is a simple pencil of quadrics, there will be four values of $\lambda \in \mathbb{P}^1$ for which the quadric Q_λ is singular. These four points on $\mathbb{P}^1$ have a nontrivial invariant; specifically, if λ is the cross-ratio of these four numbers in any order, then the j-invariant

$$j = 256 \cdot \frac{(\lambda^2 - \lambda + 1)^3}{\lambda^2(\lambda - 1)^2}$$

(first introduced in Lecture 10) is independent of the ordering and preserved by any automorphism of $\mathbb{P}^1$. Two simple pencils can thus be carried into one another only if their j-invariants coincide. The fact is that the converse is also true; the j-invariant is a complete invariant of the pencil, as we will see later.

It will not be a surprise to find that this invariant of a simple pencil may also be read off of its base locus X. Specifically, we have the following.

Proposition 22.38. *The base locus X of a simple pencil of quadrics in $\mathbb{P}^3$ with singular elements Q_{λ_i}, $i = 0, 1, 2, 3$, is isomorphic to the curve E with equation*

$$y^2 = \prod (x - \lambda_i);$$

in other words, the j-invariant of the pencil is the j-invariant of X.

PROOF. We could, of course, do this by examining the explicit equations given in the last exercise. Instead, we will argue geometrically; in fact, we will show that both X and E coincide with an a priori distinct third object. We fix a point $P \in X \subset \mathbb{P}^3$ and introduce the curve of lines through P lying in some quadric of the pencil; precisely, we set

$$Y = \{(\lambda, l) \colon P \in l \subset Q_\lambda\} \subset \mathbb{P}^1 \times \mathbb{G}(1, 3).$$

Observe first that Y is smooth, since projection of Y into the second factor maps Y onto the plane curve $\overline{X}$ viewed as a subvariety of $\Sigma_P = \{l \colon P \in l\} \cong \mathbb{P}^2 \subset \mathbb{G}(1, 3)$.

Now, to see that Y is isomorphic to X, observe that any line l lying on a quadric $Q \in L$ will meet X at two points (that is, the two points of intersection of l with any $Q' \in L$ other than Q); we can thus define a map $\varphi \colon Y \to X$ by sending (λ, l) to the point of intersection of l with X other than P (and to P if l is tangent to X at P). On the other hand, for any point $R \in X$ there is a unique quadric $Q_\lambda \in L$ containing the line $l = \overline{PR}$. For any point $S \in l$ other than P and R, a quadric $Q \in L$ will contain l if and only if it contains S; but the set of quadrics containing S is a hyperplane H in the space $\mathbb{P}^N = \mathbb{P}^9$ of all quadrics, and since the line L is not contained in H, it must meet it at one point Q_λ. Thus we may invert φ by sending R to (λ, l), where $l = \overline{PR}$ and Q_λ is the unique quadric in L containing l.

Finally, to see that $Y \cong E$, any smooth quadric $Q \in L$ will contain two lines through the point P, while a singular one will contain one. Since Y is smooth it follows that the projection on the first factor expresses Y as a double cover of $\mathbb{P}^1$ ramified over the four points $\lambda_i \in \mathbb{P}^1$ and hence that Y is isomorphic to E. □

Exercise 22.39. Show that projection of the curve $X \subset \mathbb{P}^3$ to a plane from a point P not in X either maps X birationally to a plane quartic with two nodes, a node and a cusp, two cusps or a tacnode, or two to one onto a conic curve. (*)

Exercise 22.40. Let Q be any of the four singular quadrics of the pencil containing X. Show that four tangent lines $T_P X$ to X lie in Q, and that the four points P of tangency are coplanar. Show that the 16 points obtained in this way are exactly the inflectionary points of X.

The answer to the general question about when two simple pencils are equivalent is pretty much what you might hope for on the basis of this example; we have the following.

Theorem 22.41. *Let $L = \{Q_\lambda\}$ and $M = \{R_\lambda\}$ be pencils of quadrics in $\mathbb{P}^n$, with singular elements Q_{λ_i} and R_{μ_i}, $i = 1, \ldots, n+1$. Then L and M are projectively equivalent if and only the two subsets $\{\lambda_i\}, \{\mu_i\} \subset \mathbb{P}^1$ are congruent, i.e., if there is an automorphism of $\mathbb{P}^1$ carrying $\{\lambda_i\}$ to $\{\mu_i\}$.*

This will follow in turn from the following.

Lemma 22.42 (Normal Form for Simple Pencils). *Let L be a simple pencil of quadrics in $\mathbb{P}^n$. Then there exists a set of homogeneous coordinates on $\mathbb{P}^n$ in terms of which all $Q \in L$ are diagonal; i.e., L is generated by quadrics Q and R where*

$$Q(X) = \sum X_i^2$$

and

$$R(X) = \sum \lambda_i \cdot X_i^2.$$

PROOF. Write $\mathbb{P}^n = \mathbb{P}V$ for V a vector space, let Q and R be a pair of generators of L corresponding to smooth quadrics and consider the corresponding maps

$$\tilde{Q}, \tilde{R}\colon V \to V^*.$$

The basis for V we seek will be one consisting of eigenvectors for the composition

$$A = \tilde{R}^{-1} \circ \tilde{Q} : V \to V.$$

Note that rechoosing Q and R will have the effect of replacing A by a linear combination of A and the identity; in particular, it will not change the eigenvectors of A. Note also that A cannot have two independent eigenvectors with the same eigenvalue λ, since then the linear combination $Q - \lambda R$ of Q and R would have rank $n - 1$ or less.

To see that the eigenvectors of A have the properties we want, we invoke the standard calculation: if $v, w \in V$ are eigenvectors for A with eigenvalues λ and $\mu \neq \lambda$, respectively, we have

$$\begin{aligned} Q(v, w) &= \langle \tilde{Q}v, w \rangle \\ &= \langle \tilde{R}Av, w \rangle \\ &= \lambda \cdot R(v, w) \end{aligned}$$

but at the same time

$$\begin{aligned} Q(v, w) &= \langle v, \tilde{Q}w \rangle \\ &= \langle v, \tilde{R}Aw \rangle \\ &= \mu \cdot R(v, w) \end{aligned}$$

and we conclude that $Q(v, w) = R(v, w) = 0$.

It remains to be seen that A is indeed diagonalizable. Suppose it is not; this means we have a pair of vectors $v, w \in V$ with $(A - \lambda I)v = 0$ and $(A - \lambda I)w = v$. Replacing Q by $Q - \lambda R$ we may take $\lambda = 0$, so that $\tilde{Q}v = 0$—i.e., the point $[v] \in \mathbb{P}V$ is a singular point of the quadric Q—and $\tilde{R}^{-1}\tilde{Q}w = v$. But then we have

$$\begin{aligned} R(v, v) &= \langle \tilde{R}v, v \rangle \\ &= \langle \tilde{R}\tilde{R}^{-1}\tilde{Q}w, v \rangle \\ &= \langle \tilde{Q}w, v \rangle \\ &= Q(w, v) \\ &= \langle w, \tilde{Q}v \rangle \\ &= 0 \end{aligned}$$

so that the quadric $R \subset \mathbb{P}V$ contains the point $[v]$ of Q, contradicting the hypothesis that the pencil $L = \overline{QR}$ is simple. □

Note that if we wanted to be pedantic we could refine this last lemma; specifically, in view of the fact that the locus Φ_n of singular quadrics in $\mathbb{P}^n$ has multiplicity k along Φ_{n-k+1}, for any pencil $L = \{Q_\lambda\}$ of quadrics, not all singular, we have

$$\sum_{\lambda \in \mathbb{P}^1} (n + 1 - \operatorname{rank}(Q_\lambda)) \leq n + 1$$

and the preceding argument shows that the pencil is simultaneously diagonalizable if and only if equality holds in this expression.

Proof of Theorem 22.41. Having written the two quadrics Q and $R \in L$ in the form

$$Q(X) = \sum X_i^2$$

and

$$R(X) = \sum \lambda_i \cdot X_i^2$$

we see that the set of λ for which the linear combination $Q_\lambda = Q - \lambda R$ is singular is just $\{\lambda_1, \ldots, \lambda_{n+1}\}$. Replacing Q and R by linear combinations $R' = aR + bQ$ and $Q' = cR + dQ$ thus has the effect of replacing each λ_i by $(a\lambda_i + b)/(c\lambda_i + d)$ so the pencil is determined up to projective equivalence by the subset $\{\lambda_1, \ldots, \lambda_{n+1}\} \subset \mathbb{P}^1$ up to linear fractional transformations of $\mathbb{P}^1$. $\square$

Hints for Selected Exercises

Exercise 1.3. It has to be shown that given any $q \notin \Gamma$ there exists a polynomial of degree $d - 1$ vanishing on Γ but nonzero at q. There are many ways to do this, the fastest being perhaps induction on d.

Exercise 1.11(b). For what it's worth, the line $L_{\mu,\nu}$ is given by the equations

$$(\mu_0\nu_1 - \mu_1\nu_0)Z_0 + (\mu_0\nu_2 - \mu_2\nu_0)Z_1 + (\mu_1\nu_2 - \mu_2\nu_1)Z_2 = 0$$

and

$$(\mu_0\nu_1 - \mu_1\nu_0)Z_1 + (\mu_0\nu_2 - \mu_2\nu_0)Z_2 + (\mu_1\nu_2 - \mu_2\nu_1)Z_3 = 0.$$

Exercise 1.13. To get started, observe that a homogeneous quadratic polynomial on $\mathbb{P}^3$ pulls back to a homogeneous sextic polynomial on $\mathbb{P}^1$.

Exercise 1.27. Show that the line $\alpha Z_1 - \beta Z_2$ meets the curve defined by $Z_0 Z_2^2 = Z_1^3$ exactly in the points $[1, 0, 0]$ and $\mu([\beta, \alpha])$; similarly for ν.

Exercise 1.28. Observe that homogeneous polynomials of degree 3 on $\mathbb{P}^2$ pull back via ν to homogeneous polynomials of degree 9 on $\mathbb{P}^1$, and that this pullback map cannot be an isomorphism.

Exercise 1.29. Observe that for $[Z_0, \ldots, Z_3] = \nu([X_0, X_1])$, we have $Z_1/Z_0 = Z_3/Z_2 = X_1/X_0$, giving us a quadratic polynomial vanishing on the image. Now express Z_2/Z_0 and Z_3/Z_1 in terms of these ratios to find two cubic polynomials cutting out $C_{\alpha,\beta}$.

Exercise 2.12. First, show that for a suitable choice of homogeneous coordinates $[X_1, \ldots, X_{2k}]$, any three mutually disjoint $(k-1)$-planes in $\mathbb{P}^{2k-1}$ can be taken to be the planes $(X_1 = \cdots = X_k = 0)$, $(X_{k+1} = \cdots = X_{2k} = 0)$ and $(X_1 = X_{k+1}, \ldots, X_k = X_{2k})$. For part (ii), use the fact that the Segre variety is the zero locus of quadratic polynomials.

Exercise 2.14. Probably the most efficient way to do this is simply to take P, $P' \in \mathbb{P}^1$, Q, $Q' \in \mathbb{P}^2$ and show that any linear combination of $\sigma(P, Q)$ and $\sigma(P', Q')$ lies in $\Sigma_{2,1}$ only if $P = P'$ or $Q = Q'$.

Exercise 2.19. Again, there are many ways to do this. One would be to observe that of the two families of lines on the surface S, the members of one meet the curve C exactly once each, while all but a finite number of the lines of the other family meet C three times. This finite subset of the lines of one family generally consists of four lines (if C is given by a bihomogeneous polynomial of bidegree (1, 3), this is the zero locus of the discriminant of F with respect to the second variable), and so has a cross-ratio.

Exercise 2.29 (last part). Observe that if the map φ were so given, the inverse image of most hyperplanes in $\mathbb{P}^5$ would be the zero locus of a bihomogeneous polynomial of balanced bidegree, which is not the case.

Exercise 4.12 (parts (b) and (c)). The key fact is that the rank of a skew-symmetric matrix is always even.

Exercise 4.13. One possible (and relatively elementary) way: first argue that it is sufficient to do this for the family of conics $\{X^2 + aY^2 + bZ^2\}$ parametrized by $\mathbb{A}^2$. To see that this family cannot have a rational section, show that this would entail a pair of rational functions $Y(a, b)$, $Z(a, b)$ satisfying the equation $aY^2 + bZ^2 + 1 \equiv 0$, or equivalently four polynomials P, Q, R, S in a and b satisfying $(QS)^2 + a \cdot (PS)^2 + b \cdot (RQ)^2 \equiv 0$. (There are other approaches that involve more machinery but may provide more insight; for example, you can argue that any five-dimensional subvariety of $\mathscr{X}$ will meet the general fiber of $\mathscr{X} \to \mathbb{P}^5$ in an even number of points.)

Exercise 4.14. One possible way: choose points p and $q \in \mathbb{P}^3$ such that the line $\overline{pq}$ does not meet every member of the family. For any member $X_b \subset \mathbb{P}^3$ of the family, let Q be the (unique) quadric surface containing X_b, p and q, $l \subset Q$ the line of the ruling of Q whose members meet X_b once passing through p and $r(b) = l \cap X_b \in X_b$.

Exercise 5.13. Use induction on d. For the second part, show that $d \leq 2n + 1$ points in $\mathbb{P}^2$ will fail to impose independent conditions on curves of degree n if and only if $n + 2$ of them are collinear.

Exercise 6.16. It may be helpful to know that a line $l \subset \mathbb{P}^3$ will meet the twisted cubic C only if the restriction to l of the three quadrics cutting out C are dependent; in this way we can write the equation of $\mathscr{C}_1(C)$ as a 3×3 determinant.

Exercise 7.8. This can be done in exactly the same fashion as the proof of Proposition 7.16.

Exercise 8.2. Probably the most instructive way to do this is to use the construction suggested in Exercise 6.18.

Exercise 8.4. One way to do this is to use Exercise 1.11.

Exercise 8.6. This can be done directly by calculation; alternately, look at the quadrics containing $C \cup \{p\}$.

Exercise 8.7. One way to crank this out would be to project from a point on the rational normal curve C; the image of a point $r \in \mathbb{P}^4$ will lie on a unique chord to the image of C, and we can write out the condition that r itself lie on the corresponding chord to C. A slicker way to do it would be to use the determinantal description of C suggested in Example 1.16 (with $k = 2$), bearing in mind that a sum of two rank 1 matrices can have rank at most 2.

Exercise 8.8. As in Exercise 8.7, this can be done by hand; but it is easier to use the determinantal description of the Veronese surface given in Example 2.6.

Exercise 8.12. Try for example $k = 4$, $l = 2$, and X the projection of a rational normal curve $C \subset \mathbb{P}^n$ from a point $p \in \mathbb{P}^n$ lying in a trisecant plane to C but not on a secant line.

Exercise 8.13. The union of the trisecant lines to $C_{\alpha,\beta}$ is just the (unique) quadric containing $C_{\alpha,\beta}$.

Exercise 9.18. Count parameters; that is, compare the number of coefficients in a polynomial of degree d to the number in a $d \times d$ matrix of linear forms.

Exercise 10.7. One place to look is at a projection of a rational normal curve.

Exercise 10.11. This can be done directly, but one approach is to use the determinantal description of the rational normal curve described in Example 9.3.

Exercise 10.14. The answer is C and Σ; the key ingredient in the solution is the j-function.

Exercise 10.22. One way is to look at the discussion of Zariski tangent spaces in Lecture 14.

Exercise 11.11. One answer is

$$Q(Z) = \begin{vmatrix} Z_0 & Z_1 \\ Z_1 & Z_2 \end{vmatrix} \qquad P(Z) = \begin{vmatrix} Z_0 & Z_1 & Z_2 \\ Z_1 & Z_2 & Z_3 \\ Z_2 & Z_3 & Z_0 \end{vmatrix}.$$

(Note that the Z_0 in the lower right corner of the second determinant could be any linear form.)

Exercise 11.19. To show the isomorphism of Ψ with the scroll $X_{2,2,1}$ fix distinct points $q, r \in l_0$ and a line $M \subset \mathbb{P}^3$ disjoint from l_0 and consider the images under the Segre embedding of $\mathbb{G}(1, 3) \times l_0$ of the three curves

$$\{(l_0, p)\}_{p \in l_0}$$

$$\{(\overline{qs}, q)\}_{s \in M}$$

$$\{(\overline{rs}, r)\}_{s \in M}.$$

Exercise 11.35. $m = n$ is the smallest number; this is required for polynomials of the form $x^n + x^{n-1}$ (e.g., corresponding to points on a tangent line to the rational normal curve).

Exercise 12.11. The key is to observe that the plane Λ in the construction is a general $(n - 2)$-secant plane to the rational normal curve.

Exercise 12.13. You would expect a general cubic surface S to contain a one-dimensional family of conics; it does because every conic on S is coplanar with a line contained in S.

Exercise 12.23. See [D].

Exercise 12.25. Three of the skew lines will lie on a unique smooth quadric $Q \cong \mathbb{P}^1 \times \mathbb{P}^1 \subset \mathbb{P}^3$; the fourth line will in general meet this quadric in two distinct points q, r and be determined by them. Check that the automorphism group of Q fixing the three lines and the points q and r is one-dimensional.

Exercise 12.26. The trick to look at the hyperplanes spanned by the lines pairwise. The answer to the general question is "no", but I don't know an elementary proof.

Exercise 13.8. The genera are, respectively, -1, 0, 0, and 1. The interesting point here is that the genera of three coplanar concurrent lines is different from that of three noncoplanar concurrent lines; indeed, we will see in Lecture 14 that they are not isomorphic varieties.

Exercise 13.13. For this (and the two following) use your knowledge of the Hilbert function; e.g., if you want to show $F_1, \ldots, F_k \in I(X)$ generate, show that $\dim((S/(F_1, \ldots, F_k))_m) = h_X(m)$.

Exercise 14.9. The fastest way to see that Y and Z are not isomorphic is by comparing the codimension of the conductor ideals in their local rings at the origin; alternatively, see the discussion of multiplicity in Lecture 20.

Exercise 15.5. This can be ground out directly; one shortcut might be to realize $\mathbb{P}^3$ as the space of homogeneous cubic polynomials in two variables mod scalars, the twisted cubic as the locus of cubes (as in Lecture 10) and show that the tangential surface TC is the zero locus of the discriminant.

Exercise 15.6. If a general point of a general tangent line $\mathbb{T}_p X$ to X lay on more than one tangent line, every tangent line to X would meet $\mathbb{T}_p X$.

Exercise 15.18. Observe that $\mathcal{T}'''(X)$ is the locus of lines L such that the restriction $F|_L$ of F to L has a multiple root for every polynomial $F \in I(X)$ in the ideal of X.

Exercise 15.19. The cone over an irreducible nondegenerate curve $C \subset \mathbb{P}^r$ for $r \geq 4$ will in general have all strict inequalities.

Exercise 16.11. A tangent line L to a curve X at a smooth point p will be a singular point of $\mathcal{S}(X)$ if it meets X again elsewhere (this is clear) or if it has order of contact > 2 with X at p—that is, every polynomial vanishing on X vanishes to order at least 3 on L at p (this is less clear). See Lecture 17 (up through Example 17.14) for further discussion.

Exercise 16.16. Consider the projection from the plane spanned by $\mathbb{T}_{p_1} X, \ldots, \mathbb{T}_{p_l} X$.

Exercise 17.4. See [GH1].

Exercise 17.7. Use the power series expansion of the parametrization.

Exercise 17.10. $q \in T^{(k)}(C)$ will be a smooth point if it lies on a unique osculating k-plane $\mathcal{G}^{(k)}(p)$ to C, it does not lie in the $(k-1)$-plane $\mathcal{G}^{(k-1)}(p)$, and the osculating plane $\mathcal{G}^{(k)}(p)$ has contact of order exactly $k+1$ with C at p. In this case, the tangent plane to $T^{(k)}(C)$ at q will be simply $\mathcal{G}^{(k+1)}(p)$.

Exercise 17.21. This has to be done in two stages. Show that a general surface containing C is smooth away from C (Bertini) and that it is smooth along C (Bertini applied to the blow-up of $\mathbb{P}^3$ along C, plus other considerations).

Exercise 19.7. Do this inductively with respect to k, or equivalently, apply the basic formula repeatedly.

Exercise 20.7. Yes.

Exercise 20.13. No.

Exercise 20.15. One way to do this (not the only way) is to deform the union of two conic curves having a point of contact of order 3.

Exercise 20.17. (a), (b) and (c) are all quartic curves; (d) is a sextic.

Exercise 22.1. If $\Lambda = \mathbb{P}W$ for some $W \subset V$, show that Q' is given by the composition

$$W \to V \to V^* \to W^*$$

where the middle map is $\tilde{Q}$.

Exercise 22.35. For the first part, observe that a general two-plane in $\mathbb{P}^N = \mathbb{P}^5$ will meet Φ_2 in a smooth cubic curve and miss Φ_1 altogether; take L a flex line to this cubic in such a plane. What does the corresponding family of plane conics look like?

Exercise 22.39. The quartic with a tacnode occurs when P is a smooth point of one of the four singular quadrics in the pencil containing X; the double conic when P is a vertex of one of those quadrics.

References

Basic Sources

[AM] Atiyah, M. F., and MacDonald, I. G. *Introduction to Commutative Algebra*, Addison-Wesley, 1969.
[E] Eisenbud, D. *Commutative Algebra*, in press.
[EH] Eisenbud, D., and Harris, J. *Schemes: the Language of Modern Algebraic Geometry*, Wadsworth, 1992.
[F] Fulton, W. *Algebraic Curves*, W. A. Benjamin, New York, 1969.
[GH] Griffiths, P. and Harris, J. *Principles of Algebraic Geometry*, Wiley and Sons, 1978.
[H] Hartshorne, R. *Algebraic Geometry*, Graduate Texts in Mathematics, Springer-Verlag, 1977.
[M] Mumford, D. *Algebraic Geometry I: Complex Projective Varieties*, Grundlehren der math. Wissenschaften 221, Springer-Verlag, 1976.
[M1] Mumford, D. *Red Book*, Springer-Verlag Lecture Notes in Mathematics 1358, 1988.
[SR] Semple, J., and Roth, L. *Introduction to Algebraic Geometry*, Oxford Univ. Press, 1949.
[S] Serre, J.-P. "Faisceaux Algébriques Cohérent," *Annals of Math*, 61, 1955, 197–278.
[Sh] Shararevich, I. *Basic Algebraic Geometry*, Springer-Verlag, New York, 1977.

Specific References

[A] Ahlfors, L. *Complex Analysis*, McGraw-Hill, 1953.
[BM] Borel, A., and Moore, J. "Homology theory for locally compact spaces." *Michigan Math. J.*, 7, 1960, 137–159.
[BK] Brieskorn, E., and Knörrer, H. *Plane Algebraic Curves*, Birkhäuser Boston, 1981.
[C] Coolidge, J. *A Treatise on Algebraic Plane Curves*, Dover, 1931.
[CG] Clemens, H., and Griffiths, P. "The intermediate Jacobian of a cubic threefold." *Annals of Math.*, 95, 1972, 281–356.

[D] Donagi, R. "On the geometry of Grassmannians." *Duke Math. J.*, 44, 1977, 795–837.
[Ein] Ein, L. "Varieties with small dual varieties." *Invent. Math.*, 86, 1986, 63–74.
[E1] Eisenbud, D. "Linear sections of determinantal varieties." *Am. Jour. Math.*, 110, 1988, 541–575.
[F1] Fulton, W. *Intersection Theory*, Springer-Verlag, 1984.
[FH] Fulton, W., and Harris, J. *Representation Theory: A First Course*, Springer-Verlag, 1991.
[FL] Fulton, W., and Lazarsfeld, R. "Connectivity and its applications in algebraic geometry." In *Algebraic Geometry*, A. Lidgober and P. Wagreich (Eds.), Springer-Verlag Lecture Notes in Mathematics 862, 1981, 26–92.
[GH1] Griffiths, P., and Harris, J. "Algebraic geometry and local differential geometry." *Ann. Ec. Norm. Sup.* 4ᵉ série, 12 (1979), 355–432.
[GLP] Gruson, L., Lazarsfed, R., and Peskine, C. "On a theorem of Castelnuovo, and the equations defining space curves," *Invent. Math.*, 72, 1983, 491–506.
[GM] Green, M., and Morrison, I. "The equations defining Chow varieties." *Duke Math. J.*, 53, 1986, 733–747.
[Hi] Hironaka, H. "Resolution of singularities of an algebraic variety over a field of characteristic 0," *Annals of Math.*, 79, 1964, I: 109–203; II: 205–326.
[Hi1] Hironaka, H. "Triangulations of algebraic sets." In *Algebraic Geometry, Arcata, 1974*, AMS *Proc. Symp. Pure Math*, 29, 1975, 165–184.
[IM] Iskovskih, V. A., and Manin, Ju. I. "Three-dimensional quartics and counterexamples to the Lüroth problem." *Math USSR-Sbornik*, 15, 1971, 141–166.
[JLP] Jósefiak, T., Lascoux, A., and Pragacz, P. "Classes of determinantal varieties associated with symmetric and skew-symmetric matrices." *Math. USSR Izv.*, 18, 1982, 575–586.
[K] Kleiman, S. "The Enumerative Theory of Singularities." In *Real and Complex Singularities, Oslo 1976*, P. Holm (Ed.), Sijthoff and Noordhoff, 1977, 297–396.
[K1] Kleiman, S. "The transversality of a general translate." *Composition Math.*, 28, 1974, 287–297.
[L] Lawson, H. B. "Algebraic cycles and homotopy theory." *Annals of Maths.*, 129, 1989, 253–291.
[MM] Mumford, D., and Morrison, I. "Stability of projective varieties." *l'Enseign. Math.*, 24, 1977, 39–110.
[M1] Mumford, D. *Lectures on Curves on an Algebraic Surface* Annals of Math. Studies, 59, Princeton Univ. Press, 1966.
[P] Pinkham, H. "A Castelnuovo bound for smooth surfaces," *Invent. Math.*, 83, 1986, 321–332.
[S1] Serre, J.-P. *Groupes Algébriques et Corps de Classe*, Hermann, 1959.
[S2] Serre, J.-P. "Géométrie algébrique et géométrie analytique," *Ann. Inst. Fourier*, Grenoble, 6, 1956, 1–42.
[W] Wilson, G. "Hilbert's sixteenth problem." *Topology*, 17, 1978, 53–73.

Index

Note: Where there are several entries under a given heading, the ones giving the definition of an object or the statement of a theorem are indicated by italic numbers

D

E

F

Z

Graduate Texts in Mathematics

continued from page ii

119 ROTMAN. An Introduction to Algebraic Topology.
120 ZIEMER. Weakly Differentiable Functions: Sobolev Spaces and Functions of Bounded Variation.
121 LANG. Cyclotomic Fields I and II. Combined 2nd ed.
122 REMMERT. Theory of Complex Functions. *Readings in Mathematics*
123 EBBINGHAUS/HERMES et al. Numbers. *Readings in Mathematics*
124 DUBROVIN/FOMENKO/NOVIKOV. Modern Geometry—Methods and Applications. Part III.
125 BERENSTEIN/GAY. Complex Variables: An Introduction.
126 BOREL. Linear Algebraic Groups. 2nd ed.
127 MASSEY. A Basic Course in Algebraic Topology.
128 RAUCH. Partial Differential Equations.
129 FULTON/HARRIS. Representation Theory: A First Course. *Readings in Mathematics*
130 DODSON/POSTON. Tensor Geometry.
131 LAM. A First Course in Noncommutative Rings.
132 BEARDON. Iteration of Rational Functions.
133 HARRIS. Algebraic Geometry: A First Course.
134 ROMAN. Coding and Information Theory.
135 ROMAN. Advanced Linear Algebra.
136 ADKINS/WEINTRAUB. Algebra: An Approach via Module Theory.
137 AXLER/BOURDON/RAMEY. Harmonic Function Theory.
138 COHEN. A Course in Computational Algebraic Number Theory.
139 BREDON. Topology and Geometry.
140 AUBIN. Optima and Equilibria. An Introduction to Nonlinear Analysis.
141 BECKER/WEISPFENNING/KREDEL. Gröbner Bases. A Computational Approach to Commutative Algebra.
142 LANG. Real and Functional Analysis. 3rd ed.
143 DOOB. Measure Theory.
144 DENNIS/FARB. Noncommutative Algebra.
145 VICK. Homology Theory. An Introduction to Algebraic Topology. 2nd ed.
146 BRIDGES. Computability: A Mathematical Sketchbook.
147 ROSENBERG. Algebraic *K*-Theory and Its Applications.
148 ROTMAN. An Introduction to the Theory of Groups. 4th ed.
149 RATCLIFFE. Foundations of Hyperbolic Manifolds.
150 EISENBUD. Commutative Algebra with a View Toward Algebraic Geometry.
151 SILVERMAN. Advanced Topics in the Arithmetic of Elliptic Curves.
152 ZIEGLER. Lectures on Polytopes.
153 FULTON. Algebraic Topology: A First Course.
154 BROWN/PEARCY. An Introduction to Analysis.
155 KASSEL. Quantum Groups.
156 KECHRIS. Classical Descriptive Set Theory.
157 MALLIAVIN. Integration and Probability.
158 ROMAN. Field Theory.
159 CONWAY. Functions of One Complex Variable II.
160 LANG. Differential and Riemannian Manifolds.
161 BORWEIN/ERDÉLYI. Polynomials and Polynomial Inequalities.
162 ALPERIN/BELL. Groups and Representations.
163 DIXON/MORTIMER. Permutation Groups.
164 NATHANSON. Additive Number Theory: The Classical Bases.
165 NATHANSON. Additive Number Theory: Inverse Problems and the Geometry of Sumsets.
166 SHARPE. Differential Geometry: Cartan's Generalization of Klein's Erlangen Program.
167 MORANDI. Field and Galois Theory.
168 EWALD. Combinatorial Convexity and Algebraic Geometry.
169 BHATIA. Matrix Analysis.
170 BREDON. Sheaf Theory. 2nd ed.
171 PETERSEN. Riemannian Geometry.
172 REMMERT. Classical Topics in Complex Function Theory.
173 DIESTEL. Graph Theory.
174 BRIDGES. Foundations of Real and Abstract Analysis.
175 LICKORISH. An Introduction to Knot Theory.
176 LEE. Riemannian Manifolds.
177 NEWMAN. Analytic Number Theory.
178 CLARKE/LEDYAEV/STERN/WOLENSKI. Nonsmooth Analysis and Control Theory.

9 780387 977164